THE
SENSE OF SURVIVAL

J. Allan South

Illustrated by Barbara Wilkinson
Cover by Steven D. Keele

Timpanogos Publishers

ACKNOWLEDGMENT

Sincere appreciation and gratitude is expressed to family; to friends; to knowledgeable and professional folks; to typists–who have been able to produce order from chaos; and to a kind Deity for a world full of interesting and useful information and for the endowment of whatever it was that made me write this book.

A strong effort has been made to report the information contained herein in a responsible manner. However, no liability can be assumed by the author or publisher for danger, difficulty, or harm accrued while attempting to use this information. Use it at your own risk.

The Sense of Survival
Fourth Printing

ISBN 0-935329-00-0

TABLE OF CONTENTS

INTRODUCTION

Much talk is heard of survival these days: economic survival, emotional survival, physical survival. Many are predicting gruesome occurrences for humanity and are making their preparations for nuclear holocaust, complete economic collapse, or giant earthquakes and the chaos and disorder they would cause. There is talk of hibernation, storage, sharing, and militant repulsion of outsiders.

The unprepared need to be convinced that problems will certainly come and that prudent, sensible living dictates a reasonable preparation for possible need. It doesn't have to be nuclear war (although it certainly enters our minds as a possibility); it could be a strike, loss of work, tornado, earthquake, tidal wave, civil disorder, winter storm, fire, or some other disaster. It doesn't have to engulf the country or the world; it may only affect a region, state, or county.

Disaster-vivid can also bring on disaster-insidious. Earthquakes have historically often killed more by famine and disease than by the actual earth movement. An oft-cited example is the eruption of Mt. Tambora, Indonesia, in 1815. The eruption was a killer at the outset; but the many thousands of tons of pulverized rock and dust thrown into the atmosphere altered climates causing summer frosts and ruined crops. The ensuing famine and disease (caused in great measure by underfed rodents carrying it into dwellings) were also killers. Large-scale nuclear war could produce a similar effect. Everyone can't be standing in an open field during an earthquake, but just about everyone should be able to safeguard against famine and disease—at least everyone in the "land of plenty."

Everyone may not need emergency advice; but there always have been disasters, and until the earth somehow inclines to a more ordered state there undoubtedly will continue to be. Prophets, politicians, scientists, and coffee-break rhetoricians all predict difficulties. We hear about earthquakes (try, for example, Revelations 6:12; Isaiah 13:10-13), wars, famine, pestilence, etc. See your local newspaper for the latest of these happenings.

When these cataclysmic descriptions do not involve one's own immediate situation, they seem far away. When they do, they are all too real.

Let us proceed, then, with the basic assumption that times will change, at least for some, and certain basic skills could be useful—if not in fact necessary—for comfort or perhaps even survival and when a crisis occurs, the time for preparation will be past.

In the marketplace of survival literature, there seems to be a niche for books that show or tell how to acquire some of the necessities of life that we now take for granted. The instructions given in these pages are meant to be simple and concise without losing clarity and function. While there are many approaches to some of the topics, only a few are covered, and many possible approaches have been left out. With each calamity, there is a slightly different set of preparations to make, although some basic preparations are quite universal.

To create a more visible scenario, and to place those needs before us to have some basic knowledge and/or materials at our disposal, let's use an hypothetical earthquake. This indulgence should sufficiently explain that needs do exist.

*   *   *

You are Dad.

One fine day, after hitting the snooze alarm twice and finally managing to pull the mattress off your back, you make it out of bed. It is winter. It's pitch black outside and you are still sitting on the edge of the bed when you hear what sounds like a train thundering through your living room. The house starts shaking, then pounding. Your dresser tips over and your bed slides back and forth across the room. The sound of glass breaking shears through your senses as your bedroom curtains catch flying pieces of window and allow them to fall on the floor. The thundering and bucking continue and you wonder if the giant made it down from the beanstalk and picked your house up and dropped it—or threw it—down. Suddenly you are hit on the back of the head with something very heavy and you slide off the bed onto the floor. You are aware of a suffocating weight on top of you. It itches. The 10 inches of insulation you had installed in your attic are now on your floor, along with what's left of your bedroom ceiling.

(We could stop here, but this is where the real fun begins.)

You are mostly conscious (well, maybe partly) and you hear the kids crying, screaming! You hear your wife reviling you for interrupting her sleep with this horrible trick. You hear water running and smell smoke. You get up and push aside large chunks of plasterboard. You start toward the children's rooms and step on a plasterboard nail. Where, oh, where are your shoes? (And when was your last tetanus inoculation?) The shoes are in the living room, covered with insulation. There are some others in the closet. You manage to get them on and stumble through the debris to the baby's room. You pick her up. You gather her two brothers and bring them to your bedroom. Luckily, there are no serious injuries. You have found a flashlight. The batteries are weak, but at least it's light. It is cold. The children are still hysterical. Your wife is finally convinced that this isn't a joke.

When you go downstairs, you are appalled to discover that your basement has three inches of water in it from a broken water pipe. Luckily, the water main ruptured elsewhere or it would be up to your knees by now. It is also fortunate that the main gas line ruptured because the natural gas line is broken in your house and would probably have blown the family into next week, since broken electrical wires are still arcing in the wall. (That's where that smoke is coming from.)

Suddenly, the wires stop arcing. This is a good thing, because you don't remember where to turn off the power to the house.

It's very cold.

Your wife has the children calmed by now. The two boys have to go to the bathroom and the baby needs changing. The boys go into the bathroom. Miraculously the toilet is still in one piece. They flush the toilet. you hear a splash in the basement. The sewer pipe is broken. Well, at least it won't happen again because the toilet tank isn't filling: no water.

It's still cold.

You nail a piece of plasterboard over the window in the living room after throwing some of the insulation out the window. You light a fire in the fireplace. The chimney is mostly collapsed, so the smoke goes into the open attic. You wonder how long the wood from that old cherry tree you cut up will keep you warm. It's getting light. The children are hungry. You itch from handling insulation. Your wife is staring into space and saying odd things. It is urgent that you use the bathroom, but it is cold outside. Smoke from the fire is beginning to fill the room. An aftershock knocks the plasterboard off the window and lets out some of the smoke and most of the heat.

A neighbor comes to enlist your station wagon and you to help take his wife to the hospital. She has a broken arm and an open wound and needs medical attention. You tell your wife to call your boss and tell him you won't be at work today. After consultation, you agree with her that it probably won't be necessary. You grab a piece of bread and your coat and leave, over your wife's objections. After all, she could have some tuna fish sandwiches ready for breakfast in no time.

You still have to go!

You take the neighbor and his wounded wife up the street toward the hospital. There are cracks in the road. You can only get to within about three-quarters of a mile of the hospital because the road is so damaged.

You walk and carry the hurt woman to the hospital.

You are freezing.

There is a two-block line at the hospital.

You find a secluded spot and go.

You finally get in to see a doctor and are cared for and get the neighbors back home. It is late afternoon.

Your wife has cleaned the place up a bit. She has supper ready; it's tuna fish sandwiches. The kids are itchy. They need baths. There is no water. They have to go. You chisel a hole in the backyard and make a lean-to from pieces of fence to afford privacy for a latrine.

Another neighbor comes over. He wants to bring his family over to huddle by your fire. He doesn't have a fireplace. He will furnish the remnants of his garage for firewood.

You listen to civil defense instructions on the portable radio until the batteries die. You appropriate the batteries from your son's electric baseball game. You find out that the quake has devastated a wide area.

Your blanket supply seems inadequate, but you settle down for the night.

You make it through the night.

Your wife doesn't think she can fix breakfast because you are almost out of bread and tuna fish. But there is some peanut butter.

You are thirsty. All the contents of the refrigerator were spilled and/or broken. You melt snow on the fire after jury-rigging a grill. Your wife complains about the handles melting off her stainless-steel cookware.

Mice keep getting into the bread.

Mice keep getting into your blankets.

##?!! the mice.

The kids are driving you nuts.

You are out of food. You tried to get some at the supermarket but there wasn't much left and they would only take cash for payment and you didn't have much cash. On the radio you hear that it will be at least three more weeks before food can be trucked in because of the extent of the road damage, and even then the supply will probably be inadequate to take care of everyone. You dig out the sack of wheat you stored. It is full of little white worms. You are hungry. You boil it and eat it. After three days on boiled wheat, everybody's digestive system goes haywire. You are out of toilet paper.

You are out of firewood. There are some sources of wood around (like old trees, fences, and telephone poles) but in the whole neighborhood there are only two axes and one hand saw.

It is spring and somehow you're still alive. Much has been improved and rebuilt. It becomes apparent that you need a garden. All the seeds you have are leftovers from last year; corn, beans, radishes, and carrots. It's been a long time since you had a salad. You think the weeds coming up look good. Didn't the pioneers and Indians eat weeds? You try them. At this point, some of them aren't bad.

The neighborhood gets together and decides an irrigation system is necessary to water the gardens. Water should be brought from a nearby stream. There are only four shovels in the whole neighborhood. You all dig the ditch. When you finish, the neighborhood Tonka trucks and tractors are all worn out.

Stray dogs, hungry from a lean winter, are constantly around; a couple of them begin molesting your children while they play. You chase them away with a baseball bat. The next day, while you are gone, your neighbor shoots the two disruptive animals with a .22 rifle. You hope he was careful. He says he was careful, but they were starting to be violent. You wonder if you should have a rifle. You've been a long time without meat or substantial protein. If you could find the dead dogs, you would be tempted to . . . .

You wish you knew how to raise rabbits or chickens—or *something*—for fresh meat.

You definitely wish you had some rat poison.

Your garden is growing; bugs are eating it; you have no insecticides.

Your clothes are ragged; your dress shoes and sneakers are worn out.

You wonder: if you went to a rural area, could you find some produce and beg, borrow, steal, or work for it? Then you hear of two acquaintances who have caught some shotgun pellets trying to steal from a farmer's garden.

*   *   *

Despite the flaws and exaggerations in our story, much of it (and even worse) is all too possible, and it graphically sets the stage for what follows.

And what follows?—A description of several possible disastrous events and procedures to provide protection from them (Section I), an informative nuts-and-bolts discussion of the necessities of life and their acquisition on a survival level (Section II), and a brief primer on medical care (Section III), including first aid and the use of herbs for some common maladies.

Section I is mostly about nuclear war but also includes a brief rundown on several natural disasters. The emphasis is on how to protect yourself.

Section II is about how to grow, gather, store, and use food (plant and animal); put together a "Fourteen-Day Emergency Kit"; obtain and store water; select and use equipment; obtain shelter; maintain sanitary conditions; select weapons; prepare for communication and signaling; and other miscellaneous pertinent subjects. A somewhat morally philosophical discussion of survival appears in the chapter on weapons.

Section III is about medical care. First-aid equipment and procedures are the main subjects. The growing and using of some time-tested herbs are also discussed.

Since the Table of Contents is so extensive, no index has been included.

Checklists are used where they seem relevant. They are designed to be as practical as possible without being overdone. Selected references are included to provide a (hopefully) instructive volume of sources of information and to recommend direction for further study of the subjects.

Most items of equipment are so widely available that the mention of sources is not generally necessary. However, in the case of specialty items and in some other cases sources have also been included.

From the beginning of Section I reference is made to the "Fourteen-Day Emergency Kit." It may be instructive to read about this kit at any time your curiosity prevails upon you to do so. It is Chapter 5.

I
THE LIKELY
POSSIBILITIES

There are many difficulties that we potentially face as members of human society. Many are man-made and many are products of the natural elements.

A brief resume of some of the more likely possibilities and a short, common-sense review of protective measures for them is quite in order for a survival discussion. Natural disasters: such as floods, earthquakes, winter storms, tsunamis, hurricanes, tornadoes, and volcanoes; and man-made problems of biological and chemical warfare and nuclear warfare are all included. Extensive attention is given to nuclear war, since its potential impact is so unique and far-reaching.

It is valuable to know that the National Weather Service broadcasts reports on natural phenomena that include *watches* and *warnings*.

A *watch* means that the phenomenon is *expected* in your area, and you should keep yourself alert to further information and begin preparations to protect life and property.

A *warning* means the phenomenon *is occurring* or *will occur* shortly and you should take protective measures immediately.

CHAPTER 1

Natural Disasters

FLOODS

Flood damage causes hundreds of millions of dollars in damage and costs many lives each year in the United States alone. Flooding caused by overflowing rivers, collapsing dams, and sudden heavy rainfall occur in nearly every section of the country. With some types of flooding, warning signs are gradual and sufficient to easily provide safety. Flash floods, which are usually caused by heavy rain or breaks in dams, can strike almost without warning and can be devastating.

Here are some suggested procedures for coping with flood dangers:

Before

Be especially wary of narrow canyons during rainy times; flash floods can come from miles away. If a flash flood warning has been officially given or one is otherwise imminent, stay away from arroyos, streambeds, and other waterways. Find out how far your location is above possible flood levels and where the nearest safe areas are.

Keep a portable radio close by and monitor warnings and reports.

Always keep gas in your car (at least half a tank).

Keep your "14-Day Emergency Kit" ready. If conditions are threatening, move it to your car.

If you are at home and you know authoritatively that there is sufficient time, move furniture to upper floors of your house. Disconnect all appliances. Take steps to protect food supply and other assets.

Preserve some extra "safe" water.

If you are home, secure your possessions and lock up before leaving.

During

Stay in the safest areas possible.

If your auto stalls in a wet or otherwise dangerous area, leave it immediately and go to a safer area.

Do not attempt to drive through waterways or flooded areas unless you are certain the roadway is still solid and the water level is below the middle of your wheels.

Do not attempt to walk through a waterway that is above your knees or is extremely swift.

After

Food and water supplies that have been contaminated by flood waters should be made safe before use or else discarded.

Be careful of broken utility lines and structural defects in buildings.

Report dangerous situations to authorities. Do not use open flame to examine buildings, as ruptured gas lines may be present; use a flashlight or other electric light.

Keep in touch with official instructions via radio, TV, or other means.

Follow authoritative instructions.

Use caution in driving through affected areas. Unreported and unrepaired damage may have created unsafe conditions.

Don't go "sightseeing" in disaster areas.

Don't connect electrical appliances if they are wet,and don't stand in wet areas to reconnect appliances. Turn the power off, plug in the appliance, and then stand back and turn the power back on. If something is wrong, turn the power back off immediately.

EARTHQUAKES

An earthquake is one event that is nearly inevitable for many areas. Quakes are calculated to be overdue in many locations and could easily occur in areas where they have not been predicted.

An earthquake may be large or small, may last from a few seconds to nearly a full minute, and may or may not be preceded by foreshocks. The accompanying seismic map of the United States (figure 1.1) gives a general idea of relative earthquake risk in various parts of the country.

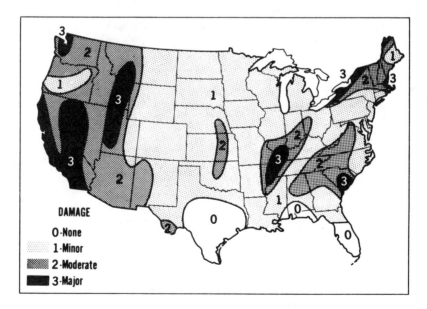

Figure 1.1   Possible Seismic Risks

Zones 0—no reasonable expectancy of earthquake; Zones 1—minor earthquake damage expected; Zones 2—moderate earthquake damage likely; Zones 3—major earthquake damage is possible. (ESSA/Coast and Geodetic Survey.)

The motion of an earthquake may be vertical or lateral or both, and the earth may be displaced many feet. This motion can reduce a building to rubble in seconds, and aftershocks can cause further damage. Even some structures built to rigid standards of earthquake resistance can, under extreme conditions, be seriously damaged or destroyed.

Without further discussion of earthquake theory or generalities, here are some precautions that can minimize risk of damage or injury:

Before

Place large and heavy objects on the floor or low shelves.

Secure gas appliances such as water heaters and dryers by bolting them down, tying them to wall studs, or otherwise fastening them into position.

China and other glassware, pictures or mirrors are potential hazards that should be secured.

Make sure your dwelling conforms to local building codes.

Keep your "14-Day Emergency Kit" ready at all times.

Store bottled items in cartons lined with plastic and padded with cardboard or newspaper.

Place restraining boards, bars or wires on front of storage shelves to prevent the goods from being shaken off. A piece of cardboard or other insulative material placed between bottles on the shelf will prevent them from striking each other and breaking.

Have an adjustable wrench or other necessary tools available for turning off utilities.

Drawn drapes can help prevent injury from flying glass.

Make sure there is gas in the car at all times.

Keep a pair of shoes for each member of your family under their beds.

Keep a flashlight for each person, and be sure each knows where his or her flashlight may be found.

During

If you are indoors, seek shelter from falling, breaking, or moving objects. A corner, under a desk or bed, or in a strong doorway or hallway are the safest places. Be cautious of masonry fixtures such as fireplaces.

If you are outdoors, avoid all buildings, high structures, or power poles. Seek open areas; if they are not available, seek a strong shelter.

If you are traveling in an auto, stop as quickly as possible in a safe area. Afterwards, proceed cautiously. Road damage may have occurred.

Be cautious of broken and other utility lines. Don't use any open flame until you are sure there are no gas leaks. Don't touch downed electrical lines.

After

Put out any fires.

Check utilities and utility lines for leaks or disconnection of any kind. Turn off any service that is ruptured. Turn off main water valve to prevent water from draining out of the house if the main line is broken.

Listen for official advice on radio and cooperate with authorities.

Inspect your own dwelling for safety. Pay close attention to brick structures such as chimneys. Check metal chimneys for continuity.

Clean up any dangerous materials that may have spilled.

Stay away from beaches. Earthquakes can cause tsunamis (tidal waves).

Be prepared for successive aftershocks. If your dwelling is unsuitable to live in, set up temporary quarters in an open area. Stay away from damaged structures.

Do not share water or food with pets if it is obvious that additional supplies necessary for survival will not be available.

Do not use water from a toilet tank that has disinfectant material in it that colors the water; it is poisonous.

WINTER STORMS

Cold, stormy weather has taken many lives and caused heavy financial losses. The greatest cause of injury and death during winter storms is accidents; next is overexertion; then exposure and freezing; and, finally, other causes such as carbon monoxide poisoning in houses or stalled automobiles, home fires, collapse of buildings, electrical injuries from downed wires and so on. Financial losses result from the above injury-related causes and other causes such as loss of livestock, paralyzed transportation and closed businesses.

Terms commonly used to describe winter storms are important to understand.

• A *blizzard* is a combination of heavy snow, cold air, and strong wind. A *blizzard warning* means the Weather Bureau expects considerable snow, winds at least thirty-five miles per hour or more, and temperatures of twenty degrees Farenheit or less. A *severe blizzard warning* means heavy snowfall, winds at least forty-five miles per hour, and temperatures ten degrees Farenheit or lower are expected.

• A *heavy snow warning* is usually given when a snowfall of four or more inches in a twelve hour period or six or more inches in a twenty-four hour period is expected. *Snow flurries, snow squalls,* and *blowing and drifting snow* may also be mentioned in a warning. These decrease visibility and generally degrade driving conditions.

• An *ice storm* is forecast when freezing rain is expected to form substantial layers of ice on roadways and other terrestrial objects. When lesser amounts of ice are expected to form, a *freezing rain* or *freezing drizzle* is forecast.

• *Sleet* forms when drops of rain freeze on their way to the ground. A sleet storm usually consists of ice particles mixed with rain.

Preparation and caution can alleviate much of the danger involved in winter storms. When you are driving along in a storm everything can be perfectly okay one minute and the next minute you can be stranded, through car failure or getting stuck, with nothing between you and freezing to death but what you have with you.

The main ideas to remember are: *be prepared* and, if an emergency occurs, *don't panic*. A friend who has spent some time up north said, "There are no angry eskimos; they all died." And panic can take away your objectivity and energy in the same way anger can.

Here are a few other things to remember:

Before

At home, keep a liberal supply of packaged foods that could be used without cooking or refrigeration. (Utilities are frequently interrupted during severe winter storms.)

Keep a backup means of heating and cooking and a liberal supply of fuel.

Maintain an adequate supply of cold weather clothing and bedding.

Stay healthy.

Stay in touch with weather information.

Keep some emergency gear in your car—jumper cables, tow strap, first aid kit, knife, fire extinguisher, flares, candles, flashlight, matches, small shovel, poncho, blanket, emergency (space) blanket, a small kit of high-energy food, and perhaps some extra clothing. Whenever traveling during weather extremes a supply of drinking water or other suitable liquid (not alcohol) should also go along.

During

At home conserve resources as much as possible. Close off areas not needed—use just one or two rooms.

Stay warm—wear hats during sleep and when awake.

Let someone know what's happening. Call a neighbor or relative if possible.

Avoid overexertion. Heart attacks and exhaustion account for a large percentage of cold weather fatalities. Also, becoming damp with perspiration makes you colder.

In your auto, travel with a companion and, if possible, during weather extremes travel in convoy with another car. Avoid travel during storms if possible. Sometimes only a few hours can make an enormous difference in travel conditions. Travel during daylight hours and plan alternate routes. Keep in touch with weather information.

Travel during daylight hours and plan alternate routes. Keep in touch with weather information.

If you become stranded stay with your car—it is probably the safest place to stay unless you are *certain* that you can easily reach nearby help. Run the motor for short periods to get heat from the heater. As soon as the car warms up, turn off the engine. Keep a bit of fresh air in the car to breathe and beware of poisoning from carbon monixide—slightly open a window for ventilation and make sure exhaust fumes are not coming in the window. At night use a flashlight or inside light to alert work crews, but be careful not to run the car battery down to the point it will not start the car. During the day tie a piece of colored cloth to your antenna. If you do decide to leave your car leave a note stating a person to contact (name, address, phone number) and the time you left and the direction you were heading.

If you have a pretty good idea that you may be getting into trouble, remember that in your car you are sitting on foam padding that is one of the best insulations around. Take a knife or other sharp object and tear the insulation out of the seats. Stuff it inside your clothes next to your skin. You can also make some makeshift booties by tying them with pieces of cloth. The price of a new seat is nothing compared to freezing to death.

When venturing out in the cold remember to avoid overworking. Dress in layered clothing with an outer layer that will prevent getting wet and keep out the wind. (For related discussion see Chapter 11, "Equipment".) Avoid drinking alcoholic beverages.

After

Repair damage to home, auto or person and replenish supplies. Another storm may be on the way.

TSUNAMI

A TSUNAMI ('soo-nah-mee), or tidal wave, is sometimes created when an earthquake or other disturbance occurs in or near the ocean. They are rare but incredibly destructive. A recent tsunami caused the loss of many lives in Japan. As these waves travel through deep water, they move rapidly and appear to be small; but as they approach shallower water, they become slower and rise to enormous heights.

If you live in a coastal area, be aware of earthquake activity. An unusual rise or fall of ocean water level can also be a warning of an approaching tsunami. If you think danger is imminent, stay close to the radio. Keep your "14-Day Emergency Kit" handy and your car filled with gas.

If time permits following a warning of tsunami, turn off all utilities before leaving home.

After a tsunami has occurred, stay away from the coast. Successive waves may occur after the major tsunami, and many lives have been lost to these successive waves.

Listen for authoritative advice.

HURRICANES

Hurricanes cause immense damage, as has often been seen in the Gulf Coast area. The extreme winds and variations in pressure found in a hurricane can crush structures and otherwise cause damage. Heavy rain and accentuated surf can cause flooding and water damage. High-speed wind propulsion of small objects causes additional damage. In addition, tornadoes spawned by hurricanes are among the most lethal natural occurences.

As with other conditions about which advisories are issued from the National Weather Service, a hurricane *watch* signals that a hurricane may occur in the area for which it is given. A hurricane *warning* means that it is *expected* to occur shortly (usually within 24 hours).

Here are some safety actions to remember:

Before

Be alert to official warnings of both hurricanes and tornadoes.

Be prepared to leave. Keep your "14-Day Emergency Kit" ready. Be sure you have plenty of batteries for your portable radio.

Beachfront and low-lying areas that may be subject to elevated surf or flooding are not safe and should be evacuated.

Secure all objects that are subject to being blown away. Tie down or bring indoors all garbage cans, furniture, toys, etc. Not only can they be lost, but they can also cause damage as they are blown into other objects.

Mobile homes are not usually substantial enough shelter unless they have been especially well anchored.

Moor boats securely.

Protect windows by boarding up, shuttering, or taping over them.

Keep your automobile fueled. It may be hard to get fuel following a storm of any magnitude.

Be sure you have a secure shelter.

If you are told to evacuate by authorities, follow their instructions. Make sure where you are to go and what facilities are there. Close up and lock your home; and if you have to leave your car, lock it up, too. Shut off utilities if you are told to do so. Travel with caution; use recommended routes and listen for instructions on your radio. Give yourself plenty of time to reach safety.

During

Stay inside. Stay away from windows. (They can be broken by wind pressure or flying debris.)

Be aware of the "eye" of the hurricane which may produce a calm for a few minutes to a half hour or more. When the storm resumes after the "eye" passes, winds will be blowing in the opposite direction from those that preceded the "eye" and are often more violent than the initial part of the storm.

After

Remain sheltered until officially informed that it is safe to leave.

Continue to listen to the radio for advice on how and where to obtain assistance.

Stay away from high-damage areas.

Be cautious with utility connections, especially when returning home. Watch for gas leaks (no open flames until you are sure there are none), broken sewer connections, and breaks in power lines. Don't connect electrical appliances if they are wet. Don't stand in wet areas to reconnect appliances. Turn off the power, plug in the appliance, and then stand back and turn the power back on. If something is wrong, turn the power back off immediately.

Report broken lines to proper responsible authorities.

TORNADOES

Tornadoes are among the most violent of all atmospheric phenomena. They appear as a grey or black whirlpool. The spinning, funnel-shaped cloud may sound like a locomotive and can destroy buildings, cars, and other property. People and animals are especially vulnerable to these storms. A tornado's path is somewhat unpredictable but is usually a narrow strip a few miles long. (In rare instances they have traveled 200 to 300 miles.) They can travel very slowly, or as fast as 60 or 70 miles per hour.

A few things to observe regarding tornadoes are:

Before

When a tornado watch is announced, keep watch in your area and stay tuned to radio or TV broadcasts. If you spot a tornado, phone the local law enforcement or Weather Service office and report it.

When a tornado warning is announced, go to a shelter immediately. An underground shelter, cave, or substantial steel-framed or reinforced concrete building is best. If you must take cover in your home, go to a corner of the basement and take cover under a sturdy table, workbench, or other shelter.

Stay away from areas where heavy appliances are directly overhead. If your home has no basement, use the center of the house as cover. An interior hallway or small room such as a closet or bathroom provide the best shelter. Stay away from windows.

Do not stay in a mobile home if a tornado is imminent.

At school, work, or other such places, a shelter may be available. If not, it is usually best to take shelter in an interior hallway in the center of the building on the lowest floor. Large rooms such as auditoriums and gymnasiums are usually unsafe and should be especially avoided.

If you are outside on foot or in your car, try to travel in a path perpendicular to the

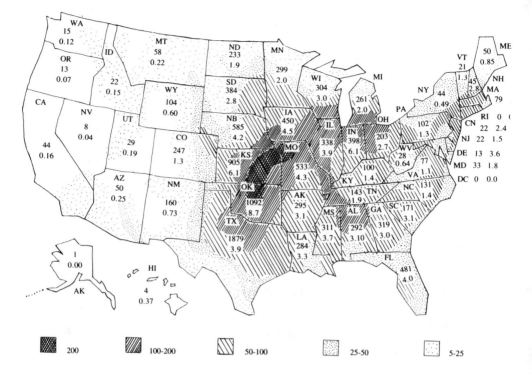

Figure 1.2   Incidence of Tornadoes (1953-1970)

The upper figure in each state is the total number of tornadoes and the lower figure is the mean annual tornadoes per 10,000 square mile zone of highest incidence. Shaded areas show total windstorms of 57.5 miles per hour or greater during the years 1955-1967 (National Weather Service).

path of the tornado. If you need to take cover, lie flat in a ditch, ravine, culvert, or other depression, against the side closest to the oncoming tornado.

During

HANG ON TO YOUR HAT!

After

Exercise great caution in entering damaged buildings which may have broken utility lines or may be unsound and collapse easily. Never use open flames where there may be leakage of natural gas or other flammable materials. Be especially careful of downed electrical wire.

Report damage to responsible parties.

Be careful using electric appliances that have become wet. Dry them out, plug them in to a circuit that has been turned off, and then turn the circuit back on. Then, if there is trouble, you won't be touching it.

Avoid disaster areas.

Continue to monitor the news media for advice on assistance.

NOTE: a good fallout shelter makes a good storm cellar, and vice versa. Be sure to take some tools with you to help clear debris from shelter entrance if necessary. Two entrances are desirable.

VOLCANOES

Volcanic eruptions have killed tens of thousands in a single eruption. The recent eruption of Mount St. Helen's is still vivid in the minds of many in the northwestern part of the United States and brings to mind what could have happened if it had taken place in a more populated area.

Nearly always there is a significant warning; sometimes there is not. Volcanic activity could accompany a large-scale earthquake and possibly even nuclear war.

Volcanoes can vent gases, heat ground water to form hot springs and geysers, flow lava and pyroclastic material, and spew ash and lava into the air. Airborne ash from volcanic activity has smothered countrysides and altered climatic conditions. There are still a number of active volcanic sites in the world.

Should you be in the area of a volcanic crisis, here are a few ideas:

Escape from the immediate devastation of an eruptive volcano would have to be immediate. Keeping to high ground is a general safety rule. Flows, as well as some toxic gases, follow lower ground. Avoid breathing ash as much as possible. A damp cloth or dust mask can help a great deal. Avoid getting the ash in your eyes or on the skin. Swimming masks or goggles could be used for this purpose. Wash off and remove contaminated clothing as soon as possible; some volcanic dust may be acidic and very irritating when dampened.

Should you actually be caught in a situation where you are in danger of being overrun by a lava flow, try to find a large rock or other object to raise yourself above the flow; or, as a final effort, attempt to find a floating object to "ride" until rescue comes. (Surfing on a lava flow may not be very recreational, but it certainly beats swimming in it!)

SELECTED REFERENCES AND NOTES

Blake, Reed H., John R. Christiansen, and Ralph L. Garrett. *Disaster Preparedness: A Family Protection Handbook.* Bountiful, Utah: Horizon Publishers and Distributors, Inc., 1984.

In Time of Emergency: A Citizen's Handbook. Washington, D.C.: Federal Emergency Management Agency, 1983. Usually available from local civil defense offices.

Thygerson, Alton L. *Disaster Survival Handbook.* Provo, Utah: Brigham Young University Press, 1979.

CHAPTER 2

Biological and Chemical Warfare

Biological and chemical (BC) warfare agents and means of delivering them have been created by the United States, United Kingdom, U.S.S.R., Japan, Germany, and many other countries. There is a distinct possibility that if war developed between the superpowers some of these BC (or CB) weapons would be used in the exchange. (This is easy to believe in view of the fact that chemical agents have been used in some conflicts in some world areas in the past few years).

Because there is a possibility of encountering some of these weapons, a brief and basic discussion of some BC agents follows:

The most important consideration in defending against BC agents is to limit or prevent intimate exposure. This can be accomplished by means of protective clothing, including gas masks, and by avoiding contaminated food and drink.

In most cases, treatment of those who are infected by a BC agent consists largely of decontamination and good care, including relief of symptoms. In those cases where more specific care is needed, the help of a medical professional should be obtained. Many cases of exposure to BC agents will not be fatal and a casualty will probably recover with the proper care and time.

Test kits and procedures have been developed which can quickly identify some chemical agents. The process of identifying biological agents is not as simple since most of them involve incubation periods for organisms to grow or produce their toxins.

BIOLOGICAL AGENTS

Biological agents chosen for warfare are pathogenic (able to produce disease), virulent (able to infect a high percentage of those exposed), and able to withstand storage and delivery (including environmental factors). Much work has been done to genetically and environmentally alter some organisms used as agents to make them resistant to traditional methods of control.

All available measures for preventing contamination by a biological warfare agent should be used, including masks, gloves, and other protective clothing.

After contact with a biological agent, rigorous hygiene should be initiated and maintained; clothing, all body parts, and objects should be thoroughly cleaned with a good germicidal cleanser. Strict measures must be adopted regarding food and water for both human and animal consumption.

Identification of the organism or toxin to which exposure has occurred is of paramount importance in dealing with it.

11

CHEMICAL AGENTS

Chemical agents include:
A.   Vessicants (blister agents)
B.   Lacrimators (irritant agents)
C.   Lung Irritants (choking agents)
D.   Sternutators (vomiting agents)
E.   Nerve Gases (anticholinesterase agents)
F.   Incapacitating Agents
G.   Psychochemical Agents
H.   Systemic Poisons (blood agents)

Vessicants

Vessicants (blister agents) are agents such as the mustards (H, HD, HN). They cause intense irritation of the eyes, skin, lungs, etc., and some can cause damage to the lymph and hematopoietic (blood-forming) organs. The onset of pneumonia is a common side effect.

Areas of contamination must be washed immediately—especially eyes—for several minutes. Remove contaminated clothing, blot excess agent from the skin. Inhalation is treated by immediate removal from the area and positive pressure oxygen.

Lacrimators

Lacrimators (tear gas or irritant agents) are agents that cause severe irritation to the eyes and skin, such as chloroacetophenone (CN) and o-chlorobensalmalononitrile (CS). These agents are frequently used for crowd control; they are very potent but not as toxic as some other agents.

Immediate removal from exposure is a logical first step when these fast-acting agents are encountered. Washing eyes out with a very mild solution of sodium bicarbonate is beneficial. Skin burns can be treated by normal first-aid procedures used to treat thermal burns.

Lung Irritants

Lung Irritants (choking agents) are agents such as phosgene (CG) which cause severe irritation to eyes and respiratory system. Chest pains, difficulty in breathing, lowering of blood pressure, palpitations, and coughing up frothy sputum are some of the symptoms. When all symptoms are present, death is imminent. Phosgene has been used in former wars but has been largely replaced in chemical arsenals by nerve agents.

Removal from contact is imperative. The use of a gas mask or oxygen can help accomplish this. The victim is usually treated for shock and administered oxygen.

Sternutators

Sternutators (vomiting agents) are usually arsine compounds (DA, DC, DM) which are used as a smoke and are very irritating. Nose and throat irritations, headache, nausea, and depression characterize the symptoms produced. These agents could be used for civil disturbances.

Victims should be removed from exposure. Irrigation of nose and throat with a mild sodium bicarbonate solution is helpful. The victim should rest; a tranquilizer may help reduce the delayed effects of depression.

Nerve Agents

Nerve Agents (anticholinesterase agents) are the big guns of the chemical warfare stockpile. There has been a move to use binary systems that use two nonlethal entities that, when mixed, produce the deadly agent. This promotes safety in handling and storage. Some of the agents on the useable list are: Sarin (GB), Soman (GO), Tabun (GA), and VX.

Nerve agents absorb readily through body surfaces and can produce respiratory difficulty including heavy secretions and bronchial spasms, central nervous disorders including convulsions, gastrointestinal upsets including vomiting and diarrhea, constricted pupils of the eye, unconsciousness, and death.

Aid to a victim of nerve gas must be prompt. Positive pressure oxygen and 2 mg. injection of atropine sulfate repeated in seven to eight minutes if necessary (usually not to exceed three doses), are effective measures. Artificial respiration is also required in severe cases. A somewhat more complex treatment involves the administration of oximes; this requires a well-trained medic.

All body surfaces that have contacted nerve agents must be washed thoroughly with mildly soapy water. Contaminated hair should be cut.

Incapacitating Agents

Incapacitating agents are those which render the victim harmless (usually unconscious or temporarily paralyzed). There is a variety of these that have been developed. One notable example is Agent BZ, which can render a victim incapacitated for up to a day or more. Treatments for these vary according to the agent used.

Psychological Agents

Psychological (or psychochemical) agents are those such as LSD (lysergic acid diethytamide) which causes abnormal mental and behavioral changes such as hallucinations, delusions, and dizziness. Usually the effects gradually dissipate after a few hours. After-disturbances occur anytime after in some cases.

A victim should be kept from excessive sensory experience and from injuring himself or others until the effects of the agent wear off. A tranquilizer may be helpful to the victim.

Systemic Poisons

Systemic poisons (blood agents) such as hydrocyanic acid (AC) and cyanogen chloride (CK) are not likely but possible to be used in warfare. (Hydrocyanic acid was used in World War I). They can cause nausea, convulsions, and severe irritation to eyes and respiratory system and can lead to respiratory failure and death.

Victims can be treated by amyl nitrate (one amyl nitrate ampule crushed and held under the victim's nose every 20 to 30 seconds up to a maximum of eight, or two crushed ampules inserted into the victim's gas mask every four minutes to a maximum of eight, and in some cases by treating for shock.

SELECTED REFERENCES

Sibley, C. Bruce. *Surviving Doomsday*. London: Shaw and Sons, Ltd., 1977.

U.S. Army Special Forces Medical Handbook. Boulder, Colorado: Paladin Press, 1982.

Harris, Robert, and Jeremy Paxman. *A Higher Form of Killing: The Secret Story of Gas and Germ Warfare.* London: Chatto and Windus, 1982.

Brow, Frederic J. *Chemical Warfare: A Study of Restraints.* Westport, Connecticut: Greenwood Press, 1968.

CHAPTER 3

Nuclear War - The Bomb

One survival situation that traverses the consciousness of the rational and irrational alike (with varying velocities) is the possibility of nuclear war.

It is important to understand that this horrifying and unthinkable event could be survived by some measure of humanity with a reasonable degree of health, sanity, and vigor. Even pessimistic "experts"predict that millions of Americans would survive a massive nuclear attack. But it is important to survival to know some of the facts on the subject.

Since, thankfully, the world's experience with nuclear war is limited, it is difficult to assess in detail what the effect of a large-scale nuclear conflict or a smaller, "blackmail" encounter would be. A great amount of study has been done, however, and many tests have been subjected to scientific scrutiny. These studies have produced a fairly accurate body of information about nuclear explosions. Many knowledgable people have attempted to publish and clarify these facts. Others have broadcast misleading arrangements of the facts. The misleading information may be well purposed and would perhaps be harmless, except that it could conceivably lead to either false preparation or lack of any preparation at all, and a nuclear incident is still a very real possibility.

The United States has made substantial efforts in weaponry to deter our antagonists and assure lethal retaliatory capability, but our current civil defense effort to insure individual protection has become little more than enough to perpetuate its own bureaucracy. Yet, civil defense is, perhaps arguably the best structure from which to approach natural disasters as well as war or warlike attack by some antagonist. The Soviets, the Swiss, and others outdo us many times over in civil defense expenditure and effectiveness. (Of course it is hard to spend money on things that *may* never be needed—like insurance policies, fire extinguishers, and civil defense.)

Our shelters are not adequately built, adequately stocked, or adequately ventilated; nor are they adequately abundant to protect us in a nuclear war. The crisis relocation program (CRP) that sends urbanites to rural America in nuclear emergencies is filled with inadequacies, and leaves many questions unsatisfactorily answered. Among the more glaring of these are: When do we decide to clear out? Who goes when we do decide, and who stays to care for necessities? Who polices the exodus (directs traffic, handles accidents, etc.)? Who convinces the country-folk to welcome and care for the many visitors? What about the bedridden and the infirm when those who were caring for them are gone? What is to be eaten? What is to be drunk? Who pays? Who supplies the gas to get where you're going? Whose field or yard do we dig up to make a fallout shelter in the event an attack does occur? If an attack occurs, what do we eat next month—or next year? If an attack doesn't occur, when do we go back, and when we get there how do we make the house payment, since we haven't worked for a couple of weeks or longer?

Of course, there are civil defense manuals printed that cover almost every one of the above questions, and they can help. But having an answer doesn't necessarily produce the results—and how many have ever seen the manuals or even know that they exist? The CRP could be useful, but in the event of war, decisions would have to be made by all individuals as to their own courses of action. Preparations should be made, and one of the first is to know the CRP plans for your area. (Contact your local civil defense office.)

It is hoped that the following information will aid in making a decision on what preparations to make and how to make some of them. The subject of nuclear war is here divided into two parts: *The Bomb* (Chapter 3), and *Protection From the Bomb* (Chapter 4). Some of the issues discussed in one chapter are discussed in greater detail in the other. If you don't want to plow through all the details, the most important part to read is "Protective Shelters" in "Protection From the Bomb". Read at least this much! (But don't feel that it is necessary to read every word of this chapter). The checklist at the end of chapter 4 may also be helpful.

The awesome power of nuclear weapons can confuse some of the individual issues associated with them. To help eliminate the confusion, the issues chosen here for explanation are titled: *The Nature of The Beast, Who Are They After?, Packages They Come In, Methods of Detonation, The Demons to be Dealt With,* and *Atmospheric and Other Indirect and Long-Term Effects of Nuclear Weapons.*

There are many "figures and facts" on nuclear war floating around that are not footnoted and do not agree with each other. Most of the tables and calculations put forth here are patterned after or are calculated in a manner similar to those in *The Effects of Nuclear Weapons* (Glasstone, Samuel, ed. *The Effects of Nuclear Weapons.* Washington, D.C.: U.S. Department of Defense, 1962, and Samuel Glasstone and P.J. Dolan, *The Effects of Nuclear Weapons.* Washington, D.C.: U.S. Department of Defense, 1977. Both are available from the Superintendent of Documents, Washington, D.C.).

THE NATURE OF THE BEAST

An understanding of what something is and how it works is always helpful in knowing how to deal with it. Following is a brief review of the workings of a nuclear bomb, along with some definitions that will be helpful in the remainder of the discussions on nuclear war.

The Chemistry of Nuclear Weapons

An atom consists of a central *nucleus* which contains most of the mass of the atom, and electrons which surround the nucleus in various configurations and are very light compared to the nucleus. The two basic kinds of particles found in the nucleus are *protons* and *neutrons*. Protons and neutrons have approximately the same mass. A proton is electrically charged (+), as is an electron (−). A neutron is electrically neutral. An atom of a given element contains a definite number of protons and, in an electrically neutral state, the same number of electrons. This gives the atom an electrical charge balance. The number of neutrons present in the nucleus can vary. When an element has the required number of protons but exists with different numbers of neutrons, the differing species are called nuclear *isotopes* of the element. Most elements occur in nature in two or more isotopic forms and many other isotopic forms of various elements have been made by man.

Many of the isotopes are unstable and are subject to *decay*, or breakdown of the isotopic form by emission of particles or electromagnetic energy or both. These emissions are called nuclear radiation and the isotopes producing them are said to be *radioactive*.

A nuclear bomb derives its energy through interaction between the nuclei of atoms and produces many, many times (literally millions of times) more energy per pound than does a conventional bomb. A conventional bomb derives its energy from interaction between atoms or groups of atoms called *molecules*. Two types of interactions between the nuclei of atoms that produce the tremendous amounts of energy required for bombs are *fission* and *fusion*. Pound for pound, fusion weapons produce more (about three times more) energy than fission.

The materials used for fission explosions are uranium and plutonium. Uranium is among the heaviest naturally occurring elements. Plutonium is artificially produced.

Fission occurs when a speeding neutron strikes the nucleus of another atom and the nucleus breaks apart. This process produces enormous quantities of energy and the fragments of the "broken-apart" nucleus become the nuclei of other elements, many of which are radioactive. This is one of the sources of radio-active material from a nuclear bomb and the primary source of radioactivity in fallout.

Fusion occurs under conditions that cause two nuclei to fuse together. This process also produces vast quantities of energy and high-energy particles, of which the most notable are neutrons.

The materials used in fusion are at the lightest end of the periodic table of elements rather than the heaviest, as with fission. An isotope of hydrogen is commonly used. Hydrogen normally has one proton for a nucleus with no accompanying neutrons. Deuterium is an isotope of hydrogen which has a nucleus of one proton and one neutron, and the hydrogen isotope tritium has a nucleus of two neutrons and one proton. When two deuterium atoms are fused together they create the nucleus of a helium atom. The use of isotopic hydrogen as a fusion material in bombs led to the term *hydrogen bomb*.

Figure 3.1   Fission and Fusion

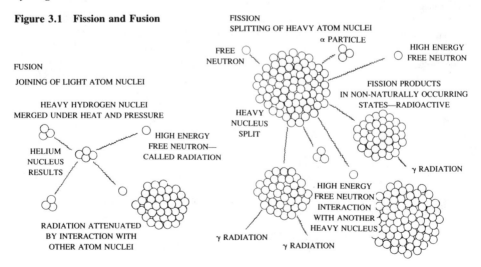

The condition used in nuclear weapons under which fusion occurs is extremely high heat. The source of the heat in these weapons is a fission explosion. The term *thermonuclear bomb* is thus derived from the process of high heat (thermo) producing a nuclear reaction. The fission nuclear reaction causes the additional fusion nuclear reaction. (Either fision or fusion may be properly called "nuclear"). This fusion process liberates highly energized neutrons which can then cause additional fission if other fissionable (fissile) material is also present in the bomb. This additional fissionable material is usually present as an encasing layer covering the bomb.

This, in simple terms, is the process involved in modern nuclear weapons. A central fission explosion causes a peripheral fusion reaction in fusible material, which causes additional fission in an encasing layer of fissionable materials. The bombs dropped in Japan at the end of World War II were fission bombs only.

The Neutron Bomb

A neutron bomb, then, is not the exotic freak that it may have been labeled, but rather a thermonuclear (fission-fusion) bomb which has been arranged in such a way that it has a lesser amount of fissionable material than would be present in a "traditional" hydrogen bomb (there is no encasing layer of fissionable material). The "neutron" weapon still produces a blast and heat, but—and this is the important point—it is able to kill by way of the "shower" of high-energy neutrons without also producing the larger quantitites of radioactive fission products and the total destruction necessarily present in a similar killing zone for a traditional thermonuclear device. For example, neutron weapons could possibly counter an offensive on a sizeable and populous area such as Europe and still leave most of it standing. A conflict in such an area with traditional nuclear weapons could largely destroy it. Neutron weapons are small and hence have limited killing range. One of the most forceful arguments against such weapons is that they become all too usable (especially on a small scale) and subsequently a great temptation for use by aggressive and ambitious leaders.

Measuring Explosive Power

As a means of describing the amount of energy released by a nuclear weapon, it is compared to TNT, a well-known non-nuclear explosive. Due to a large difference in yields of energy between the TNT and a nuclear weapon the terms kiloton (1000 tons) and megaton (1 million tons) are used, so that a one-kiloton (1 KT) nuclear weapon can produce the same amount of energy as one kiloton (1,000 tons) of TNT. A one-megaton (1 MT) explosion would be equivalent to 1 million tons of TNT. The nuclear weapons detonated in Japan in World War II were reportedly less than 20 kilotons each.

PACKAGES THEY COME IN

Many sizes of nuclear weapons exist, as well as many methods of delivering them. It is unlikely that the smaller tactical nuclear weapons would be used by an antagonistic power to invade North America unless there had been a previous exchange of large strategic weapons (although some have proposed this as a possible "blackmail"-type invasion). For this reason the discussion is limited here to strategic weapons and foreseeable problems involved in a large-scale nuclear exchange. These large weapons can be launched from mobile and stationary land-based launchers, submarines, ships, and airplanes.

Weapon Sizes

There is a limit at which the size of the weapon and its capability for damage do not justify the cost and the problems involved in its construction and delivery system. It could be thought that one big bomb could do it all. Conceivably it could, but it would be outrageously impractical if not impossible to build and deliver.

Besides this, the radius of damage of a nuclear weapon does not directly enlarge with its size. That is, a bomb delivers its power in a spherical pattern, and hence the damage done by the blast of a ten-megaton bomb is roughly twice that of a one-megaton bomb, not ten times. A twenty-megaton bomb would have a damage radius only slightly larger than a ten-megaton weapon, and so on. Although there may have been some few larger weapons made, the largest of the standard Soviet delivery systems carries approximately a twenty-to-twenty-five megaton warhead. A large-scale weapons exchange would probably consist of a few twenty-to-twenty-five megaton, ten-megaton, and five-megaton weapons and many more one-megaton and smaller weapons coming this way. U.S. missile systems mostly carry smaller submegaton warheads and are (reportedly) very accurate. This kind of missile force is designed to produce maximum damage per ton of explosive at multiple target sites.

Multiple Warheads

One of the innovations that broadens a missile's capability to destroy is the loading of several small warheads on one missile in place of a single large one. Each one can then be independently targeted from very high altitude on reentry into the atmosphere. These are called multiple independently targetable reentry vehicles (MIRVs). More blast damage can be initiated from several well-placed smaller warheads than from one larger warhead equal in TNT equivalent to the sum of the smaller ones. In addition to damaging a wider area, they are also more difficult to intercept. For most industrial,

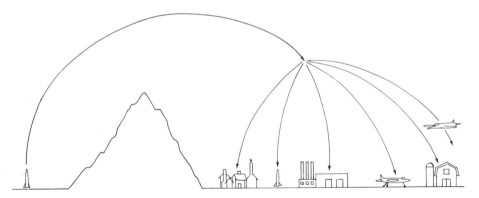

Figure 3.2   MIRV's

Multiple independently Targetable Re-entry Vehicles (MIRV's) are missiles containing more than one warhead. After the missile has reached the general location of the target from very high altitude and then starts down, the warheads separate and travel toward their separate destinations.

residential, and even many military installations, MIRVed missiles are considerably more effective than single-warhead missiles. Underground missile silos, control centers, and other "very hard" targets are more effectively destroyed by large warheads which detonate at approximately ground level. Much of the current Soviet strategic missile force is equipped with MIRVs.

The Soviet Arsenal

The Soviet arsenal is still largely in tact and is composed of bombers capable of carrying free-fall bombs and air-to-surface missiles; mobile and stationary bomb-laden, continent-hopping intercontinental ballistic missiles (ICBMs); submarines with submarine-launched ballistic missiles (SLBMs) which can be launched from under water; and surface ships with long-range and low-altitude, highly accurate ground-hugging cruise missiles. For comparison, figures 3.3 through 3.7 give some of the specifics. These diagrams are taken from "Soviet Military Power" (prepared by the United States Department of Defense—see references at the end of Chapter 4) and compare some U.S., Soviet, and NATO weapons and deployments. Current Soviet weapons are largely fission weapons which are "dirtier" in that they produce dangerous fallout, but they are cheaper and easier to make.

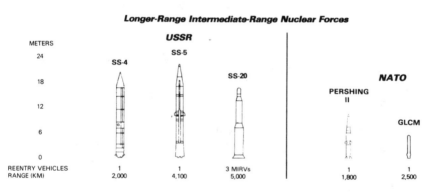

Longer-Range Intermediate-Range Nuclear Forces

USSR

METERS				NATO	
REENTRY VEHICLES	1	1	3 MIRVs	1	1
RANGE (KM)	2,000	4,100	5,000	1,800	2,500

Total deployed on 1 March 1983: 581

Total deployed on 1 March 1983: None

Missile Production
USSR and NATO

Missile Type	1978[1]	1979[1]	1980[1]	1981	1982	1981 NATO[2]
ICBMs	225	225	250	200	175	0
LRINF	100	100	100	100	100	0
SRBMs	250	300	300	300	300	0
SLCMs	600	700	750	750	800	700
SLBMs	250	200	200	175	175	90
SAMs	53,000	53,000	53,000	53,000	53,000	6,900

[1] Revised to reflect current information.
[2] Includes that produced by the United States; excludes France.

Figure 3.3　NATO and Soviet Intermediate Range Missiles

(U.S. Department of Defense)

USSR Fourth Generation ICBMs

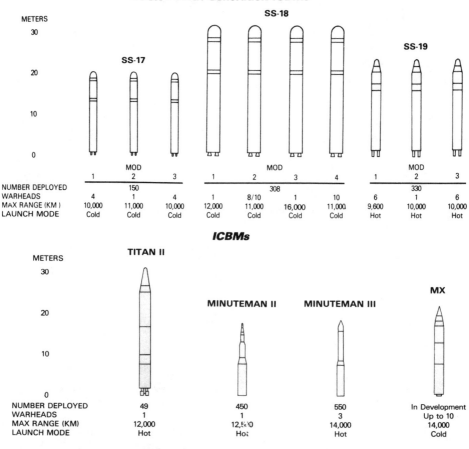

	MOD 1	MOD 2	MOD 3	MOD 1	MOD 2	MOD 3	MOD 4	MOD 1	MOD 2	MOD 3
NUMBER DEPLOYED		150			30	8			330	
WARHEADS	4	1	4	1	8/10	1	10	6	1	6
MAX RANGE (KM)	10,000	11,000	10,000	12,000	11,000	16,000	11,000	9,600	10,000	10,000
LAUNCH MODE	Cold	Cold	Cold	Cold	Cold	Cold	Cold	Hot	Hot	Hot

ICBMs

	TITAN II	MINUTEMAN II	MINUTEMAN III	MX
NUMBER DEPLOYED	49	450	550	In Development
WARHEADS	1	1	3	Up to 10
MAX RANGE (KM)	12,000	12,500	14,000	14,000
LAUNCH MODE	Hot	Hot	Hot	Cold

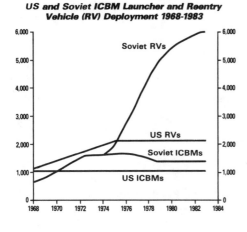

US and Soviet ICBM Launcher and Reentry Vehicle (RV) Deployment 1968-1983

Soviet RVs
US RVs
Soviet ICBMs
US ICBMs

Figure 3.4   U.S. and Soviet Land Based Strategic Missiles

(U.S. Department of Defense)

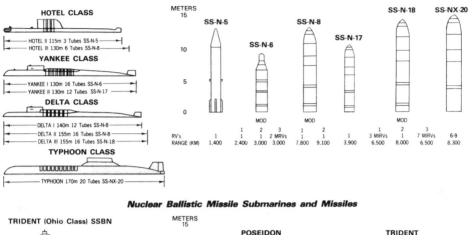

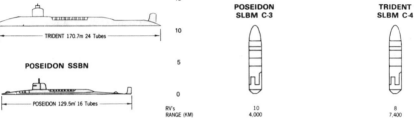

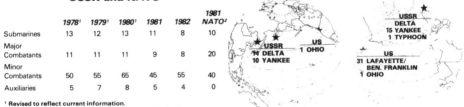

Naval Ship Construction USSR and NATO

	1978[1]	1979[1]	1980[1]	1981	1982	1981 NATO[2]
Submarines	13	12	13	11	8	10
Major Combatants	11	11	11	9	8	20
Minor Combatants	50	55	65	45	55	40
Auxiliaries	5	7	8	5	4	0

[1] Revised to reflect current information.
[2] Includes that produced by the United States; excludes France.

Figure 3.5   Comparison of Some Soviet and U.S. Sea-Going Forces

(U.S. Department of Defense)

Figure 3.6   Some U.S. and U.S.S.R. Interceptor Aircraft and Missiles

(U.S. Department of Defense)

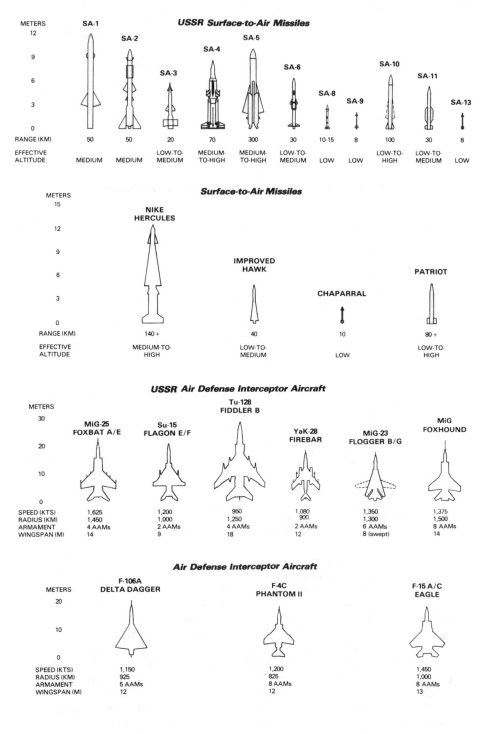

USSR Surface-to-Air Missiles

	SA-1	SA-2	SA-3	SA-4	SA-5	SA-6	SA-8	SA-9	SA-10	SA-11	SA-13
RANGE (KM)	50	50	20	70	300	30	10-15	8	100	30	8
EFFECTIVE ALTITUDE	MEDIUM	MEDIUM	LOW-TO-MEDIUM	MEDIUM-TO-HIGH	MEDIUM-TO-HIGH	LOW-TO-MEDIUM	LOW	LOW	LOW-TO-HIGH	LOW-TO-MEDIUM	LOW

Surface-to-Air Missiles

	NIKE HERCULES	IMPROVED HAWK	CHAPARRAL	PATRIOT
RANGE (KM)	140 +	40	10	80 +
EFFECTIVE ALTITUDE	MEDIUM-TO-HIGH	LOW-TO-MEDIUM	LOW	LOW-TO-HIGH

USSR Air Defense Interceptor Aircraft

	MiG-25 FOXBAT A/E	Su-15 FLAGON E/F	Tu-128 FIDDLER B	YaK-28 FIREBAR	MiG-23 FLOGGER B/G	MiG FOXHOUND
SPEED (KTS)	1,625	1,200	950	1,080	1,350	1,375
RADIUS (KM)	1,450	1,000	1,250	900	1,300	1,500
ARMAMENT	4 AAMs	2 AAMs	4 AAMs	2 AAMs	6 AAMs	8 AAMs
WINGSPAN (M)	14	9	18	12	8 (swept)	14

Air Defense Interceptor Aircraft

	F-106A DELTA DAGGER	F-4C PHANTOM II	F-15 A/C EAGLE
SPEED (KTS)	1,150	1,200	1,450
RADIUS (KM)	925	825	1,000
ARMAMENT	5 AAMs	8 AAMs	8 AAMs
WINGSPAN (M)	12	12	13

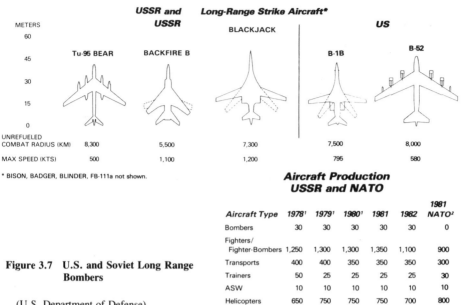

USSR and **Long-Range Strike Aircraft***

Figure 3.7   **U.S. and Soviet Long Range Bombers**

(U.S. Department of Defense)

* BISON, BADGER, BLINDER, FB-111a not shown.

Aircraft Production USSR and NATO

Aircraft Type	1978[1]	1979[1]	1980[1]	1981	1982	1981 NATO[2]
Bombers	30	30	30	30	30	0
Fighters/ Fighter-Bombers	1,250	1,300	1,300	1,350	1,100	900
Transports	400	400	350	350	350	300
Trainers	50	25	25	25	25	30
ASW	10	10	10	10	10	10
Helicopters	650	750	750	750	700	800
Utility	100	100	50	25	25	100
Total	2,490	2,615	2,515	2,540	2,240	2,140

[1]Revised to reflect current information.
[2] Includes that produced by the United States; excludes France.

WHO ARE THEY AFTER?

In order to get a feel for what we are up against at a given location in the event of nuclear war, it is sensible to consider the likely targets. Although it would be difficult to exactly predict where every bomb would land in such a conflict, it is practical to make some judgments according to the best information available and prepare accordingly.

According to reports published by government agencies and others (see the selected bibliography and notes, end of Chapter 4), the targets of greatest interest to an antagonist in an initial first strike a decade ago would likely be missile silos, air bases, other runways 10,000 feet or longer, and missile submarines and their bases.

Other military bases, seaports, and large airfields would be next on their list, followed by electrical generating facilities and heavy industrial areas. Population centers would have been targeted to some degree depending on changing strategies. However, the situation today is somewhat different. With the break up of the Soviet Union, the control of many weapons has become less centralized. Further, there is danger of some of these weapons falling into the hands of ambitious and dangerous people. This creates a whole new series of potential dangers. Blackmail, revenge, or a deranged "holy war" become all too possible. And the targets in these instances could likely be population centers.

With the high degree of accuracy which has now been achieved in missile technology by the United States, Russia, and others, it is conceivable that the majority of the missiles fired in a war would land reasonably close to their targets.

Fallout, on the other hand, would endanger communities large and small, especially those nearby and downwind from the hard targets hit during a first strike.

This information may assist in deciding what preparations are likely to be practical and most helpful in the event preparations are in fact needed.

METHODS OF DETONATION

A bomb conveyed by standard delivery systems may be exploded at or slightly below ground level, at low altitudes above ground, or at very high altitudes. Different targets require different methods of exloding a weapon to maximize damage to them, and weapons are set to detonate accordingly.

As mentioned, an underground silo containing a missile may best be destroyed by the weapon being detonated at the surface of the earth at or near the silo. On the other hand, a low-altitude (a few thousand feet above ground) airburst maximizes damage to most industrial and residential areas, and even most military installations, and is therefore a likely choice at such targets. A high-altitude (air) burst (100,000 feet or more) causes little ground damage but can cause severe damage to electrical devices, especially radio equipment (see EMP).

Ground-level detonations cause less initial damage to most civilized areas than low-altitude airbursts but cause much larger amounts of radioactive residues. A maximized airburst can cause substantial blast damage to an area roughly twice the size of that caused by the same size groundburst.

Lower level air bursts display more dangerous initial radiation and blast damage than ground-level detonations but produce little residual radioactive material (fallout) that reaches the ground while still dangerous if the fireball of the explosion remains largely above the ground. Since the greatest damage is done to cities by this type of detonation, it is reasonable to assume that an antagonist would be likely to use this method on urban areas with the result that many cities may not be contaminated with a large amount of fallout, except as a result of proximity to other "hard" targets. For example, the atomic bombs dropped on Japan in World War II were airburst and produced essentially no radioactive fallout. (There was, however, substantial initial nuclear radiation.)

THE DEMONS TO BE DEALT WITH

One of the first things to understand about a nuclear explosion is that an explosion by any other name is still an explosion. Shock waves in the form of high pressure and high winds from a large blast cause severe damage.

The other effects of a nuclear blast are: intense light and heat (thermal radiation), initial and residual nuclear radiation, and electromagnetic pulse (EMP). The size, shape, point of detonation, distance from it, terrain, weather, and prevailing winds all dictate what the effects will be on a given locality. In addition to these direct effects, there are other indirect effects, such as climatic and atmospheric changes, that would necessarily attend the detonation of a large number of nuclear weapons.

Earlier estimates of the total likely explosive power of an initial nuclear exchange between the United States and the Soviet Union were about 10,000 megatons, and many calculations and predictions have been made on the basis of this volume of explosive power. Current information points to a significantly smaller initial exchange.

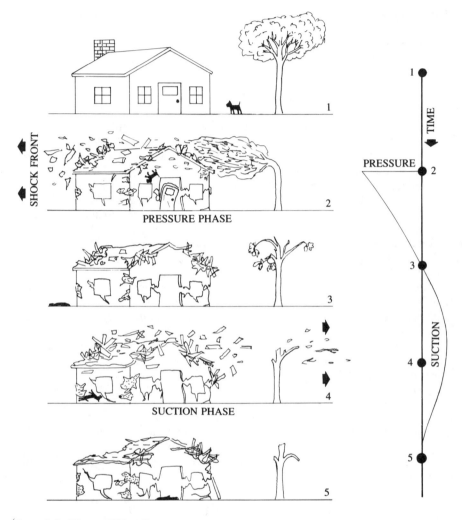

Figure 3.8   Phases of Blast Pressures

The blast pressure varies with time as the wave passes.

Having A Blast

The blast of a large nuclear weapon causes a powerful shockwave which travels at a high rate of speed in all directions from the blast. The wave, as normally described, has two destructive components. The extremely rapid expansion that occurs in a blast causes a compression and a movement of the air. The compression is described in pounds per square inch (psi) over normal and is called *overpressure*. Hence, a one psi overpressure may be thought of as being one pound per square inch above the normal atmospheric pressure.

The other component of the blast wave is the *dynamic pressure* and is manifest largely as a wind initially moving away from the center of the explosion. This wind, or dynamic pressure, would be in the neighborhood of one-hundred miles per hour at a distance of six miles from an airburst one-megaton blast—enough to "blow the man down" if he were standing upright.

The overpressure exerts collective force on surfaces and thus in higher magnitude tends to crush hollow objects such as cars and buildings. An overpressure of 0.5 psi, which can break windows and cause other damage in residential areas, could occur at a distance of twenty miles from a one-megaton blast. An overpressure of about 4.5 psi is enough to flatten the average single-family dwelling. This overpressure can occur at a distance of approximately two miles from a twenty-kiloton detonation, five miles from a one-megaton, or ten miles from a ten-megaton blast.

The above approximate distances are given for a weapon burst in the air at an altitude that would produce the maximum damage, henceforth called *maximized air burst*.

The combination of overpressure and high winds near a nuclear blast causes extreme damage. The speed at which the destructive wave front travels and its duration vary with the size of the weapon, distances, and, to a slight degree, some other factors. Normally a person in an unprotected area, but at a distance where other factors would not be lethal, could survive the effects of the blast wave by lying on the ground. (For a more complete idea of damage effects and time of arrival after detonation, see tables in "Protective Shelters" in Chapter 4).

The direct effect of overpressure alone on an unprotected human is estimated to cause a one percent fatality rate at an overpressure of 35 to 45 psi, and a 99 percent fatality rate at 55 to 65 psi (Glastone, 1962, p. 537). However, the winds of the dynamic pressure would be over 1,000 miles per hour at 35 psi from a maximized airburst—certain death to the unprotected. Indirect effects such as flying debris and collapsing buildings would also constitute a great hazard.

After the shock wave passes a given area, the pressure continues falling until an underpressure develops and a reverse wind begins blowing toward the explosion. This also may cause damage, but the power of the reverse or negative phase of the blast wave is much lower than the positive phase going away from the blast.

The blast wave from a nuclear explosion generally causes damage at lower pressures than conventional chemical explosions because they are maintained for longer periods of time.

There are phenomena that bring a bit of unpredictability to blast damage severity at some locations. Two of these are the refraction (or bending) and reflection of the blast wave. In these cases the refracted or reflected waves meet other wave patterns to give an additive effect. This may be illustrated by imagining yourself in the middle of an intersection in your car. If one car hits you from the left side, that is bad enough; but if another hits you at the same time from the right side, you have double trouble.

There are precedents for this bouncing around of blast waves in the testing of nuclear weapons. Glass has been broken literally hundreds of miles from a test blast where it would definitely not have been expected. In a time of multiple nuclear detonations, it would be difficult indeed to predict where and how much this "bending" and "bouncing" of blast waves would cause damage.

In addition to refraction and reflection, hills and other obstructions can alter the expected effects of blast waves.

Thermal Radiation

Thermal radiation represents a significant portion of the energy expended in a nuclear detonation. (For a one-kiloton explosion this is roughly equivalent to 400,000 kilowatt-hours of power; see Glasstone and Dolan, 1977, p. 277). At the time of detonation the intense flash of the rapidly expanding fireball can start fires miles away, superheat the air causing chemical smog, cause burns, and cause severe damage to the eyes. (Retinal spot burns, which are permanent eye injury but which may not be severely debilitating except in extreme cases have occurred at a distance as great as one or two hundred miles from test detonations. In cases of flashblindness, which is temporary eye injury, vision returns to normal after a few minutes to a few days). The extreme heat of the detonation vaporizes everything nearby.

In a large weapon detonation, as the fireball expands to its maximum proportions it continues its thermal radiation. It may last several seconds. A one megaton bomb takes about nine seconds to radiate eighty percent of its thermal energy, and a twenty-megaton bomb takes considerably longer. (This is a good reason to secure protection immediately when a blast occurs). In a smaller, fission detonation, such as those used over Japan during World War II, the thermal pulse is essentially instantaneous.

Atmospheric conditions can greatly alter the thermal effects of a bomb. Water (clouds) or dust can absorb and scatter much of the heat between the blast and the measuring point. On the other hand, if the blast occurred *beneath* cloud cover, the reflected heat from the clouds could significantly add to the heat at other points below.

Other large surfaces can similarly alter the intensity of the thermal radiation at a given location.

It has been estimated that the temperature directly below the nuclear blasts in Japan in World War II was in the neighborhood of 6,500°F. and nearly 3,000°F. some 4,000 feet away. The detonations in Japan occurred at approximately 1,850 feet above the ground.

Light colors tend to absorb less heat than darker colors and moist items tend to be less damaged than dry items. Bare skin can be severely burned while that covered with even light clothing remains unhurt. Table 3.1 describes some of the possibilities of damage to different objects.

The detonation of many weapons would undoubtedly trigger large-scale fires over wide areas. The poisonous fumes from these fires would probably fill a considerable portion of the Northern Hemisphere with air that would make the EPA's red lights go wild. The smoke would gradually dissipate and in the process would undoubtedly cause a considerable amount of "acid rain."

Nuclear Radiation

Nuclear radiation is produced at the time of the explosion and is harmful as it initially travels from the explosion (initial or prompt radiation) and as it emanates from particles which have become radioactive through the processes of the detonation (residual radiation or fallout). Nuclear radiation consists of particles and electromagnetic energy. The measurably dangerous nuclear particles radiated are neutrons, alpha particles, and beta particles. Neutrons represent by far the most significant danger from particles.

The electromagnetic waves produced in a nuclear explosion are of a wide range of wave lengths initially giving visible light of all colors plus invisible radiations beyond the ends of the visible spectrum. In radioactive decay, many isotopes give off high-energy gamma rays. Gamma rays are the primary concern in residual nuclear radiation.

TABLE 3.1

APPROXIMATE DISTANCES IN MILES FROM GROUND ZERO FOR THERMAL EFFECTS

Property	35 KT	1.4 MT
Ignition of blue cotton denim (10 oz.)	1.9	5.2
Ignition of newspaper (printed area)	2.6	8.0
Ignition of dry grass	2.6	7.5
Ignition of dry leaves	2.5	9.0
Ignition of pine needles (Ponderosa)	1.8	6.5
Ignition of new white bond paper	1.1	4.9
Ignition of cotton-nylon fabric (5 oz., olive color)	1.9	5.0
Ignition of rayon-acetate drapery (wine color)	1.7	5.5
Ignition of black cotton rags	1.8	6.4
Ignition of mineral-surface roll roofing	—	4.8

Personal	1 MT	10 MT
First-degree skin burns	12	25
Second-degree skin burns	11	22
Third-degree skin burns	7	17

Note: In addition to skin burns, retinal eye burns can be experienced for 60 mi. from a low-altitude (10,000 ft.) 1-MT burst and 95 mi. from a higher altitude (50,000 ft.) burst at night and 35 and 55 mi. (respectively) during daylight (clear day). For a 10-MT burst the distances are about 85 mi. for low-altitude and 155 mi. for high-altitude bursts at night and 55 and 95 mi. during daylight. In some circumstances, retinal spot burns can occur at 200 mi. and more.

U.S. Department of Defence and U.S. Department of Energy (Dolan and Glastone)

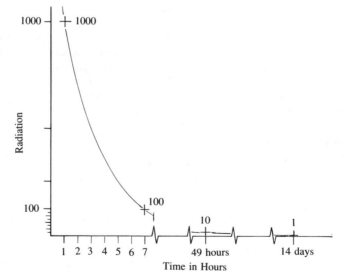

Figure 3.9   Radioactive Decay as Described by the Rule of Seven

Total radioactivity decays quite rapidly and generally follows the graph described above.

Rule of Seven   During a nuclear explosion a complex mixture of radioactive materials is produced. Each radioactive particle thus created decays at a different rate, making it difficult to predict the half-life (the time necessary for one-half of the radiation to dissipate) of the radioactive material as a whole. However, the total decrease in radiation intensity from the fission products of the bomb is generally described by the *rule of seven*. This general rule is that the total radiation intensity will decrease to 1/10 its value for every 7 times the time. For example, if the radiation intensity were called 100 at 1 hour after a detonation then it would be 10 (1/10 of 100) at 7 times 1 hour, or seven hours. At 49 hours (7 times 7 hours, or about 2 days) the radiation would be about 1 (1/10 of 10) and at about two weeks (7 times 2 days) the radiation would further decrease to about 0.1, or 1/1000, of the original value of 100. With multiple detonations at varying times it would become more difficult to calculate the rate of decay, but in simple circumstances the rule is generally reliable for about six months after a detonation, after which the radioactive decay becomes more rapid than predicted by the rule.

Units of Measurement   Units of measurement that are commonly used in describing radiation are roentgen, rad, and rem. All three represent units of energy. Roentgen is a measure of electromagnetic radiation present, a rad is a measure of the radiation energy absorbed by body tissue, and a rem (roentgen equivalent in man) is a measure of the biological effect in man from the energy absorbed. These terms have sometimes been used interchangeably in places where more definitive meanings were unnecessary. The rem is used in most of the descriptions stated here.

Initial Nuclear Radiation   Initial (or prompt) nuclear radiation consisting of neutrons, alpha and beta particles, x-rays, and gamma rays is usually defined as that radiation occurring during the first minute after the onset of a detonation. It is very powerful at close ranges and is reduced by mass—that is, the more mass between you and it, the more of it will be absorbed by the mass before it gets to you. The wrinkle in this story of the dangers of prompt radiation is that in order to be close enough to a *large* nuclear blast to receive a lethal dose of initial radiation it would normally be necessary to be so close that the blast and thermal radiation would be much more of a threat to survival than would the initial nuclear radiation. The exception to this is a weapon such as a neutron bomb, but at present only small neutron bombs are constructed. For this reason, protection against initial nuclear radiation would probably be much less important for survival than protection against blast and thermal radiation in a large-scale strategic nuclear war. In the case of smaller nuclear weapons, however, the relative danger of initial nuclear radiation becomes much greater.

Animals would be affected by initial nuclear radiation in a manner very similar to people. Some plants are very resistant to radiation, but a nuclear blast would be biocidal for some distance.

Induced Radiation   In the immediate vicinity of a ground-level or very low altitude nuclear blast, the earth and other adjacent materials are induced to a radioactive state by the neutrons and other radioactive materials emanating from the blast. As a result, these areas may remain radioactive to a greater extent and for a longer period of time than areas receiving only fallout. This kind of radioactivity is, naturally, called "induced radiation."

TABLE 3.2

APPROXIMATE DISTANCES IN MILES FOR INITIAL RADIATION EXPOSURE

Radiation Dose (in rems)	100 KT	1 MT	10 MT	
100	1.3	1.8	2.4	
500	1.1	1.5	2.1	U.S. DOD, DOE
1,000	1.0	1.4	2.0	Dolan and Glastone

One additional factor that could influence the amount of induced radiation present is the striking of nuclear power facilities by nuclear weapons. This would raise radiation levels in adjacent areas.

Fallout (Residual Nuclear Radiation)   Residual nuclear radiation comes in many varieties. The fission products of a nuclear explosion include more than two-hundred different isotopes of thirty-six different elements—most of which are radioactive. These products are mixed with other materials by the detonation and become part of the bomb residues. Approximately two ounces of fission products per kiloton equivalent of *fission* in the blast are formed. (This equals about 125 pounds per fission-produced megaton).

At the instant of detonation the temperature near the center of the explosion is millions of degrees. This high heat vaporizes the radioactive fission products of the explosion and surrounding materials. As the force of a surface blast pushes or sucks earth and other debris pulverized and/or vaporized by the blast up into the mushroom cloud that is formed, some of the gaseous and molten radioactive material becomes part of it. Much of the resultant material is a glassy or cinder-like substance that does not readily break down or dissolve by natural process. This can limit to a degree the radioactive elements that are "taken up" by plants and animals. As this dust and sand and other radioactive products of the blast fall back to the earth they constitute what is called "fallout". (Though the figures differ a little according to who is giving them and which of several variables are in effect, a one-megaton groundburst can remove a few million cubic meters of earth—and that's a lot of particles!)

IMMEDIATE FALLOUT   Heavy immediate fallout should be expected in those areas adjacent to (and where prevailing winds carry it from) locations where groundbursts would be used to destroy missile silos, control centers, or other such targets. Figure 3.10 shows locations of probable targets as described in "Who Are They After" (this chapter) on a U.S. map. Figure 3.11 shows those targets most likely to receive groundbursts and what may be realistic approximate dangerous fallout patterns., These fallout areas are roughly calculated by taking into account prevailing winds and anticipated explosive activity but are only approximations.

SEASONAL AND WEATHER EFFECTS ON FALLOUT   Figures 3.12 and 3.13 show possible fallout areas as calculated by a government agency if many target areas were hit with surface bursts in spring and fall. These are probably unlikely fallout

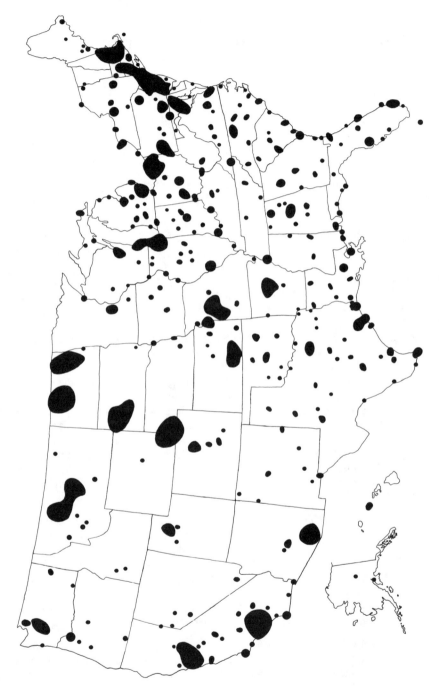

Figure 3.10   Probable U.S. Targets

The dots here indicate high-priority targets and danger areas, including military, industrial and population.

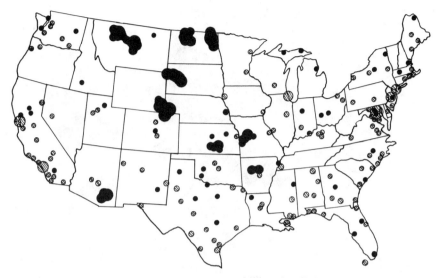

● Missile Silos, Military Bases, and Command Centers   ◎ Military installations and population centers with substantial military value.

Figure 3.11   Military and Other High-Priority Targets

Hard Targets and Likely Areas of Heavy Fallout

This map shows areas that would be very likely to receive groundburst weapons and probable areas of fallout.

Figure 3.12   Possible Fallout Conditions After An Attack On A Spring Day

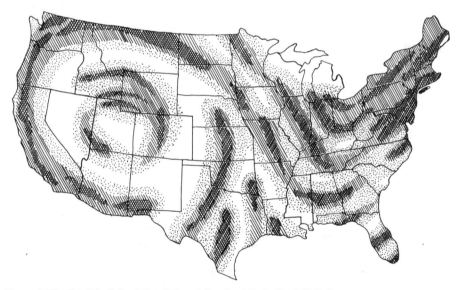

Figure 3.13   Possible Fallout Conditions After An Attack On A Fall Day

Survival Actions

☐   No Shelter required.

▨   Occupy shelter up to 2 days.

▨   Occupy shelter 2 days to 1 week.

■   1 week to 2 weeks of shelter may be needed followed by decontamination of ''hot'' areas.

These fallout patterns may or may not be correct, but they do illustrate the effect of seasons on fallout distribution. (U.S. Department of Defense)

patterns but illustrate the effect of seasonal prevailing winds. Note that the prevailing winds over the United States are, however, west to east. The winds affecting fallout are mostly high-altitude wind streams (see figure 3.14).

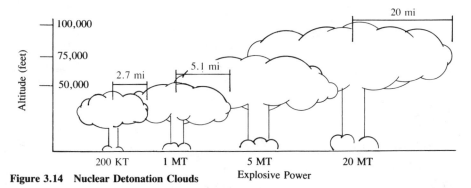

Figure 3.14   Nuclear Detonation Clouds

There are many variables that affect the amount of fallout produced and the area of distribution, however. For example, fallout can be "rained out" of the air by normal precipitation, causing radiation "hot spots."

Current probable military strategies indicate that many targets would be targeted with airbursts, thus reducing the fallout threat. Nevertheless, fallout would still be a major concern.

DELAYED AND LOCAL FALLOUT   The finer particles of radioactive fallout may take days or weeks to reach the earth and may only then be brought down by atmospheric precipitation. Some of it may circle the earth a few times before settling. This is sometimes referred to as "delayed fallout", and it constitutes a much less significant danger than do the larger particles. The larger particles reach the ground rather quickly, and that which falls in approximately the first twenty-four hours may be referred to as "local fallout". In most areas where dangerous amounts of fallout would accumulate after a detonation, the bulk of the accumulation would occur within a few hours. After this time the cumulative radioactive decay would exceed the accumulations of additional fallout (unless additional detonations occurred). In other words, the highest level of dangerous radioactivity would occur, in most areas, within a few hours of the explosion.

OTHER FACTORS   Many factors affect the process of fallout reaching the surface of the earth: makeup (i.e., the type of earth and debris of which the fallout is made), terrain, windspeed, humidity, "layering" of different atmospheric conditions, and so on. For this reason fallout "hot spots" and "cold spots" may occur in some areas where they may not be expected. To give a rough idea of how fallout would travel in an idealized situation without including any of the above-mentioned variations, figures 3.15 and 3.16 are included. They give, respectively, the radiation dose rate at the given times, and the total cumulative dose at the given times for a three-megaton surface burst with a one megaton fission yield. The diagrams are made reflecting a steady wind of fifteen miles per hour blowing at all altitudes in the direction indicated. Higher-altitude prevailing winds, which would have great effect on fallout patterns, are generally much swifter than this. The greater the wind, the wider the area of distribution and the greater the dilution of the fallout except in areas of overlapping

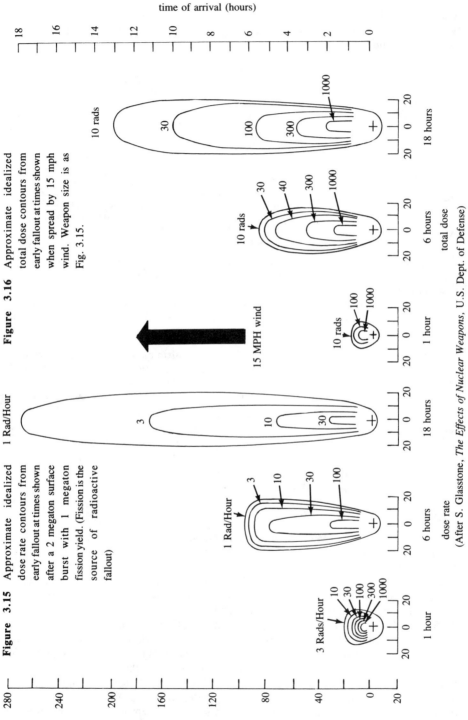

Figure 3.15 Approximate idealized dose rate contours from early fallout at times shown after a 2 megaton surface burst with 1 megaton fission yield. (Fission is the source of radioactive fallout)

Figure 3.16 Approximate idealized total dose contours from early fallout at times shown when spread by 15 mph wind. Weapon size is as Fig. 3.15.

time of arrival (hours)

15 MPH wind

total dose

dose rate

distance from ground zero (miles)

(After S. Glasstone, *The Effects of Nuclear Weapons*, U.S. Dept. of Defense)

fallout from separate blasts. The figures assume idealized conditions. The doses are given in roentgens and roentgens per hour.

A close look at these figures (3.15 and 3.16) can arm a person bent on surviving a condition of heavy fallout with some important facts. One point to be made is that in a downwind location there is a time of delay before the fallout begins reaching the surface of earth. Early dose rates are low, but accumulation of additional fallout rapidly increases the radioactivity to a peak usually occurring within a few hours of the blast. The rapid decay of the fallout then reduces the dose rate.

This means two things: one, that when you need protection you need it fast; and two, that in some cases if you can secure protection for the initial period of peak radiation, you could then, after a day or two or even perhaps after a few hours, leave that protection for a short period of time to obtain supplies or to travel to a less contaminated area or more secure shelter without receiving a lethal or even a disabling dose of radiation. (Radiation measuring equipment is necessary to determine radiation levels.)

AN IDEALIZED EXAMPLE   To make this principle more vivid, consider the following example: John and Mable Smith live 25 miles downwind from a detonation. A little over an hour after the detonation a dose rate of about 10 roentgens per hour was observable. This rate rapidly increased to over 1000 roentgens per hour before the second hour after detonation. Then the rate decreased to about 300 roentgens per hour by the sixth hour while some fallout was still arriving and to about 80 by the eighteenth hour. (See figure 3.15).

Now, turning to figure 3.16 consider the total cumulative dose that was present at the Smith's house. This total dose was about zero at one hour because the fallout had just begun arriving. The total cumulative dose was about 3,000 roentgens after 6 hours and about 4,500 roentgens after 18 hours.

These doses are given for a completely unprotected person, but John's and Mable's simple fallout shelter reduced the dose to a level that caused no immediate illness necessitating medical care. (An expedient trench shelter with three feet of earth covering it would reduce the eighteen-hour cumulative dose of 4,500 roentgens to approximately 4.5).

Obtaining the proper protection, however, may require staying in an undesirable shelter. This could be at least unpleasant and at most absolutely impossible for very long. With this in mind, consider the possibility of staying in the shelter for four or five days until the fallout has decayed to a large extent. (After two days the dose rate outside the Smith's house in the open would be about 10 roentgens per hour.) By then it would be possible to leave for "greener pastures" (better shelter, supplies, and/or a less contaminated area) and take about an hour to get there without receiving an excessively harmful radiation dose. A radiation anticontamination suit is desirable at this point (see "Radiation Anticontamination Suits" in chapter 4).

It may be necessary to leave after a day or less, but taking advantage of maximum protection during the time of highest dose rate is important to survival. Any material or distance placed between you and fallout will help to some degree.

At this point questions could arise such as how it could be known where "greener pastures" were or how one could get there if he knew where they were. Answers to these questions become complex and uncertain. There are a few things to keep in mind, however. In some areas where fallout could be extensive, actual damage to structures and vehicles could be minimal; you *may* be able to use your automobile and some roadways. Radio bulletins and other communications from responsible parties could be of great value in securing information about where to go and where to get what is needed. It is also important to plan more than one possible escape route. A radiological monitoring device is absolutely essential to make accurate judgments in this matter. (See also "Avenues of Retreat" in chapter 4.)

Up to this point only the general effects of fallout have been considered. Other, more specific aspects of fallout are important to consider, including what is to be expected from it in the long term.

EMISSIONS FROM FALLOUT   The radioactive properties of fallout "decay" by emission of alpha and beta particles and gamma rays. Beta particles are electrons; alpha particles are the same as the nuclei of helium atoms, consisting of two protons and two neutrons.

In air, alpha and beta particles are low in power. An alpha particle emitted from a radiation source can be stopped by as little as a sheet of paper or a very short distance in air (one or two inches), and a beta particle emitted from a radiation source can be stopped by heavy clothing or approximately ten feet of air (see figure 3.17). Simply keeping fallout from touching the skin can prevent alpha and beta burns. Alpha burns from skin contact with fallout affect only the surface layers of the skin, and although they are uncomfortable and unsightly the skin will normally heal to its original or near-original state within a matter of weeks.

When fallout enters the body through an orifice or an open cut, however, alpha and beta burns can be very harmful. Internally, the emission energy is absorbed in a local area by the body, and many of the substances that emit these radiations also collect in specific locations in the body. In addition, the internal tissues are generally more sensitive to damage than external tissues. Avoiding inhalation is imperative.

If radioactive materials are ingested or otherwise enter the body, the rate at which they decay and the length of time they are held in the body before elimination by normal biological processes both affect the damage the radioactivity can do. (The time held in biological process is sometimes described in terms of the biological half-life, which is the time it takes for the body to remove half of the material specified.)

Every precaution should be used to prevent fallout from coming in contact with any body surface. Precautions include protective clothing, dust masks, goggles, and immediate washing of any contaminated body area. It is possible that in many cases

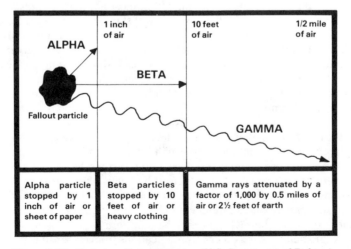

Figure 3.17   Emissions From Fallout   (U.S. Department of Defense)

inhalation of fallout may not be a large problem due to the fact that most immediate local fallout would be quite large in size and difficult to inhale.

Gamma radiation is very powerful and is absorbed by body tissues as it travels through the body. Gamma radiation requires much greater shielding mass and/or distance for attenuation (or reduction) than alpha and beta radiation and, hence, instead of affecting only the area of particle contact it irradiates the whole body. Five feet of water, one-half mile of air, or about three feet of packed earth between the gamma radiation source and the point of measurement will reduce the radiation to about 1/1000 of its value.

Neutrons, which would be radiated essentially only as initial nuclear radiation, are somewhat more difficult to attenuate. Sufficient mass will do the job, but it is more effective to slow them down and then absorb them. When neutrons are scattered in the slowing process and when they are absorbed, they generally produce gamma rays in the process. Iron, boron, and barium are good materials to slow neutrons; and water can further reduce their velocity and capture them. Practically speaking, damp earth or concrete would make a satisfactory neutron shield at most any area where the blast would not require even greater protection.

FISSION PRODUCTS IN FALLOUT   The radioactive products are judged to be too numerous and many of them not significant enough to be mentioned specifically here. Those discussed here probably represent the most significant immediate and long-term health hazards; these are Iodine-131, Strontium-90, and Cesium-137.

Iodine-131 can be inhaled as a gas or ingested in contaminated water or other liquids. Iodine congregates in the thyroid gland and if it is radioactive it can cause severe damage—especially in young children. Growth and mental and physical processes can all be affected.

Iodine-131 has a short half-life and a short biological half-life. For this reason it would represent a significant health hazard for only a short time (a few weeks at the most).

Flooding the system with nonradioactive iodine is an effective measure against biological damage from the radioactive I-131 (see also "Medical Effects of Radiation").

Strontium-90 (Sr-90) is one of the significant long-term villains in a nuclear war. It is absorbed through the digestive system and, because it is chemically similar to calcium, it moves to the bones. (It is thus called a "bone-seeker.") Due to its long biological life and long half-life it can stay in the body, emitting low-energy radiation (beta), for long periods of time. Since it is a beta emitter it would irradiate only the immediate areas around the area of deposit. Possibly the greatest factor in preventing damage from the effects of Sr-90 comes from its rather circuitous route to the body. The Sr-90 deposited on the ground would tend to stay on the surface and not be incorporated into plants with deep roots. Even then it would have to compete with calcium and other clean elements for incorporation into the plants. If a person eats a contaminated plant, the radioactive Sr-90 will compete for digestion and metabolism in the body.

Some cereal grains tend to incorporate Sr-90 quite readily and hence may have elevated concentrations in them at some time after the fallout is deposited.

If an animal eats a contaminated plant the Sr-90 likewise may or may not be metabolized. If it were metabolized by, for example, a cow, it may be deposited in the cow's milk or it may become part of the cow's bones. If it were deposited in the milk it may or may not be metabolized by a man drinking the milk. That Sr-90 in the cow's bones would probably have little or no effect on the animal's meat.

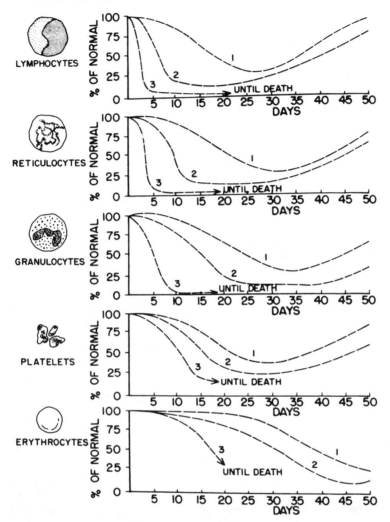

Figure 3.18   Typical Response of Blood Cells to Whole Blood Irradiation

Curves numbered 1 represent the response to a low radiation dose which would require hospitalization but would probably not be lethal. Curves numbered 2 represent the response to a radiation dose that would be lethal to about half those exposed. Curves numbered 3 represent a dose that would be lethal to 95 to 100 percent of those exposed. (*Emergency War Surgery*, U.S. Department of Defense)

The maximum contamination of Sr-90 in edible foodstuffs may not occur for two to three years after a nuclear war due to the foregoing processes. Extremes of contamination could cause blood and bone disorders in animals.

Another factor limiting the danger of Sr-90 and some other isotopes is the fact that much of the radioactive material incorporated in fallout is "locked" into the fallout as the superheated material cools and is, hence, not very available for breakdown and incorporation into plant or animal. As mentioned, much of the fallout is a glassy sand that is not readily soluble. (There is still significant danger from it, however.)

Cesium-137 (Cs-137) is chemically similar to sodium and potassium and, hence, concentrates in the meaty tissues of an animal ingesting and metabolizing it. The Cs-137 is a gamma-ray emitter and therefore produces a whole-body irradiation instead of local irradiation.

Plants do not readily incorporate cesium. For this reason the significant contamination likely to occur would be through direct ingestion of fallout material containing the Cs-137 or eating meat from a contaminated animal.

Animals are also affected by fallout and other residual radiation in a manner similar to man. Animals should be prevented from grazing during at least the first week following the deposit of fallout and should be fed uncontaminated food and drink.

Medical Effects of Nuclear Radiation   It is important to have some idea of what medical effects could be expected from nuclear radiation.

Most biological functions seem to be affected greatly by the rate at which the radiation is received. Radiation doses received over a short period of time (acute) are more harmful than the same dose received over a long period of time (chronic). (We are constantly subjected to small amounts of radiation from cosmic sources, naturally occurring radioactive elements, and other terrestrial sources.) In other words, a dose of five-hundred-rems received in one hour would very likely be lethal, while a five-hundred-rem dose received evenly over a year's time may not even require medical treatment (although it may have serious long-term effects). It is also true that exposure to only a part of the body is not as serious as whole-body exposure of the same magnitude. The body seems to repair the affected area more quickly when other parts of it are healthy.

Genetic damage is recognized as being cumulative but is still somewhat dependent on the rate of exposure.

An acute whole-body exposure to a 100-to-200-rem radiation dose would result in illness but would rarely be fatal. Illness consists of fatigue, discomfort, nausea, and possibly vomiting during the first day or so; and some may experience no illness at all. This is followed by a "latent period" of about two weeks or so after which some symptoms may reappear. During the latent period changes in the blood and bloodforming (hematopoietic) tissues occur that are harmful, and it may take weeks or months to reestablish healthy blood.

Probably the most important consideration in radiation damage expected in a widespread contamination is the damage to the blood-forming organs of the body. This is manifest by a reduction in the number of white blood cells and alteration of other blood components (see figure 3.18). This means that an individual may not initially become greatly incapacitated by a radiation dose but may become acutely susceptible to infection. In simple terms, we might say that one may live through the immediate effects of a substantial radiation dose only to be killed afterward by an infection acquired from stepping on a nail or from a burn or blister.

In the process of time the body has power to regain its infection-fighting capacity.

In addition to the above symptoms, about two or three weeks after receiving an acute dose of three-hundred rems or more, a loss of hair (epilation) will become apparent. This malady also heals with time.

Large acute exposures from two-hundred to one-thousand rems begin producing other symptoms. Bleeding into various organs of the body and under the skin are common. These begin appearing two to three weeks after exposure or earlier. Spontaneous bleeding from the mouth, intestinal tract, and kidneys (blood in urine) occur. These result from the reduction in numbers of blood platelets which reduces the ability of the blood to clot properly. At these dose rates medical care, including the administration of antibiotics, is essential. The critical period in this exposure range is four to six weeks. Death is almost certain to occur from a thousand-rem acute exposure within two months.

Exposure in the range of five-thousand rems has rapid effects on the central nervous system. Symptoms include respiratory distress, incapacitation, and stupor. Death is almost certain to occur within a few hours to a week after this large an exposure. Tables 3.3 and 3.4 describe relative sensitivities of body organs and approximate effects of ionizing radiations.

Some Precautions Against Radiation Injury   No known medicine is active against radiation sickness itself. Exposure weakens the body and makes it more susceptible to infections and disease. These, of course, may be treated by antibiotics and by other medical measures which help ward off infections and strengthen the body. Radiation sickness is not contagious. It cannot be "caught" from an individual who has it or even the dead body of one who has received a lethal dose, provided the person or body is not contaminated with radioactive particles. Radiation sickness preys first on the weak. The very young, the very old, and the ill or injured are generally most susceptible.

TABLE 3.3

RELATIVE SENSITIVITIES OF VARIOUS BODY PARTS TO IONIZING RADIATIONS

Most Sensitive
Lymphoid tissues, bone marrow, spleen, reproductive organs, gastrointestinal tract.

Less Sensitive
Skin, lungs, liver.

Least Sensitive
Muscle, nerves, adult bones.

When the presence of radioactive residues is imminent, potassium iodide may be taken to prevent thyroid contamination with radioactive iodine isotopes. Radioiodine can be ingested in water and in milk from cows that have eaten contaminated feed and can be inhaled as a gas and absorbed into the blood through the lungs. It is difficult to "filter" the gas from the air, so treatment with potassium iodide is very likely one of the most effective medical measures that can be taken to prevent ill health from nuclear weapon residues. When the thyroid gland is flooded with clean iodine the absorption of radioactive iodine is statistically and physically much less probable. The dose recommended by Creson Kearny in his book *Nuclear War Survival Skills* (see selected bibliography at the end of chapter 4), is 130 mg per day for adults and 65 mg per day for children. These dosages may be approximated by using a saturated solution of potassium iodide, which may be made by filling a small bottle about sixty percent full with potassium iodide crystals and then filling it to about ninety percent full with clean water. Some of the crystals should remain undissolved at the bottom.

With a standard-size household medicine dropper the above-recommended adult dose is contained in approximately four drops. The infant dose is approximately two

TABLE 3.4

APPROXIMATE CLINICAL EFFECTS OF IONIZING RADIATION DOSES

Whole-Body Dose in Rems	Dose Received Over	Expected Acute Clinical Effects	Therapy/ Recuperation Periods
100	1 day	A small percentage will experience nausea and vomiting a few hours after exposure lasting less than a day.	Reassurance
	1 month	None	Reassurance
	3 months	None	Reassurance
300	1 day	Nausea, vomiting within a few hours, damage to blood forming or hematopoietic tissues. Death may occur in a small percentage from hemorrhage and/or infection. Epilation (loss of hair) is also likely to occur.	Reassurance, antibiotics, blood transfusion/6 weeks
	1 month	Malaise, damage to hematopoietic tissues.	Antibiotics / 4-6 weeks
	3 months	Probably none depending on peak dose rates.	Reassurance
600	1 day	Severe illness within a few minutes to a few hours; severe damage to hematopoietic tissues; hemorrhaging; gastrointestinal disturbances; reproductive damage; more than half will die within 2 to 6 weeks.	Blood transfusions; Antibiotics; bone marrow transplants could be considered. / Up to 1 year
	1 month	Malaise; damage to hematopoietic tissues; possible damage to reproductive organs and gastrointestinal tract.	Antibiotics/months
	3 months	As at 1 month only damage would be significantly reduced.	Antibiotics/months
5000	1 day	Death within a day or two; severe gastrointestinal disturbances and bleeding, severe damage to central nervous system.	Comfort

An individual surviving 6 weeks after an acute radiation dose is generally expected to live. Genetic mutations have been shown to be greatly reduced in mice by providing a considerable time lapse between acute ionizing radiation doses and conceptions. Effective genetic recuperative period was less for females than males, but a period of several months would probably be advisable. Man would probably be somewhat similar to mice in this respect. Other late somatic effects include reduced fertility; various forms of cancer, including leukemia; cataracts; fetal injury; and other health degradation leading to early death.

drops. The solution is bitter but may be mixed with food or drink to avoid the unpleasant taste.

Because of the relatively short half-life of the most dangerous radioisotope of iodine, the maximum expected time for protection necessary would be eighty to one-hundred days, and probably much less. The standard recommended length of time for taking this prophylactic dose is fourteen days.

A much more convenient and practical way of taking this clean iodine is to procure some tablets. They are available packaged for the purpose in 130 milligram tablets and are now widely available. (See "Sources" at the end of Chapter 4.)

Possible side effects from taking potassium iodide are allergic reaction (including fever, soreness, swelling), skin rash, swelling of salivary glands, and malfunction of the thyroid gland. The thyroid may be induced to overactivity and enlargement (goiter). If serious or allergic reactions occur, stop taking it immediately and seek medical assistance. Do not exceed the recommended dose.

In the case of those already affected by some degree of radiation injury, restful, clean quarters and a good diet aid greatly in recovery. If radiation exposure sufficient to cause mild internal bleeding occurs, fluids should be given. At any injury level, fluids are important. Boullion, fruit juices, herb teas, or even a mixture of 1/2 teaspoon baking soda and 1 teaspoon salt per quart of water will help.

Diarrhea can be checked by taking kaolin or pectin; some appropriate herb teas such as raspberry leaf tea, may also help.

If radiation particles are ingested, some have suggested that taking pectin or eating pectin-rich foods such as apples may help "encapsulate" the particles and aid in their removal from the digestive system.

Finally, since the disease-fighting mechanisms of the body are seriously interrupted with exposure, the taking of antibiotics can assist an exposed person in fighting infections.

EMP (Electromagnetic Pulse)

Electromagnetic pulse (EMP) becomes a significant factor in nuclear war when a large weapon is detonated at very high altitude (about 100,000 feet). This pulse of electromagnetic energy in the range of radar and radio waves can induce electrical current in conductors such as wires, cars, and pipes. The surge of electrical current thus entering conductive channels can cause damage to electrical devices. Telephones (although many telephone systems are protected against EMP), radios, televisions, computers, and other appliances can be put out of service by a strong EMP. Currently made solid-state devices, including the ignition systems of most automobiles, are generally more susceptible to EMP than the older types. (This means that a strong surge of EMP could ruin the electronic ignition on your car.)

During a large high-altitude nuclear test in the Pacific, extensive power outages and communications disruptions occurred in the Hawaiian Islands about 750 miles away. In fact, it has been estimated that electrical services and communications could be damaged or disrupted in the whole United States of America by the high altitude detonation of one large nuclear weapon over the geographic center of the country. Low-altitude and ground bursts do not normally distribute sufficient EMP to cause much electrical damage at distances where the electrical equipment is still in one piece after the explosion.

It is likely that a warring antagonist would begin an attack with an EMP surge sufficient to cripple communications. Of course, there would be ways to reestablish

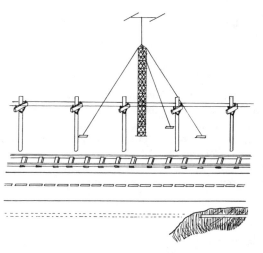

Figure 3.19   EMP

An electrical pulse is created by nuclear detonations—especially detonations at high altitude. This electromagnetic pulse (EMP) is picked up by conductive materials and can damage connected electrical devices.

authoritative broadcasting and other communications after a short period of time, but to count on immediate advice from your local civil defense office could be a mistake. Plans need to be made in advance. (It becomes evident at this point that some CB or other communications equipment may also be desirable survival equipment.) If there is a power outage it may be wise, however, to go immediately to your portable radio.

A radio may survive EMP by being placed in a refrigerator, oven, or some other metal box or by being wrapped in foil. According to the experts, a small, battery-powered transistor radio with a short antenna (about thirty inches or less) should survive any probable wave of EMP without being so protected. Table 3.5 lists the relative susceptibility of various devices to EMP.

All electrical appliances and devices should be disconnected from antennas and power outlets to prevent damage from EMP. (Of course it may be hard to know when a wave of EMP is about to happen by.)

An EMP warning device can be easily constructed that will sound an alarm in the event of any power interruption. (Plans for one can be acquired for $ 5.00 from Paul Seyfried, c/o Packers, 1950 East Forest Creek Lane, Salt Lake City, UT 84121.)

TABLE 3.5

SUSCEPTIBILITY OF VARIOUS DEVICES TO EMP

Very Susceptible:
    Low-power, high-speed digital computers
    Computers and power supplies
    Intercom systems
    Alarm systems
    Life-support systems
    Transistorized transmitters and receivers

Some types of telephone equipment
Power system controls and communication
Communication junctions

Less Susceptible:
Equipment such as above using vacuum-tube instead of solid state construction and
having no semiconductor rectifiers; also, those using low-current switches, relays, and
meters
Hazardous equipment such as detonators, explosive mixtures, rocket fuels,
pyrotechnical devices, and squibs
Long power cable runs using dielectric insulation
High-energy storage capacitors and related equipment
Inductors

Least Susceptible:
Regular 60-cycle-per-second high-voltage equipment such as motors, transformers,
lamps, heaters, relays, circuit breakers, rotary converters, and air-insulated cable runs.

ATMOSPHERIC AND OTHER INDIRECT AND LONG-TERM EFFECTS OF NUCLEAR WEAPONS

The large-scale detonation of nuclear weapons would decrease the ozone layer in the
atmosphere and inject large quantities of dust and smoke into the atmosphere. There is
also a possibility that landslides, earthquakes, avalanches, and/or volcanic activity
could be induced by such disturbances. In addition, there would probably be a series
of events leading to buildup of radioactive materials in some plants and some areas
with attendant effects on agriculture and other elements of the ecosystem.

The Ozone Problem

The extreme temperatures of nuclear detonations heat the surrounding air to the
extent that it changes the chemistry of the air. Oxides of nitrogen, particularly nitric
oxide (NO), are formed. These form in such large quantities that they contaminate the
atmosphere over a wide area.

The ozone in the upper atmosphere shields the earth from the high-energy
untraviolet (UV) part of the light spectrum. Much of the UV light energy is absorbed by
the ozone and the energy is converted to heat. The results are a warming of the
atmosphere and a shielding from UV radiation at ground level. The reactive chemical
species, mentioned above, that are produced in a nuclear explosion mingle with the
ozone and alter large quantities of it chemically so that considerably less protection
from the UV light is given. There is then a subsequent lowering of atmospheric
temperature and a reduction in UV shielding.

Consequences of even a slightly lower temperature on agriculture can be dramatic
(see "Dust and Smoke in the Air" below), as can a more intense UV radiation. Many
plants and animals are very sensitive to UV radiation. Common crops such as peas,
beans, tomatoes, sugar beets, lettuce, and onions are very sensitive to UV light and
could be easily damaged or killed by UV increases. Some grasses are moderately
sensitive to UV light; but, fortunately, many of the very important agricultural staples,
such as wheat, barley, corn, soy, alfalfa, many succulent species, and some kinds of

trees, show high resistance to UV radiation. These facts may be important to keep in mind when storing seeds for a ''survival garden'' and deciding what to plant.

It is estimated that ultraviolet radiation could increase to five or six times its present level (*Long-term Worldwide Effects of Multiple Nuclear Weapons Detonations*, p. 42) before returning to near-normal levels after three to five years. It would take several years for the ozone layer to be completely restored to normal.

Animals, including humans, are very susceptible to burns of the skin and eyes and there is a potential of a small percentage of genetic damage from greatly magnified UV radiation. Realistically speaking, individuals would protect themselves by ''dressing for the occasion'' and by becoming more nocturnal. When severe sunburn occurred, within minutes very few would leave themselves open to additional unnecessary exposure. It would likewise be necessary to protect domestic animals from the abundant UV light to the greatest extent possible by such actions as providing cover for them.

Other animals would suffer more but would probably also become increasingly nocturnal. Birds would be particularly vulnerable. Most larger mammals would not be able to be as nocturnal in winter, when foraging is a difficult task and less cover is available to them. They could be devastated by the effects of ''snow-blindness'' from the increased high-energy light.

Dust and Smoke in the Air

Massive injections of dust into the atmosphere have been recorded from several volcanic eruptions. The National Academy of Sciences study mentioned above and entitled *Long-term Worldwide Effects of Multiple Nuclear Weapons Detonations* suggests that the eruption of Krakatoa in 1883 placed about the same amount of dust in the atmosphere as would a 10,000-megaton nuclear war. That figure amounts to tens of millions of tons of dust. Others estimate that this eruption produced less than a 10,000 megaton exchange would. The 1815 eruption of Mt. Tambora probably produced considerably more dust than would a 10,000 megaton exchange. It shot an estimated 100 billion cubic yards of ash into the atmosphere.

The former projected effect of this dust is a reduction in solar radiation by a few percentage points for a one to three year period and a drop in average temperature of about one degree celsius (depending on how widely the dust spread). The exact amount of cooling of the atmosphere attendant to this atmospheric dust may be uncertain, but a cooling would occur in large areas. There are recently reported calculations (see below) indicating a much greater cooling effect.

Even a slight cooling can have a dramatic effect on agriculture. For example, with an average one-degree (Celsius) cooling of northern North America, it is estimated that the growing of wheat would become impossible in Canada (*Long-Term Worldwide Effects of Multiple Nuclear Weapons Detonations*, p. 3). As the dust causing a temperature reduction slowly settled out of the atmosphere over one to three years, the temperatures could be expected to return to normal and agricultural possibilities restored.

Information reported by scientists at the conference ''The World after Nuclear War,'' held in Washington, D.C., October 30 through November 1, 1985, suggests that, in addition to the dust from nuclear blasts, the smoke and the toxins from fires started by a large-scale nuclear war (5,000 megatons or more) would have a devastating effect on much of the flora and fauna of the earth, especially in the Northern Hemisphere where most or all of the detonations would be likely to occur. It was also suggested that a temperature drop of thirty degrees to fifty degrees Celsius could be initiated by the blocking out of the sun's light and heat by the gigantic cloud cover; and that this condition, which has been dubbed "nuclear winter," could last several weeks. At the same time, the heavy cloud cover would prevent or diminish the process of photosynthesis in plants, causing heavy disruption of their life cycles. The "rain-out" of the toxic clouds would also create harsh acid environments in widespread areas.

The Southern Hemisphere could receive more smoke, dust, and radioactive fallout than was formerly believed due to changes in prevailing atmospheric winds. Several reports of computer modeling and other projections have been published since the conference and seem to have various levels of credibility.

The most pessimistic projections seem to be based on figures which compound "worst case" data. However, there are just a few things to keep in mind if you would plan to survive these projected difficulties: better shelter, warmer clothes, more fuel, more food storage, seeds, and preservation of quality animal breeding stock.

Other projections predict that much of the smoke produced by widespread fires would return quickly to earth and the moisture involved in the explosion would cause a raining out of much of the smoke almost immediately. These predictions, if true, would "warm" the "nuclear winter" considerably.

Other Calamities

The blast of a large weapon, especially when detonated at or near ground level, produces a ground shock wave similar to a moderate earthquake. There are many precedents to illustrate what kinds of things happen when an earthquake occurs. Avalanches, earthslides, and possibly even volcanic and seismic activity could be induced by multiple detonations of large weapons. (Having this fact called to your attention may not change your mind about what to have for breakfast or even where to build a shelter, but it's something to keep in mind.) Other difficulties might include excessive erosion due to loss of vegetation, unexpected departures from normal weather patterns, water availability, and supermarket prices.

Long-term Radiation Buildups

It has been suggested that a 10,000-megaton nuclear weapons exchange would not produce large or even serious *direct long-term nuclear radiation* damage to animals or crops (*Long-term Worldwide Effect of Multiple Nuclear Weapons Detonations*, p. 10). There may be some buildups of radionuclides in some plants with uncertain effects. (For example, mosses and lichens tend to concentrate some fallout contaminants. This could cause damage to these plants, the animals which forage on them, and so on through the food chain.)

There is also a possibility of pathogenic organisms being induced to mutation by nuclear and/or ultraviolet radiation. Such mutant organisms may be resistant to current methods of control for crop or personal protection. The extent and difficulty of this kind of event would be difficult to predict. Long-term contamination of water supplies is not expected to be a problem. (*Long-term Worldwide Effects of Multiple Nuclear Weapons Detonations*, p. 10).

The best estimates that can be scientifically made are that a nuclear war, even of considerable magnitude, would still allow most of surviving humanity and most areas of the world to exist at "safe" levels of nuclear radiation after a period of dissipation. There would be "bad" areas but most nonparticipating and nonadjacent countries would probably be induced to little more long-term radiation liability than is now extant.

If the recent projections of probable extreme temperature drops and toxic environments are correct, the first few months after such a war would be difficult at best for survivors. After that, it is likely that some climatic changes could occur and most certainly some other environmental changes would occur.

(Selected references, some notes and sources of equipment relating to nuclear war appear at the end of chapter 4.)

Nuclear War—Protection From the Bomb

We do not know how many would survive a nuclear war or what life would really be like in the aftermath of one. We do know that the best protection from the immediate effects of a nuclear blast is to be far, far away from it when it occurs. And this, of course, would not be possible for everyone.

Practically speaking, however, protection from nuclear war can entail a long list of measures and devices. Some of them may seem simple and prudent, and some may seem ridiculously complicated. But if you can consider extensive nuclear war a possibility, then you must think through the consequences and the available protective measures to enable you to rationally select the actions you desire to take. There *are* steps that can be taken, some of them very simple, which can greatly increase your chances of surviving a nuclear war.

In the former USSR, Switzerland or much of Scandanavia, as examples, the government would make the decisions for you. They would build the deep underground tunnels and shelters (USSR). They would require a certain size of shelter with all attendant facilities for all new construction or substantial remodeling of older existing structures (Switzerland). They would require drills and place public attention on protective programs.

In the United States of America we are left much to our own devices. The free enterprise system works very well to establish devices known to be needed, but in the case of nuclear incident and its attendant difficulties most people are too busy, too preoccupied, and would rather spend time and money on other things. This thinking cannot necessarily be faulted if we never need protection; but if we ever find that we do, it could be difficult at that stage to change our minds.

Much has been written through official civil defense channels and by other authors about expedient shelters that may be constructed when nuclear war seems imminent. This has made survival from strange and awesome weapons seem more possible. One of the flaws in this kind of preparation (or lack of it) is that in the event of war there may simply not be time enough to build a very substantial expedient shelter. Even a simple trench shelter would take many long hours of hard work to make. If you were trying to avoid the effects of a bomb that had already fallen when you started building one, it would be a little too late.

The good news about expedient shelters is that any able-bodied person can construct one at very little (or no) cost and in almost any general location. They can provide substantial blast protection and very good fallout protection if you can survive staying in one for the required length of time (and it is certainly better than the alternative). And there is a chance that sufficient warnings—such as intensely strained foreign relations or impending blackmail by some radical could occur to alert us to build expedient shelters. (Honestly consider for a moment just what kind of stimulus it would take to get you to "abandon ship" and start digging an expedient shelter or hang up the "out to lunch" sign and report to your pre-built shelter.)

Because many of us would probably not have access to a pre-built shelter, and because we don't know where we would be or what we would be doing in the event of nuclear attack, it is important for every able-bodied, responsible person to know how to build an expedient shelter.

More substantial shelters can be constructed, at a widely varying range of personal cost, that would provide greater protection from both blast and fallout.

It is always important at any given point in difficulties to take the most effective protective action. Consequently, it seems prudent to make some previous preparations according to how "real" you think the problem is and according to the limits of your own resources. We proceed, then, by describing warnings, immediate protective measures, protective shelters, special problems of shelter living (including radiation monitoring and radiation anti-contamination suits), avenues of retreat, and means of sheltering animals from fallout.

WARNINGS

A nuclear war may not be the one-or two-day event that is frequently imagined. It could be days or even weeks before the last bomb explodes. At any point, a warning becomes one of the most important protective measures.

Although many public alarm systems have been found to be nonfunctional, most will still work. The trouble is that most people would not know what one meant if it sounded. Thus, a review is in order. The standard signal is a three to five minute blast on a siren or a series of shorter blasts on horns or whistles. Learn what the signal is in your area!

One of the "must have" items in everyone's emergency gear is a portable radio and a supply of fresh batteries. If an emergency occurred or were imminent, the radio could be an invaluable source of authoritative information. Many radio stations run weekly tests of the Emergency Broadcast System to familiarize listeners with the distinctive warning signal and explain how emergency instructions would be given. (See also EMP in Chapter 3.)

IMMEDIATE PROTECTIVE MEASURES

Some protective measures may be taken in advance, such as building a shelter or buying some radiological monitoring equipment, but other immediate measures must be taken if little or no previous warning is given. The very least warning of nuclear attack likely to occur would be to see the flash of the explosion. This could be your first warning or it could take place while you are en route to a shelter after a previous warning of attack has been given.

Figure 4.1   Measures For Immediate Protection

If this occurs and you feel a pervading warmth at the time you see the light, take shelter instantly and avoid looking toward the light. In smaller nuclear weapons the thermal pulse is essentially instantaneous. In larger megaton-size weapons with large portions being fusion-produced the thermal pulse lasts for several seconds, which would allow time for some protective action such as taking cover, or at least falling to the ground face down.

This intense heat can cause severe burns to eyes and exposed skin and can start fires in flammable material, but at significant distances from the explosion you may shield yourself against it by relatively simple means, such as placing a layer of fire-resistant material between you and it. The thermal pulse is a flash, like intensified rays of sunlight; it is not a wave of hot air. (A so-called "popcorning effect" may occur, however, mainly as a product of airburst weapons. This phenomenon occurs when grains of sand and other particles near the blast which are heated to extremely high temperatures, explode like popcorn, and rise in the air, being further heated. The winds of the blast then carry these superheated particles away from the explosion. These could be of danger to those in the open or in open shelters—without doors—such as many of the expedient shelters. Much of the danger could be alleviated by having in place doors on the shelter or by at least covering exposed skin surfaces with a substantial cloth at the onset of the flash.

If you are more than a few steps from cover when you observe the flash of a nuclear weapon, fall on the ground immediately. Some U.S. Army literature recommends falling face down, protecting your face and hands, and keeping your head covered and your eyes closed when there is no time to reach cover. Civil defense literature recommends lying on your side in a partially curled-up position and covering your head with your hands and arms. Do not remain standing.

Within seconds after the heat is felt, the blast wave will pass. This type of wave accompanies any explosion; the larger the explosion, the larger the blast. Identifying some of the flying objects that these blast waves can produce, if close enough to them, will quickly explain some of the dangers: glass, trees, dirt, fences, people, parts of buildings, animals, vehicles, miscellaneous debris, and so on. If you can very quickly make it to some type of cover before the blast wave reaches you, do so. A ditch, highway underpass, excavation, basement, culvert, borrow pit at the side of the road, or heavy object (rocks, machinery, or car for instance) would provide some protection. A depression is preferable over an object that may be "moved" by the blast wave. Get below ground level if possible but do not take more than a second or two to get to cover. Once there, stay secured until the heat and blast waves are past and objects have stopped falling—*at least two or three minutes!* Wait until the reverse phase of the blast wave, which blows toward the center of the explosion has subsided. Don't become part of the flying debris. Also remember that additional blasts could hit the area after the first.

TABLE 4.1
ARRIVAL TIMES OF BLAST WAVE

Size of Explosion	Arrival Time (in seconds) at one mile distance	Arrival Time (in seconds) at two miles distance
20 KT	3	7.5
1 MT	1.4	4.5

If you are at home or in a building, get away from windows immediately. Pieces of glass can be like a shotgun blast. Stairways and windows are natural channels that can conduct blast waves into a building. Stay away from them. A basement or below-ground area will generally provide greater blast protection than areas above ground. A lean-to against a wall, made with even a mattress, in the most secure area may afford some additional protection.

Once the blast wave subsides, new problems arise. Areas close to a nuclear explosion receive radioactive fallout within minutes, so don't take longer than ten or fifteen minutes to get to a more permanent shelter. For other areas farther away from the blast center it takes longer for fallout to begin. No predictable pattern can be dictated that would tell exactly where or how much fallout would fall at a given location. (See "Who Are They After" in Chapter 3.) The winds, rains, land forms, and other geographical factors have much to say about fallout distribution, but the point is: don't let the "blasted" stuff fall on you. Get into a sheltered area where you have supplies (including your radio), and stay there until it's safe to leave.

The most substantial protection from the initial blast pressures and fires is provided by strong, fire-resistant materials outside the plane of wind and concussion waves (e.g., below ground level). Protection from nuclear radiation is provided by having distance and/or mass between you and the radiation source.

PROTECTIVE SHELTERS

The likelihood of your locality being exposed to heavy blast may influence where and what kind of shelter you could entertain the thought of building. If you live next door to a missile silo or a nuclear submarine base, you may opt for an out-of-town retreat as your only reasonable location to secure shelter. On the other hand, if you live in Hurricane, Utah, or Smith River, California, your straw hat may be all the protection you'll ever need (although it would certainly be recommendable to have more available).

Some Considerations In Selecting General Locations to Build Shelters

A shelter may be a minimal structure designed to preserve life with very little comfort, an elaborate structure with many amenities, or something in between. An expedient shelter could probably be safely built at an appropriate time of alarm at or near your home if you live twenty miles or more from a smaller city or industrial area. (Some have suggested that Soviet strategy included targeting cities as small as 50,000 population, although unlikely in an initial exchange. However, current world conditions include such a wide variety of possibilities, predictions may be more difficult. A margin of 40 miles or more from a large city, military base, or other major target is a recommendable safety factor for building an appropriate expedient shelter. If you live closer you should probably either build a more elaborate permanent shelter or be prepared to leave home with sufficient equipment and supplies to set up your shelter at a safe distance from the expected target areas and hope you have time to do so.

The above distances are chosen because of the damage radius of the size of weapons expected to be thrown at the specified targets. The actual distance at which a small expedient shelter could provide safety is somewhat shorter, but in consideration of large targets being attacked at multiple locations, and missile inaccuracy, these distances are given to provide a greater safety buffer zone.

For example, a 4.5 psi overpressure will flatten most residential dwellings. A twenty-megaton groundburst would produce a 4.5 psi overpressure at about 8.3 miles from ground zero. (Ground zero is the point at ground level directly under a bomb.) A twenty-megaton maximized airburst would produce the 4.5 psi at about thirteen miles. A one-megaton ground burst would produce a 4.5 psi overpressure at about three miles (about 4.7 miles for an optimized airburst). At these distances and pressures, even a simple but well-built expedient fallout shelter could provide sufficient protection from blast and thermal effects. (The one problem found in testing this kind of shelter at these pressures was that when dry, loose sand was used to cover trench shelters the high winds of the blast blew so much of the sand covering away that what was left was insufficient to provide adequate fallout protection.)

Fallout can also be a reason to locate an expedient shelter a significant distance from home. If you live close to a target likely to be hit with a surface burst, the local fallout, especially downwind, would be a *great* hazard. (You would probably want to go upwind to build a shelter. Upwind is generally west, but don't go upwind toward another nearby target.)

The actual distance you might choose to remove yourself from a target may be further dictated by the number of large targets in your general area. For example, if you live near a large town or a small military base and it is the only target for two hundred miles in any direction, you may be safe in concluding that it would be targeted with a small weapon, and consequently eight or ten miles may be a safe distance at which to build shelter, but the farther the better. On the other hand, if you live in an area where several large cities or industrial or military areas are separated by only a few miles, the direction you take for shelter could be more critical than the distance. You would want to build away from all the target areas, not away from one but toward another. An additional factor to consider is that if you live near a Strategic Air Command base where large bombers could be taking off at the onset of a nuclear conflict, the enemy may airburst nuclear weapons in a wide perimeter around the base to destroy airborne aircraft. This could enlarge the danger radius considerably.

Table 4.2 gives some approximate blast pressures and approximate maximum accompanying wind velocities at distances from one hundred kiloton, one megaton, and twenty megaton maximized airbursts and groundbursts. A quick look at this chart clearly illustrates the rapidly enhanced safety provided by distance between you and a nuclear blast.

The approximate damage occuring to some structures at various overpressures is described in Table 4.3.

Considerations in Selecting
Immediate Shelter Locations

The immediate location of a shelter can also have significant bearing on its value in a crisis. One of the very real dangers accompanying a nuclear blast is fire. In a

TABLE 4.2
BLAST PRESSURES AND WINDS

Bomb Size		Maximized Airburst Distance in Miles					Groundburst Distance in Miles				
		1	5	10	15	20	1	5	10	15	20
100KT	overpressure (psi)	15.9	1.3	0.2	—	—	7.8	0.5	—	—	—
	wind velocity (miles/hr)	360	46	<20	—	—	240	19	8.0	—	—
1MT	overpressure (psi)	42	4.0	1.4	<1.0	<0.6	36	2	0.6	<0.4	0.2
	wind velocity (miles/hr.)	1660	135	52	<30	<20	790	68	22	12	8.5
20MT	overpressure (psi)	1.5 mi 168	19	6.5	3.6	2.4	1.5 mi 160	10.8	3.2	1.6	1.0
	wind velocity (miles/hr.)	1.5 mi 2000	440	200	115	76	1.5 mi 1820	300	110	59	36

Department of Defense, Department of Energy.

TABLE 4.3
DAMAGE FROM OVERPRESSURE

Overpressure	Distance in Miles from Optimized Airburst		Damage
	1 MT	10 MT	
1 psi	13	29	Windows broken, parked airplanes damaged, house siding damaged.
5 psi	4.4	9.5	Motor vehicles damaged, most houses collapse, 1% eardrums rupture.
10 psi	2.8	6.0	Human bodies in the open can be blown hard enough to cause nearly 100% fatality rate.
15 psi	2.3	5.0	Lung injuries begin, large multistory building severely damaged or collapse.
25 psi	1.4	3.1	Smaller reinforced concrete dwellings fail and collapse.
35 psi	1.1	2.4	1% fatality from lung damage, heavy steel and concrete structures severely damaged.
45 psi	1.0	2.1	99% eardrum rupture.
65 psi	.83	1.8	99% lung damage fatalities.
200 psi	.51	1.1	8''-thick buried concrete arch collapses (16-ft. span with 4-ft. earth covering.

large-scale nuclear war, a large percentage of the homes and other flammable materials in or near target areas would be damaged. Fires and smoldering would occur; and the products of the combustion, including carbon monoxide and other noxious entities, could be deadly to occupants trapped in a shelter inside a smoldering home. Even if the home itself were not smoldering, smoke from other nearby fires could be hard to exchange for cleaner air. A collapsed house could also make shelter entry and exit difficult. For these reasons, it may be preferable to locate a shelter at least a short distance away from the home, or else to provide a fail-safe entrance/exit *and* a substantial method of ventilation from an area away from the home. Fires started in homes far enough away from a blast to remain structurally undamaged could possibly be extinguished before the occupants needed to take shelter from fallout. Fire initiated in closer structures are frequently extinguished by winds and pressures of the blast although they may continue to smolder.

A permanent shelter could double as a storage room or root cellar or serve some other useful purpose if the room were constructed in a manner such that it would give fallout protection if needed. This could ease the pain of putting substantial expense into a structure that may never be used and may be of little monetary value to someone buying your residence if you move. In a permanent shelter U.S. Civil Defense literature recommends minimum floor space of twenty-five square feet for each individual if two-tier bunks are used. A height of at least six and one-half feet is also recommended.

Shelters that will provide acceptable protection from *blasts* at fairly close ranges must be sturdy, usually underground structures with heavy blast doors, ventilation equipment with valves suitable for blocking out the blast, and so on. Some sources of information and equipment for this type of shelter are given at the end of the chapter. However, radiation shelters would be more universally needed, they are simpler to build, and most of them also afford some degree of blast protection. Instructions for building expedient fallout shelters and an expedient blast shelter are included in Appendix A along with some other shelter ideas.

The story of the three pigs may be applied here. One made his house of straw, one of sticks, and one of bricks. The objective of radiation protection is to keep mass between you and the radiation. As in the case of the pigs keeping the wolf out, the heavier, more solid material keeps radiation out better. Distance and time also reduce the radiation (figure 4.2). Radiation, however, travels in a straight line when there is no interference. This means that giving it a corner to go around will reduce it considerably. The exceptions are those rays that make it through the wall and those that scatter. Scattering refers to its deflection by molecules of material that it is passing through. Skyshine—radiation reflected by molecules of air—is one form of scattering. Figure 4.3 illustrates this way in which radiation can enter and shows that just sitting in an open trench is not enough.

Measurement of Shelter Effectiveness

Before discussing radiation shelters further, some standard terms should be defined. The first is *protection factor* (PF). The amount by which a shelter will reduce radiation is the PF. For example, if a shelter will cut the radiation to 1/30, its PF is 30. Current minimum standards recommended by the U.S. Government for fallout shelters is a PF of 40. By doubling the amount of mass of the shielding material a PF of about 1000 can be obtained, and this is a much more recommendable value.

The second term is closely related to the first and is called *half-thickness*. The half-thickness is the thickness of a material required to reduce the radiation transmitted

FORMS OF NUCLEAR RADIATION

NEUTRON	η	ENERGIZED NEUTRON	LARGE MASS	NEUTRAL CHARGE	HIGH ENERGY
ALPHA	α	ENERGIZED HELIUM NUCLEUS	LARGER MASS	POSITIVE CHARGE	LOW ENERGY
BETA	β	ENERGIZED ELECTRON	SMALL MASS	NEGATIVE CHARGE	LOW/MEDIUM ENERGY
GAMMA	γ	ELECTROMAGNETIC ENERGY	NO MASS	NO CHARGE	VERY HIGH ENERGY PARTICLE

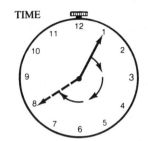

TIME

7-10 RULE FOR RADIATION DECAY

AT TIME T_1 (ONE HOUR AFTER DETONATION)—ONE O'CLOCK—RADIATION INTENSITY, R, IS 100 ROENTGENS

AT TIME T_28 O'CLOCK (A 7-FACTOR TIME INCREASE(—RADIATION INTENSITY, R, IS REDUCED TO 10 ROENTGENS (A 10-FACTOR INTENSITY DECREASE)

DISTANCE FROM SOURCE

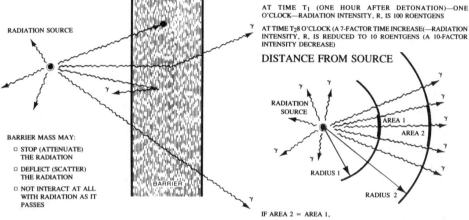

RADIATION SOURCE

BARRIER MASS MAY:

☐ STOP (ATTENUATE) THE RADIATION

☐ DEFLECT (SCATTER) THE RADIATION

☐ NOT INTERACT AT ALL WITH RADIATION AS IT PASSES

IF AREA 2 = AREA 1,

THEN THE RADIATION FLUX ON AREA 2 IS LESS THAN ON AREA 1, AND IS INVERSELY PROPORTIONAL TO THE SQUARE OF THE DISTANCES OF THE TWO AREAS FROM THE RADIATION SOURCE.

Figure 4.2   Ways Nuclear Radiation is Reduced

(U.S. Department of Defense)

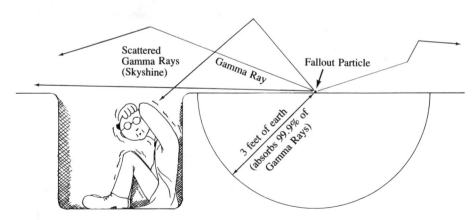

Scattered Gamma Rays (Skyshine)

Gamma Ray

Fallout Particle

3 feet of earth (absorbs 99.9% of Gamma Rays)

Figure 4.3   Scattering of Gamma Rays by Skyshine

An open trench does not provide adequate shielding from heavy fallout.

through it to one-half of its strength. The half-thickness of concrete is 2.2 inches. This means that it would take 2.2 inches of concrete to give a PF of 2 or cut the radiation passing through it in half. Thus, it would take five times this amount, or 11 inches of concrete, to give a PF of 32 (1/2, 1/4, 1/8, 1/16, 1/32). A thickness of 22 inches of concrete gives a PF of about 1024. These figures are in keeping with commonly adopted values in civil defense and other literature.

As can be seen from Table 4.4 there needs to be about 28 pounds of shielding material distributed evenly over one square foot for each halving layer (half-thickness) of shielding so that a protection factor of 16 would be provided by about 112 pounds per square foot, a PF of 64 by 168 pounds per square foot, a PF of 256 by 224 pounds per square foot, a PF of 1024 by 280 pounds per square foot, and so on. (The actual figures may vary slightly from these.)

TABLE 4.4
HALF-THICKNESSES OF SOME MATERIALS

Material	Approx. Density (lbs/cu. ft.)	Half-Thickness	PF*40	PF*1000
Steel	490	0.7 inches	3.7"	7"
Concrete	146	2.2 inches	11.8"	22"
Earth (Packed)	100	3.3 inches	17.5"	34"
Water	62.4	4.8 inches	25.4"	48"
Wood	40	8.8 inches	46.5"	88"

*Approximate thickness (in.) needed to achieve this protection factor (PF)

TABLE 4.5
SOME MATERIALS APPROXIMATELY EQUIVALENT IN SHIELDING CAPABILITY

Acccording to the Department of Defense publication *In Times of Emergency* (1977, p. 20), the following materials would provide approximately equal shielding (a PF of about 4) from fallout radiation:

    4 in. concrete
    10 in. water
    18 in. wood
    14 in. books or magazines
    8 in. hollow concrete blocks
    6 in. hollow concrete blocks filled with sand
    7 in. earth (this, of course, could be contained in sacks, boxes, or other containers)

In the event of nuclear war there is a possibility of little or no warning, or a warning of a day or even more of a probable attack. A person could have a prebuilt shelter, a preconceived but unbuilt expedient shelter, or no idea at all about a shelter. Therefore, the remaining relevant subjects examined here are: the prebuilt shelter; the preplanned expedient shelter; what we shall call "immediate expedient shelter"; some other shelter alternatives, and when to leave a shelter. Another possibility, already mentioned, that can be used in conjunction with one or more of the above is changing location, which is also discussed in "Avenues of Retreat".

Prebuilt Shelter

Prebuilt protection may come in the form of public shelters, privately built shelters, and/or commercially produced shelters.

Figure 4.4   Prebuilt Shelters

One of the options to prepare against nuclear attack is to prebuild a shelter.

Public Shelters   Public shelters do exist in many communities, usually in the form of existing buildings, mines, tunnels, subways, and other such facilities. These shelters should be marked by the standard yellow and black fallout shelter sign (if you can still see the paint). Even if you have no plan for using them you and your children should know where these shelters are in your community in case by some circumstance it were the only available alternative. Family members may be away at school, work, or other places when the need would arise (or fall) to use a shelter. They should be instructed on where to find shelter and what to do if an attack should occur. The civil defense crisis relocation program may not have time to be implemented; and even if it were, the available shelters should be valuable assets to the communities to which relocation would be made.

These public facilities may not be adequate for all members of a community, but some would want to use them while others would choose to stay at home or may, out of necessity, take shelter in some other suitable place of refuge. There are also those for whom building or otherwise acquiring a personal shelter would be nearly impossible. Some of them live where it would be necessary to have substantial shelter and may not have facility for retreating. If you are one of them, do three things right away: (1) get a fourteen day emergency kit together as described in this book, (2) see your local civil defense people about where the nearest adequate shelter is and how much protection it will give, and (3) if possible plan an avenue of retreat with suitable alternatives.

Public fallout shelters will probably have poor to no ventilation, little or no light, little or no food and water, and few if any sanitary facilities. If any of these would be items of interest to you it is suggested that you take them along with you when you go. They may, however, have radiation monitoring equipment and personnel trained to use it.

Privately Built Shelters   There are some fairly reasonable shelter designs published by the U.S. Department of Defense. Some representative types of these are included to give a general, or maybe even a specific, idea of one that you may wish to use. These are included in Appendix A. Remember needs for ventilation, water, heat, and light. A private source for a good shelter plan is included at the end of the chapter.

With any shelter built around home, landscaping can also be done with provision of shelter in mind. Planter boxes, high-mass walls, or even a swimming pool can provide good radiation barriers. Remember that gamma radiation is still harmful after traveling hundreds of feet through air, so that there is an additive effect not only from fallout that is close but also from that which lies at a distance. Any barrier that stops some of this radiation could be a protective asset.

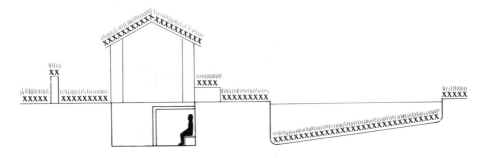

Figure 4.5   Use of Landscape Features for Radiation Shielding

High mass walls and fences, planter boxes, and swimming pools can all aid in blocking radiation.

If you are interested in improving your home to include a shelter, or if you are building a new home and wish to include some increased sheltering capability, or if you are just plain interested in the subject, I recommend finding a copy of Joel Skousen's book *Survival Homes*. It tells what to do and gives sources for materials and equipment, but, unfortunately, it is out of print.

Commercially Produced Shelters   At this writing there are a few commercial manufacturers and/or suppliers of bomb shelters and equipment. Some of them are listed at the end of this chapter.

Some building contractors have built shelters and have the expertise to do so. If you are interested, contact your local contractors' board at city or state level for an idea of who could build one for you.

Some Preplanned Expedient Shelters

If you don't want to build an elaborate or even simple permanent shelter, it is

Figure 4.6   Preplanned Protective Measures

Planning realistic measures of escape and/or shelter could be a
vital preparation.

definitely good to have a plan in mind for at least some shelter, just in case the ''wolf''
comes knocking at your door. There is also the possibility of being caught away from
home when a nuclear exchange began, in which case you may have to extemporize.

Many materials around the house could be fashioned into a makeshift shelter to
provide adequate protection in some circumstances: clothes, mattresses, appliances,
doors (taken off their hinges), tables, books, magazines, bedding, boards, rugs, earth,
furniture, sand, firewood, luggage, containers of water, boxes, bricks, rocks, snow,
benches, tools, automobiles, and so on.

A shelter could be made in a variety of places, depending on where one lives and
what the circumstances are. Let's name a few: the basement, the crawl-space under the
house, a hole or trench, an underground utility area (access from manhole), a lean-to
against a heavy solid structure, a cave, the center of the house (furthest from outside
walls). Shelter would not necessarily have to be endured forever—perhaps just a few
days or maybe a couple of weeks, and that is within the endurance of just about
everyone.

In building the shelter at the time of need, the object is to place as much shielding
around you as possible in the time available. With some preplanning and preparation, a
shelter could be constructed quite quickly. Even lying on the floor of a basement next to
an outside buried wall could provide substantial protection. Since radiation does
damage to blood-forming tissues, by shielding even your legs you can increase your

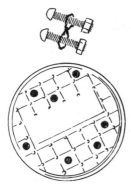

Figure 4.7   Manhole Shelter and Key

Underground utility areas (many power lines and other utilities are run
in underground concrete ''hallways'' in large cities) can provide
adequate shelter from radiation. Caution should be used to avoid toxic
fumes. Two bolts wired together can serve as a key by slipping one bolt
through a hole and lifting the lid with the other.

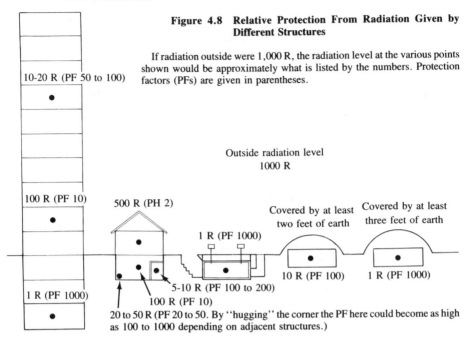

Figure 4.8   Relative Protection From Radiation Given by Different Structures

If radiation outside were 1,000 R, the radiation level at the various points shown would be approximately what is listed by the numbers. Protection factors (PFs) are given in parentheses.

10-20 R (PF 50 to 100)

Outside radiation level
1000 R

100 R (PF 10)

500 R (PH 2)

Covered by at least two feet of earth

Covered by at least three feet of earth

1 R (PF 1000)

10 R (PF 100)

1 R (PF 1000)

5-10 R (PF 100 to 200)

100 R (PF 10)

1 R (PF 1000)

20 to 50 R (PF 20 to 50. By "hugging" the corner the PF here could become as high as 100 to 1000 depending on adjacent structures.)

chances of recovery from radiation sickness. The large bones can help renew the blood entities necessary for good health.

Basement Shelters   If the shelter is to be in the corner of the basement, a work bench or table, for example, could be the basic structure, with boxes, sandbags, drawers, and so on stacked on it and appliances, boxes, tools, books, etc., pushed up against it. The outside walls of a concrete or concrete block basement would be safe to use for shelter walls if they are below ground. The most underground part of a basement or cellar should be used for the shelter. A basement used for a shelter should be a sturdy masonry structure with a well-built, well-supported floor above it.

Crawl Space Shelters   If your home has a crawl space and not a basement, a fallout shelter could be improvised by first providing an accessible entrance such as a door through the floor or foundation wall. Then select an area for the shelter. The center of the house would normally be best for a fallout shelter since it is farthest from the outside walls. If the home is subjected to significant blast pressures, however, the floor joists tend to collapse between supported areas making it generally safer to be near an outside wall. Construct walls of shielding material from the ground to the floor of the house. These walls could be sandbags, blocks, bricks, or whatever you have. The space could be dug deeper to provide greater comfort by allowing a sitting or even standing position to be assumed. It should also have enough area to allow all occupants to assume a prone position, since this may become a popular position if the occupants need to remain there for very long. If it became necessary to occupy the shelter, a roof for the shelter area could be quickly constructed by placing shielding material (books, furniture, earth, etc.), on the floor of the house directly above the shelter. Provision should also be made for ventilation (see "Ventilation," this chapter).

Shelter in Buildings and Other Protective Spaces   Basements of large buildings

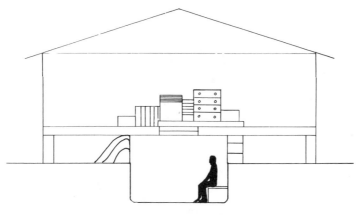

Figure 4.9   Shelter In A Crawlspace

can provide a high degree of radiation shielding. Some of these structures have been officially labeled as fallout shelters in days past. There are drawbacks to these places, however. The first is that the basement of a large building is probably not a very desirable place to be if there is a significant possibility of heavy blast damage. The second is that the possibility of obtaining adequate ventilation in such a place is not good unless previous preparations have been made.

The same reservations should be considered in selecting mines, tunnels, caves, or subways for radiation shelter. These places could obviously provide some very good radiation shielding, but if you would use one consider the ventilation problem as well as shielding at entrances, heat, light, water, etc.

Some Federal Emergency Management Agency (FEMA) literature suggests that a shelter could be built under the edge of the concrete slab floor of a house. This is

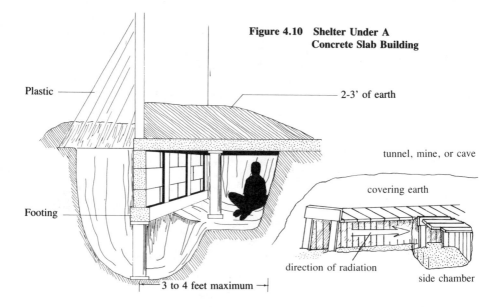

**Figure 4.10   Shelter Under A
Concrete Slab Building**

Plastic

2-3' of earth

tunnel, mine, or cave

covering earth

Footing

direction of radiation

side chamber

3 to 4 feet maximum

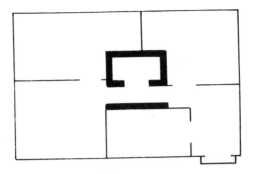

Figure 4.11   Use of High Mass Interior Walls for Radiation Shielding in Houses Without Basements

An interior room can be constructed of thick reinforced concrete walls and ceiling, and also with a high-mass wall shielding the entrace.

possible, but the amount of work involved and problems with obtaining adequate ventilation could be prohibitive. The concrete floor should be braced and space should not be dug more than four feet from an outside supporting wall.

Shelters can also be built at ground level in the interior of homes with concrete slab floors by making the walls and ceiling of an interior room out of reinforced concrete and providing a shielded entryway (Figure 4.11).

Trench Shelters   A trench shelter may be constructed by digging a straight or an L-shaped trench; covering the trench with doors, logs, or boards; and then covering the doors, logs, or boards with shielding material. The shielding material would probably be earth, in which case it should be at least eighteen inches thick (PF 40), but preferably three feet (PF 1000). The trench should not be so wide that the roofing material (boards or doors) would collapse under the weight of the shielding material piled on top of it (see Appendix A). It is interesting to note, however, that even hollow-core doors have been found to be sturdy enough if proper spacing as described in Appendix A is observed. Trench shelters are practical and effective.

Arching the earth over the trench shelter provides added strength for blast protection by "loading" stress onto the sides of the trench. This is amazingly effective, even when flimsy doors are used. The arching should extend beyond each side of the shelter 5 to 6 feet (see Appendix A). Covering the interior walls with sheets can help preserve

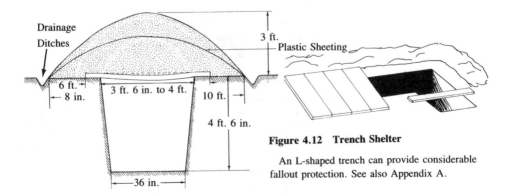

Figure 4.12   Trench Shelter

An L-shaped trench can provide considerable fallout protection. See also Appendix A.

cleanliness and can help keep the dirt from losing its integrity. A variation to the trench is to dig approximately the same trench, drive the car over it, bank the sides with earth and then fill the floor of the auto with shielding material (probably dirt).

The trench shelter is about as simple and effective as any that can be quickly built from almost nothing. All that is needed is a shovel, some doors that have been removed from their hinges or similar material, plastic sheeting, and a place to dig.

Since the covered trench or a "small-pole shelter" seem to be very practical approaches to expedient shelter, plans for their construction are included in Appendix A. The small-pole shelter is included because it gives not only fallout protection but also a very substantial protection from blast when properly constructed. It is basically a trench shelter made more structurally sound by the use of interior shoring with poles.

Lean-to and Existing-Structure Shelters   A lean-to shelter is also simple to construct. Doors, boards, plywood, or other construction material are "leaned" against a substantial existing wall. A six to eight inch deep groove or trench is dug to receive the bottom end of the door to prevent it from sliding away from the wall. Cover the lean-to roof with eighteen inches to thirty-six inches of soil or sand as a shielding material. Be sure to use sturdy enough material for the roof to hold up the heavy shielding material. If doors are used, use the heaviest ones available. The inside may be dug out for extra space. The ends should be covered with a sufficient wall to provide shielding; this could be made of earth or any other satisfactory shielding material.

Another alternative for sheltering structures is to preplan a more elaborate arrangement using existing walls or new walls incorporated into living spaces (such as a bar made of block or brick or some other good shielding material). The roofing material for such a shelter could be boards that are stored nearby, a bookshelf, or other furniture that would provide shielding or hold sufficient shielding materials. Convenient shielding materials include sandbags, blocks and bricks.

One note to stick in the back of your mind is that packed snow can also provide fallout protection. (A thickness of at least five feet is recommended.)

Immediate-Expedient Shelter

Immediate expedient is my own phrase to describe actions that must be taken immediately when one is caught away from shelter and a nuclear weapon is detonated.

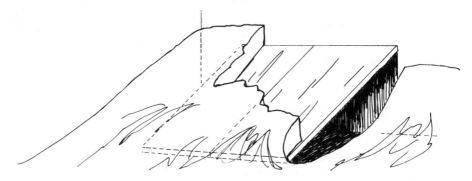

Figure 4.13   Lean-to Shelter

A lean-to can be quickly made against a solid shielding wall.

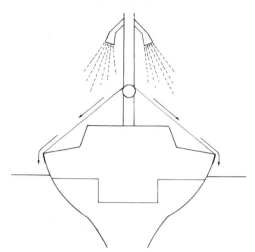

Figure 4.14 Boat Shelter

A boat that is in at least five or six feet of water and at least three thousand feet from shore can easily provide a PF of 1000. Provision should be made to keep fallout off the boat.

This is the subject of discussion in "Immediate Protective Measures" at the beginning of this chapter and is mentioned here to reinforce the awareness that these measures are very much a part of a protective shelter system.

Other Shelter Alternatives

The substantial distance between *fallout* on the ground and an airplane in flight provides a significant shield against radiation. A distance (or altitude) of 3,000 feet provides a protection factor (PF) of approximately 1,000. Of course, it may be hard to stay in flight for two or three weeks while the fallout dissipates, but an undamaged plane could be a good escape device from fallout areas after the bombs have stopped falling.

Boats can also provide good shelter from radiation. Even in relatively shallow water (about six feet) the effects at the water's surface of fallout on the bottom of a lake or pond could be essentially nullified. This in itself would provide a great reduction in radiation exposure to those aboard the craft. By keeping fallout washed off the boat's decking and other surfaces and/or using a waterproof rain fly over the boat, the radiation exposure could be further reduced (see Figure 4.14). Simply head for the middle of the lake with plenty of supplies and stay there until the radioactivity has dissipated, but remember that you must be at least 3,000 feet from any shore of the body of water.

When to Leave a Shelter

The subject of when to leave a shelter and for how long is a matter of interest. Even when the radiation dose rate is very low it may still be advisable to minimize one's exposure to it by sleeping in the shelter, for example. Some reasonable approximate limits are included in Table 4.6.

The official U.S. guidelines allow restricted outside activity when the outside radiation level is two R/hr, with a maximum allowable exposure for any one day of six R. This would then be about three hours at two R/hr., neglecting any dose aquired from inside the shelter.

Remember that with only another day or two the dose rate will probably diminish

TABLE 4.6
ALLOWABLE RADIATION EXPOSURES

Outside Dose Rate (R/hr.)	Reasonable Allowable Activity Limits
0.5	Sleep in shelter. Work and other activity may be conducted outside. Limit exposure when possible and avoid radiation "hot spots".
0.5 to 2	At 2 R/hr a maximum of 2-3 hours exposure should be observed. At 1 R/hr perhaps 5-6 hours could be tolerated. Limit exposure as much as possible.
2 to 25	At 10 R/hr no more than 10-15 min. per day of outside exposure should be allowed. Only essential outside tasks should be performed.
25 to 100	At 100 R/hr only absolutely essential duties should be performed outside and then for only a very few minutes.
100+	Only life-threatening circumstances or the promise of acquiring better shelter within a short distance should entice you to leave.

considerably. Rotate any essential outside tasks so that all outside exposure isn't given to one person. Also, it may be possible to limit exposure by conducting some activities such as exercise for short periods of time in a partially sheltered area if one is available. Older persons may be willing to assume a slightly increased proportion of the low level exposure, since low level exposure problems tend to appear several years after the exposure.

Another strategem to consider is the possibility of making a quick return trip to your home after radiation levels have decreased through decay in order to recover additional food supplies and equipment. Your 14-day Emergency Kit should be designed to handle a shelter stay. Be advised to take as much water with you as possible. If you have a stocked shelter at a distance, of course, a simple escape kit would be sufficient to leave home with, as some writers suggest.

SPECIAL PROBLEMS OF SHELTER LIVING

Staying in a shelter for even a short duration requires some special equipment. Ventilation, temperature control, light, sanitation, radiation monitoring, radiation anticontamination suiting, food, and water can all pose somewhat different problems than at other times and in other places.

Ventilation

Shelter living could be truly "breathtaking" without some substantial form of ventilation. U.S. civil defense literature specifies that three cubic feet of air per minute (CFM) be provided for each person occupying a shelter in order to keep oxygen coming in and carbon dioxide going out and to keep heat down in hot weather. If a heating

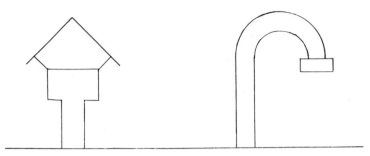

Figure 4.15   Shelter Vent-Pipe Configurations

device were used in cold weather it would also be very important to provide good ventilation to prevent buildup of carbon monoxide and carbon dioxide. Realistically speaking, you could live on less, but this is an accepted minimum standard for air. Cresson Kearny, in his book *Nuclear War Survival Skills* (p. 47), recommends 40 CFM per person for shelters in hot humid weather to keep temperatures down. This is based on substantial and realistic experimentation.

There are several ways to accomplish this ventilation. Commercial units powered by hand or electricity can be installed in a shelter. Or, with a little good old-fashioned ingenuity, a fan system could be made from a 12-volt battery and an appropriately sized fan. This, of course, would require that batteries or some other power source be stored. A pipe or conduit at least four inches in diameter should be used to bring air into the shelter if a fan is used. A filter should be used over the air inlet, to remove airborne radioactive debris. A double layer of standard furnace filter could work for this purpose. The intake should be at least twelve inches off the ground. Also, a hood should be placed over the inlet or a curved configuration should be used to prevent fallout from entering the air inlet (Fig. 4.15). Valves and baffles are necessary to prevent shock waves from entering a shelter through air vents in a closed blast shelter.

Matching blower capacity with intake and outlet capacities to create slight pressurizations of the shelter can help prevent stale air from from drifting back in. Further information can be obtained from equipment vendors.

For an expedient shelter, a prebuilt Kearny Air Pump (KAP) is a practical device for ventilation and cooling. Plans to build one are shown in Appendix B. If you don't have some other plan for ventilation, prebuild a KAP or some other alternative of a proper size to match your shelter plans.

Remember that air inlets to basement shelters should be located away from the house far enough to prevent intake of smoke and other combustion products.

Temperature Control, Light, and Sanitation

A sophisticated shelter might include power generation and a heating/air conditioning unit; simple shelters must rely on simpler methods. In some climates and at some times of year, shelter living could be tolerated quite well without any supplemental light and heat. In other areas and at other times of year, living at all without heat or very warm clothing could be nearly impossible. The possibilities for solving these problems are many, and it seems sufficient to mention the problem and offer a few ideas on the subject. Candles, lanterns, lamps, fires, heaters, and so on all give light (and some of them heat) but also use fuel and oxygen and give off noxious fumes. If your shelter and supplies could provide fuel, oxygen, and adequate ventilation, then fine; but if you have little oxygen supply (a tightly closed shelter), you must be careful not to use it up with some type of fire. Heavy clothing and bedding should be counted on as the first line of defense against cold in a shelter.

Virtually any heater could be used to warm food. A backpacking or camping stove could handle the cooking chores for the amount of time it would be necessary to stay in a shelter, however foods that need no cooking should be used in most shelter situations.

A flashlight or electric lantern gives light but also uses batteries. With judicious use and a spare set of batteries, one of these could provide some sanity-saving light for a couple of weeks, which may be long enough. A large multicell flashlight with a small amperage light bulb will shine for several days on a set of good batteries (see also "Heat and Light" in chapter 11).

Sanitation is covered in chapter 13 of this book, but a brief review seems in order here. An inexpensive and useful emergency toilet is a five-gallon plastic bucket with a tight-fitting lid, a supply of plastic sacks, toilet paper (at least 2 rolls per person), some household disinfectant (add a little after each use to control odor), and, if possible, a large garbage can with lid to store the wastes held in the plastic sacks as they are filled. When it is safe to leave the shelter, the contents of the garbage can could then be buried under at least two feet of soil to prevent transmission of disease by insects or vermin and prevent problem odors. The chemicals used in portable flush toilets may also be used in a plastic bucket toilet.

There are other options for toilet facilities, such as a portable flush toilet or a seat with attached plastic bag. If money is a problem, a plastic bucket with a four-pack of toilet paper and a bottle of disinfectant could be acquired for very little money. If money is not a concern, a portable toilet and supply of water and chemicals would be very nice. But the problem of having an emergency toilet is one that must be dealt with not only for a nuclear war situation but also for any one of a long list of other emergencies.

Other basic sanitation supplies should be on hand. Feminine hygiene items (if applicable), detergent, soap, water purification equipment, disposable diapers and plastic pants (if applicable), plastic or rubber sheeting, old newspapers, trash containers (buckets and plastic sacks), and insecticides could all be useful. In at least one assessment made after a major disaster, those involved were asked which of their supplies gave out and which they most wished they had more of. None other than diapers topped the list for those who had babies.

Control of vermin could also improve comfort and hygiene for shelter living. Rodents and some insects transmit diseases readily. If they are controlled before a disaster, they would not constitute quite such a threat during an emergency. There are many good commercial insecticides available. If you have a shelter or a specified area for a shelter it may be a matter of interest to spray these areas regularly to keep the bugs out. At least you should have a can of spray, a flyswatter, and screen around the house just in case. A supply of mouse and rat traps and/or poison could also be a great boon to shelter living—especially expedient shelter living. A good air gun could also help.

One last factor of sanitation is care of the dead. There is a possibility of death from radiation, injury, disease, or natural causes while confined to a shelter. Of course, certainty of death should be established—not always as easily as may be thought. The body or bodies would not only be undesirable to leave in the shelter but could quickly attract vermin. Probably the best approach to disposal is to completely enclose the body in plastic (sacks or sheeting), taping all joints tightly closed. Then quickly set the body outside the shelter, and throw or roll it away as far as possible from the shelter entrance without "wading" in the fallout and being sure not to carry any fallout back into the shelter. As soon as radiation levels permit, the body should be given a proper burial.

Radiation Monitoring

In order to be able to successfully defend against radiation, it is important to know what the *dose rate* is inside and/or outside a shelter (by using a survey meter) and also what the *total dose* is that you have received by using a dose meter (dosimeter). It is also important to record the data obtained.

Survey Meters   Radiation Survey Meters are used for measuring the radiation dose rate at a given time. A survey meter should measure gamma radiation from zero to about the 500-roentgen-per-hour (R/hr.) range. Geiger counters and other low-range meters that measure in the milliroentgen (1/1000 of a roentgen) range are of no use in a fallout shelter. In fact, they can be dangerously misleading. The CDV-715 ionization-can meter (a standard war-survey meter) and the CDV-710 meter are both acceptable; these were made for civil defense purposes and are presently available from some surplus outlets and other sources. Others are also available. See Sources at the end of the chapter.

Civil defense offices will sometimes give out equipment to volunteers who are willing to meet their requirements.

All radiological monitoring equipment should be tested to make certain it is accurate. If you invest in some of this hardware, make the effort to get it checked by a state or independant lab. There are some sad stories about spending considerable money on meters and finding out they are grossly inaccurate. And you could be depending on it in life-threatening circumstances.

Several monitoring situations are important to understand. For example, after the fallout stopped falling you would observe, through monitoring inside your shelter, a maximum reading and then a slow decline. At about the time of maximum radiation a reading can be taken out of doors three feet above the ground in the open. This should

Figure 4.16   CD V-715 Gamma Ray Survey Meter

Figure 4.17   CD V-720 Gamma Ray Survey Meter

be done as rapidly as possible, an anticontamination suit should be used, and precautions should be taken to prevent any fallout from entering the shelter when the one taking the reading returns to the shelter. The reading outside the shelter divided by the reading inside the shelter at about the same time will give the approximate PF of the shelter. For example, if the reading inside were 5 R/hr. and the reading outside were 500 R/hr., then the PF of the shelter would be 500/5 or about 100. Knowing this, the approximate reading outdoors could be calculated at any time after by taking a reading inside the shelter and multiplying by the PF.

The problems become more complex with multiple detonations and other factors and a good set of radiological monitoring instructions is important to have and to understand. The instructions included with the purchase of a radiological monitoring device should be sufficient to describe procedures for most situations. If they are not, seek others from the manufacturer or obtain some civil defense manuals.

Instructions for making a homemade fallout meter, the KFM (Kearny Fallout Meter), are available from the National Technical Information Service, U.S. Department of Commerce, 5285 Port Royal Road, Springfield, VA 22161 for about eight dollars. Send for *The KFM, A Homemade Yet Accurate and Dependable Fallout Meter* (ORNL-5040). This is inexpensive but does take an investment of time and effort to make and understand how to use. Plans for the KFM are also included in the book *"Nuclear War Survival Skills."* I highly recommend aquiring a copy of this book, which can be purchased for under ten dollars. Another book that is very helpful in understanding the problems of radiological monitoring is Bruce Clayton's *Fallout Survival: A Guide to Radiological Defense* (see end of chapter).

Dosimeters   It is desirable to have a dosimeter for each person to monitor his or her total exposure and a charger for a small group. Those worn by X-ray technicians measure in the milliroentgen range and are so sensitive and have such a small deflection range that they would essentially useless in a fallout situation. Dosimeters with a

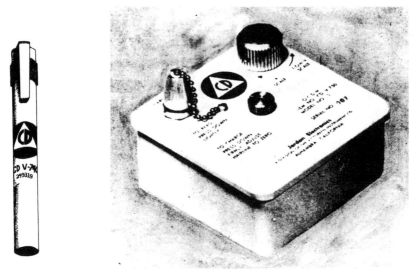

Figure 4.18   CD V-740 Dosimeter   **Figure 4.19   CD V-750 Dosimeter Charger**

Dosimeters measure cumulative radiation doses. They need to be recharged when they reach a maximum reading.

measure in the milliroentgen range and are so sensitive and have such a small deflection range that they would be essentially useless in a fallout situation. Dosimeters with a scale of zero to about 200 roentgens, such as the civil defense CDV-740, are acceptable. They are usually pen-like in shape, and as they receive radiation the device is proportionally discharged. It is generally necessary to hold it up to a light to see the reading. Doses should be recorded accurately for each individual. When the dosimeter reaches full deflection—its maximum of measurement—the dose and the time it took to acquire it should be written down and the measuring device reset or recharged to zero again using a charger. One charger will do for several dosimeters.

Periodic checks should be made for leakage when dosimeters are not in use. A charged dosimeter should show less than a 20 R deflection in a year's time. Sometimes it is necessary to recharge a dosimeter a second time within a few minutes of the first charge. If the second one doesn't hold, the instrument is probably defective.

Radiation Anticontamination Suits

A protective anticontamination suit is clothing to keep fallout off you—nothing more, nothing less. It will not stop gamma radiation but can help prevent alpha and beta burns and aid in prevention of carrying fallout into a shelter. This anticontamination suit must be worn at any time fallout is arriving or still highly active and the wearer is out in it.

Commercial anticontamination suits are available at reasonable cost, or you can invent one of your own. The commercial suits are usually made of Tyvek, which is a synthetic that resembles a tough paper. Homemade suits could be made from plastic (including garbage sacks and Halloween costumes) or rainwear. Shoes should be covered and all other openings closed. This includes rubber gloves and face protection—preferably a dust or gas mask. Goggles are also included in most commercial kits and are a good idea in a homemade job. Whole-face protection is best.

In using the suit, remember that fallout must be thoroughly washed off the suit before the wearer enters the shelter, or else the suit should be taken off and left at the entrance. Most fallout could probably also be brushed off. Remember also that beta burns cannot be totally prevented by a thin layer of plastic on the skin. Heavy clothing worn underneath the suit could greatly reduce them, however.

Food and Water

The 14-day Emergency Kit would be sufficient for many instances of shelter living. If it were necessary to stay in a shelter longer than this, the chances are good that the shelter could be left for long enough to obtain food from (hopefully nearby) larger storage supplies without great risk to health.

Water is taken for granted a lot these days but it could easily become a premium item. A minimum of seven gallons of water for each shelter occupant should go with you to a shelter (2 quarts per day for 14 days) and preferably at least twice as much. A prebuilt shelter should have a suitable container or reservoir filled with water and kept in good condition at all times. If you go to a shelter, take all the water you can practically take with you. (See ''Water'', Chapter 10, for additional ideas.)

Other Considerations

Maintaining emotional stability is a matter of prime concern in the circumstances created by a catastrophic situation, nuclear or non-nuclear. The miasma accompanying some of these potential scenes could be horrendous. Keeping people busy fixing meals and performing other chores, giving each other ''positive'' comments, reading, writing, doing exercises, engaging in personal soul searching, and playing games can all be very helpful to occupants of nuclear fallout shelters, or survivors in other possible disaster situations. Remember that any person can only think of one thing at a time; if that one thing is uplifting, or useful, or entertaining, then discouragement, lethargy, anger, and other uglies have no room to enter. What has happened has happened, and no amount of hysteria can help it.

Some suggested duties to put shelter occupants in charge of are: fixing meals, clean-up, medical, communications, recreation, water, radiation monitoring, security (fire watch and guard duty), defense, training, and cultural refinement.

How hard would it be to find a doctor? Would there be enough local, state or national government left (including police, fire, courts, EPA, equal opportunity, and the building inspector) to be of assistance? Would there be order or disorder? These questions, of course, are not possible to exactly answer.

Figure 4.20   Water For Expedient Shelters

AVENUES OF RETREAT

Where and how to escape high-risk areas is another important consideration. If it is evident that you are in a prime target area, it is advisable that you select a site for a retreat or shelter area far enough away to escape probable damage and yet close enough to be possible to get to. This effort may be a joint effort with a friend living in an outlying area, or a solo venture. Some survival writers recommend buying, leasing, renting, or bartering for space in an acceptable area. Others maintain that the only reasonable approach to surviving is to move (permanently) to an appropriate rural area. Another practical approach would be to combine the shelter with a recreation site. It may be desirable to select a site that is aligned with the local civil defense crisis relocation plans. Otherwise, it is possible that you could be pushed to go where your retreat is not.

Routes and alternate routes of travel for departure from home (if you live in an obvious target area) to a retreat should be predesignated. If you don't have some pretty firm plans, confusion could easily arise.

The Soviet civil defense plans call for their citizens to leave the large urban areas for shelter or retreat. If there are obvious international tensions and the Soviets send their countrymen scurrying to their shelters, it would probably be time to think about leaving high-risk areas for your own retreat—a preconstructed shelter or a suitable area for building an expedient shelter. Materials and tools for building an expedient shelter should be part of the preplanning. A check list of tools and supplies for this purpose is included at the end of this chapter.

Very high risk aeas—such as intercontinental ballistic missile (ICBM) silos, Strategic Air Command bases, Navy submarine bases, and perhaps the largest cities—would be so dangerous to be around that it would be advisable (as mentioned in "Protective Shelters") to be at least forty miles away from them (and not downwind) to be able to receive adequate protection from a simple expedient shelter. A distance of at least twenty miles from smaller, lower-priority targets would be necessary for similar protection. If conditions prevail that would require building shelter closer than these distances, the shelter would also need to be correspondingly more resistant to blast.

SHELTERING ANIMALS FROM FALLOUT

Like humans, animals require shielding, uncontaminated food and water, and proper ventilation. In case of nuclear war it is recommended that a nucleus of breeding stock be protected to the maximum extent possible since it may not be possible to give protection to all animals. Keeping animals in a building would help reduce radiation exposure to a great extent. Other things that can be done to help animals is to wash or brush fallout off them and keep them corralled for a few days after fallout has landed. The animals at the outside of the herd will shield those at the center of the herd from the outlying radiation (see page 156 for information on use of animals exposed to fallout).

Animals should be prevented from grazing on or otherwise ingesting contaminated feed or water. Most stocked feed and water supplies could be largely protected by simply covering them with plastic to prevent the fallout from getting in them. Animals can survive without water for about forty-eight hours. Figure 4.21 gives some additional ideas for protection of farm animals from nuclear radiation.

The following feed, water and ventilation recommendations for animals are

Figure 4.21   Sheltering Animals from Fallout

High-mass shielding can also be used to protect animals from fallout. Water should be guarded against contamination.

published in *Protecting Family and Livestock From Nuclear Fallout*, Agricultural Extension Circular No. 330.

TABLE 4.7
WATER REQUIREMENTS FOR ANIMALS

Animal	Ample Supply Gal./Day	Limited Supply Gal./Day
Cattle	17	7
Hogs	2½	1¼
Poultry		
Layers and Broilers	¹⁄₁₆	¹⁄₂₀
Turkeys	⅓	⅛
Sheep	1½	1

TABLE 4.8
MINIMUM FEED AND SPACE REQUIREMENTS OF ANIMALS

Animal	Feed/Day	Space Sq./Ft.
Cattle		
Cow	1 lb. hay/cwt. body wt.	20
Calf	1 lb. hay/cwt. body wt. + ¼ lb. 40% protein supplement	12
Sheep		
Ewe	1 lb. alfalfa hay/cwt. body wt.	10
Lamb, 60 lb.	1 lb. alfalfa hay/cwt. body wt.	4

TABLE 4.8 (CONTINUED)
MINIMUM FEED AND SPACE REQUIREMENTS OF ANIMALS

Animal	Feed/Day	Space Sq./Ft.
Swine		
Lactating sow	5 lbs. corn + ½ lb. 35% protein supplement	32
Hog, 100 bls.	3 lbs. corn + ½ lb. 35% protein supplement	4
200 lbs.	4 lbs. corn + ½ lb. 35% protein supplement	6
Poultry		
Laying hen	¼ lb. mash	0.7
10-lb. turkey	0.4 lb. mash	1.5
25-lb. turkey	0.7 lb. mash	2.0

Equivalent feeds may be substituted. Hay should be at least ½ legume.

TABLE 4.9
VENTILATION REQUIRED IN ANIMAL SHELTERS

Animal	Cfm/Animal Winter	Cfm/Animal Summer
Cattle		
400-lb. calf	30	80
800-lb. dairy cow	70	200
1000-lb.	100	225
1600-lb.	130	300
Hen	½	6
Sheep		
Nursing Ewe	10	30
60-lb. lamb	7	20
Swine		
Sow and litter	50	100
100-lb. hog	15	40
200-lb. hog	25	75

TABLE 4.10
DOSES EXPECTED TO KILL 50 PERCENT OF LIVESTOCK WITHIN 60 DAYS (LD 50/60)
Total Gamma Exposure in Roentgens

Animal	Barn	Corral	Pasture
		Area	
Cattle	500	450	180
Sheep	400	350	240
Swine	640	600*	550*
Horses	670	600*	350*
Poultry	900	850*	800*

*Approximations; not based on experimental data.
(Gladstone & Dolan, 1977, p.622)

NUCLEAR WAR PREPARATION CHECKLIST

☐ 1. Learn civil defense crisis relocation plans for your area—where to go and what you could expect the civil defense system to provide for your when you get there.

☐ 2. Learn location of fallout shelters in your area and what facilities and supplies they can provide.

☐ 3. Obtain a 14-day Emergency Kit.

☐ 4. Make a permanent shelter and/or select a plan and gather materials for an expedient shelter. Include a suitable method of ventilation for either shelter. A ventilation device can be made ahead of time for an expedient shelter.

☐ 5. Obtain or otherwise make provision for the use of radiological monitoring equipment (survey meter, dosimeter, and dosimeter recharger). A Kearny Fallout Meter could be made ahead of time.

☐ 6. Obtain an anticontamination suit or gather materials to jury-rig one of your own design.

☐ 7. Secure an extended storage supply, including food, water, and other supplies. Also include potassium iodide.

☐ 8. Make provisions for reasonable protection for at least some of your farm animals (if applicable).

☐ 9. Keep a likely form of transportation in good working order and plan escape routes and alternates from your area, keeping in mind how likely a target area you live in.

☐ 10. Be sure all members of your family are knowledgable concerning the previous points of preparation, including how to take immediate protective measures and how to turn off utilities to your house.

EXPEDIENT FALLOUT SHELTER CHECKLIST

☐ 14-day Emergency Kit—including tools
☐ Ventilation equipment such as a premade KAP
☐ Radiological monitoring equipment
☐ Instructions for building a shelter and other nuclear survival information
☐ One 4' x 8' sheet of ½'' or thicker plywood or 2 doors from the house for each person to be sheltered (or other materials for a comparable shelter)
☐ polyethylene sheeting, canvas, or poly tarp
☐ Potassium iodide
☐ Nails, wire, and whatever else you need for the shelter design you would choose to build: Take this book and a copy of Kearny's *Nuclear War Survival Skills*.

SELECTED REFERENCES AND NOTES

Clayton, Bruce D., PhD. *Fallout Survival: A Guide to Radiological Defense.* Boulder, Colorado: Paladin Press, 1984.

————. *Life After Doomsday.* New York: The Dial Press, 1980; Boulder, Colorado: Paladin Press, 1980.
This is a very good book by a sensible and interesting author.

The Committee for the Compilation of Materials in Damage Caused by the Atomic Bombs in Hiroshima and Nagasaki. *Hiroshima and Nagasaki: The Physical, Medical, and Social Effects of the Atomic Bombings*. Eisei Ishikawa and David L. Swain, translators. New York: Basic Books, Inc., 1981.

Douglass, Joseph D., Jr., and Amoretta M. Hoeber. *Soviet Strategy for Nuclear War*. Stanford, California: Hoover Institution Press, 1979.

In Time of Emergency: A Citizen's Handbook. Washington, D.C.: Federal Emergency Management Agency, 1983.

American Survival Guide, published by McMullen Publishing, Inc., 2145 West La Palma Ave., Anaheim, CA, 92801-1785
A very good source of information for all kinds of survival and everyday living.

"High Risk Areas, for Civil Preparedness, Nuclear Defense Planning Purposes," TR-82. Washington, D.C.: U.S. Defense Civil Preparedness Agency, April 1975.

Committee to Study the Long-Term Worldwide Effects of Multiple Nuclear-Weapons Detonations, Assembly of Mathematical and Physical Sciences, National Research Council. *Long-Term Worldwide Effects of Multiple Nuclear-Weapons Detonations*. Washington, D.C.: National Academy of Sciences, 1975.

Kearny, Cresson H. *Nuclear War Survival Skills*. Cave Junction, Oregon: Oregon Institute of Science and Medicine (P.O. Box 1279, Cave Jct., OR 97523), 1987.
This book is a must for everyone.

Office of Civil Defense. *Shelter Management Textbook*. SM-16.1, July 1971—D119.8/8:16.1.

McCarthy, Walton W. *The Nuclear Shelterist*. Great Neck, New York: Todd and Honeywell, Inc., 1986.

Robinson, Arthur with Gary North. *A Fighting Chance*. Cave Junction, Oregon: Oregon Institute of Science and Medicine (P.O. Box 1279, Cave Jct., OR 97523), 1986.

Survival Directory: A Guide to Personal Survival. Lakehurst, New Jersey: Mutual Association for Security and Survival.
Names of high risk cities and possible escape routes.

U.S. Congress. Senate. Committee on Foreign Relations. *Analysis of Effects of Limited Nuclear Warfare*. 1975.

U.S. Department of Defense. *Soviet Military Power*. Washington, D.C.: The Department, 1985, 1988.

U.S. Department of Defense and the U.S. Department of Energy. *The Effects of Nuclear Weapons*. 3d ed. [Philip J. Dolan and Samuel Gladstone, eds.]. [Washington]: The Departments, 1977.

U.S. Department of Defense and the U.S. Department of Energy. *The Effects of Nuclear Weapons*. 2nd ed. [Samuel Glasstone, eds.]. [Washington]: The Departments, 1962.

Zuckerman, Edward. *The Day After World War III*. New York: The Viking Press, 1984.
What our civil defense system would do for us after...

The United States Government has published a plethora of information on nuclear war. Much of it is very technical and useless to most of us, but some is very helpful. Check with your local civil defense office and your local library.

SOURCES FOR NUCLEAR WAR EQUIPMENT

Radiation Monitoring

Nitro-Pak Survival Foods and Supplies, 325 West 600 South, Heber City, UT 84032, (801) 654-0099, (800) 866-4876.

Major Surplus and Survival, 435 West Alondra Blvd., Gardena, CA 90248, (800) 441-8855, (310) 324-8855.

Alnor Nuclear, 2585 Washington Road, Suite 120, Pittsburg, PA 15241. Distributors of the RD10 dose rate meter, EMP protected, made in Finland very good.

Guillory and Associates, P.O. Box 591184, Houston, TX 77259-1184. Distributors of a British-made, EMP protected, high quality dose rate meter.

Dosimeter Corporation, P.O. Box 42377, Cincinnati, OH 45242.

Atomic Products Corporation, P.O. Box 1157, Center Moriches, NY 11934.

Kearny, Cresson H. *Nuclear War Survival Skills*. Cave Junction, Oregon: Oregon Institute of Science and Medicine, 1987. Includes plans for a Kearny Fallout Meter, an inexpensive and effective dose rate meter. The institute's address is P.O. Box 1279, Cave Junction, Oregon 97523, (503) 592-4142.

Nuclear Defense Shelters and Equipment, P.O. Box 31662, Lafayette, LA 70593-1662. Contractors and distributors of all kinds of equipment.

Fans and Ventilation Equipment

Texas Air Handlers Spec Air, Inc., 13999 Goldmark Dr., Suite 401, Dallas, TX 75240, Phone (214) 644-6806. Distributors of hand crank, electric, and combination air blowers.

Nitro-Pak Survival Foods and Supplies, 325 West 600 South, Heber City, UT 84032, (801) 654-0099, (800) 866-4876.

Major Surplus and Survival, 435 West Alondra Blvd., Gardena, CA 90248, (800) 441-8855, (310) 324-8855.

Kearny, Cresson, H. Nuclear War Survival Skills. Cave Junction, Oregon: Oregon Institute of Science and Medicine, 1987. Includes plans for building and using a directional fan, Kearney air pump, and plywood double-action piston air pump. The institute's address is P.O.. Box 1279, Cave Junction, Oregon 97523, (503) 592-4142.

Nuclear Defense Shelters and Equipment, P.O. Box 31662, Lafayette, LA 70593-1662. Contractors and distributors of all kinds of equipment.

Potassium Iodide

Nitro-Pak Survival Foods and Supplies, 325 West 600 South, Heber City, UT 84032, (801) 654-0099, (800) 866-4876.

Emergency Essentials, 165 South Mountain Way Dr., Orem, UT 84058. (801) 222-9596, (800) 999-1863.

Intertech Trading Company, Limited, 170 South Mountain Way Drive, No. 110, Orem, UT 84058. Bulk U.S.P. grade ¼ lb. for $ 6.80, 1 lb. for $ 18.90 as of this writing. Phone (801) 226-6992.

Shelters and Other Equipment

EMP Survival System, P.O. Box 73448, Houston, TX 77273. They sell an emergency electronic ignition bypass device that is said to get your car going if it is stopped by EMP.

Nitro-Pak Survival Foods and Supplies, 325 West 600 South, Heber City, UT 84032, (801) 654-0099, (800) 866-4876.

Subtech, Inc., 138 I, Northwood, NH 03261. Prebuilt fiberglass shelters, equipment, and information. Literature packet is $5.00.

Plans for an 8 foot by 40 foot shelter including blast door and valve, and a list of materials can be aquired from Paul Seyfried and Sharon Packer for $30. These good folks have been helpful in updating this book and are informed in many aspects of nuclear war survival. They and their families built this shelter for about $6,000. For the plans write to Paul Seyfried, c/o Packers, 1950 East Forest Creek Lane, Salt Lake City, UT 84121.

Marcel M. Barbier, Inc. Shelter Design, Construction, 3003 Rayjohn Lane, Herndon, VA 22071.

Major Surplus and Survival, 435 West Alondra, Gardena, CA 90248, (800) 441-8855, (310) 324-8855.

Sirius Concrete Shelters, Rt. 38, Box 2174, Livingston, MT 59047.

The book *Principles of Protection, The U.S. Handbook of NBC Weapons Fundamentals and Shelter Engineering Standards, 5th ed.,* is available from TACDA, P.O. Box 1057, Starke, FL 32091. (See ''Civil Defense'' below.)

The book *Nuclear Defense Issues,* by Sharon Packer and Paul Seyfried is available through: Bill Seyfried, Civil Defense Volunteers, P.O. Box 2355, Sandy, UT 84091-2355. Cost is $15.00 post paid and this practical, informative book is well worth the cost.

Civil Defense

The American Civil Defense Association (TACDA), P.O. Box 1057, Starke, FL 32091. They publish the Journal of Civil Defense and TACDA Alert, conduct national seminars, and generally work to promote responsible civil defense. Supporters include an impressive list of scientists, polititians, educators, analysts and other authorities.

Strategic Defense

High Frontier, 1010 Vermont Ave., N.W., Washington D.C. 20005. Lt. Gen. Daniel O. Graham founded High Frontier. He is also chairman of the American Space Frontier Committee and the Coalition for the Strategic Defense Initiative (SDI). This team of strategists, scientists and engineers are proponents of very realistic and practical solutions to our strategic defense. Unfortunately the politics of this rather solvable problem border on the insolvable.

II
THE NECESSITIES
OF LIFE

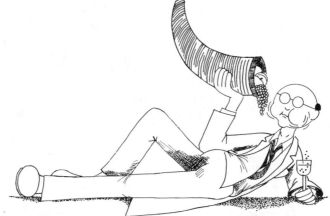

   Necessities of life include what is necessary to keep healthy and properly equipped. Emergency kits; nutrition; storage; finding, growing and using plants and animals; water; equipment; shelter; sanitation; weapons; communication and signaling; transportation; barter; care of infants; keeping cool; knots; and sanity-saving games are all items of interest.

CHAPTER 5

"14-Day Emergency Kit"

Within the civil defense system of America is provision to relocate city dwellers to rural areas in case of emergency—especially nuclear war. The Crisis Relocation Program (CRP) recommends a seventy two-hour emergency kit consisting of prepackaged, easily portable, readily accessible supply of food (the food should not require refrigeration or cooking) and other necessities that can serve your family's needs under any conditions that may occur in your area until you can relocate and be supplied by the Civil Defense System or by some other means. In most disasters the first seventy-two hours are the most critical to survival and many adverse situations develop so quickly there is little or no time for preparations.

Of course a seventy two-hour kit would take care of things in many foreseeable difficulties. Even in a nuclear war it may be reasonably safe in many areas to leave shelter after seventy-two hours for long enough to retrieve additional nearby supplies. But in some instances a seventy two-hour kit would not be enough. The question is: If the emergency lasts more than 72 hours, who is going to feed you? In all candor it is unlikely that the Civil Defense System is sufficiently strong to feed very many people in very many areas. (Call your local Civil Defense Office and ask them!)

Here are some examples where a seventy two-hour supply may not do the job:
1. A nuclear war, in which case it is reasonable to be prepared to survive for at least two weeks in a shelter with only minimal cooking.
2. A widespread earthquake which may not allow supplies to be forwarded to your area for longer than seventy two hours.
3. The *threat* of nuclear war, which could bring about widespread relocation to rural areas.

With very little more preparation, a seventy two-hour kit can contain enough food and water for two weeks and become a Fourteen-Day Emergency Kit. The kit could then handle a two-week fallout shelter stay or a relocation. The foods can be replaced and rotated into normal use every half-year or full year to prevent waste and to assure a useable supply.

The kit should include provisions for food, shelter, heat, sleeping gear, clothing, light, tools, sanitation, personal items, first aid, valuables, water, and speciality and miscellaneous items. The most important factor to consider in a crisis is protection from the extremes of the crisis, including the weather. After that, water is most essential to survival; food is next; and then the other equipment.

The following list gives several possibilities under some headings. These are only meant to be suggestions of various possibilities and not an all-inclusive list of essentials. For instance, under shelter are listed "tube tent, lightweight nylon tent, family-size tent, and/or motor home.'' You obviously do not need to buy them all; just obtain the best you can practically afford that will do the job. Or, you simply may not be interested in having some of the items listed in your kit. For related discussion, see the individual subjects elsewhere in this book. The discussion on equipment in Chapter 11 may be especially helpful.

FOOD

The food supply should consist of foods you and your family like that provide reasonably balanced nutrition. Special dietary needs, such as those for babies and diabetics, should be considered. Your short-term emergency supply should not require refrigeration and should need little or no cooking. Energy-rich food is helpful in keeping up body energy. Salty foods increase need for water intake and should be minimized.

Although the food should be able to be used without cooking, it would be desirable to provide a means of cooking or at least heating it. Canned heat and heat-tab stoves are very inexpensive and would do the job, as would a regular camping or backpacking stove—which is preferred. A camping cook kit, kitchen cookware, or some shortening cans could be used to cook in.

One list (*Essentials of Home Production and Storage*, The Church of Jesus Christ of Latter-day Saints, p. 11) estimates the following to supply one person with essential nutrients and 2100 calories per day for three days (seventy two-hours):
—1/2 lb. canned tuna fish or pork and beans
—1/2 lb. nonfat dry milk
—1 lb. graham crackers
—1 lb. dried apricots
—46 oz. canned orange or tomato juice
—1/2 lb. peanut butter
By multiplying these amounts by four and adding some hardtack candy and a couple of cans of stew or freeze-dried meals, this would easily become a fourteen-day rather than a seventy two-hour supply.

The story of the farmer who told his wife to add another cup of water to the soup whenever a visitor was present for a meal gives some further "food for thought" here. In a hard-press situation you could add some salt, boullion and hot water to a "meal" that is made to feed four and feed twenty-four. The "soup" might be a little short on calories but it could still be filling and refreshing.

Another concentrated food item that is very convenient and attractive for an emergency supply is "food tabs". These tasty tablets are nutritionally balanced, easily assimilated, and can be stored for several years with no significant deterioration. Several varieties are available. (See Sources at end of chapter.)

Dehydrated and freeze-dried foods designed for backpacking and camping are ideally suited for emergency supplies, but adequate water supplies must accompany them.

Meals Ready to Eat (MREs) are very practical and easy to use. These retort foods are used as field rations by the U.S. military and are essentially canned meals in "soft cans"—foil and plastic. They are lighter weight than canned foods, need no cooking, and are widely available (some sources are listed at the end of this chapter). MREs should be heated to make them more palatable. Heating in a water bath is usually most convenient. A variety of entrees are available. Try them first and select the ones you like. Some are rather strong-flavored and may not suit everyone's tastebuds. In fact it could be worse—members of at least one family I know refuse to eat some flavors.

If an MRE pouch is bulging or swollen do not eat it! This is similar to a bulging can and means the contents are spoiled and can cause extreme illness. The soft pouches are more easily perforated than cans—a frequent cause of spoilage.

Dehydrated fruits that are available with MREs are tasty and make good snack food, but eat them slowly and drink water with them to prevent dehydrating your digestive system. This has been a recurring problem and can be fairly serious.

Other suggestions for food include nuts, nut butters, crackers, cereals, canned fruits and vegetables, boullion, canned juices, cookies, freeze-dried meals, canned meats and stews, cheese spreads, hardtack candy, other candy, dried fruits and vegetables, canned milk, powdered milk, vitamin pills, and at least one pound of table salt per person (many uses).

Food preparation items may be only a sierra cup and a spoon; or paper plates and plastic utensils; or a good camping cook kit, including pans, plates, and cups; or something improvised from the home cupboard for your own needs. Cooking or heating can also be done with empty shortening (or similar) cans or aluminum foil. Dish soap, paper and cloth towels, a pitcher for mixing and pouring, and something to use for a dish pan should also be included. And don't forget the can opener!

SHELTER

Shelter could be plastic tube tents (2 per person), lightweight nylon tent, large family-size tent, and/or motor home. In addition, a poncho is essential; a piece of polyethylene sheeting, a space blanket, and/or a tarp are very useful.

HEAT

Heating needs may be met with a campfire, camp stove, kerosene heater, propane heater, portable metal stove (some of these fold up and can burn wood or charcoal), heat tab stove, canned heat and stove, matches, lighters and/or other means of lighting fire, and fuel for whatever is used.

SLEEPING GEAR

Include warm dry clothes; some combination of wool blankets, other blankets, and/or sleeping bags; a "space blanket;" and an insulating pad.

CLOTHING

Adequate, sturdy, comfortable clothing is a must. Include sturdy footwear (which may not be available after the disaster) poncho, coat, hat, gloves, socks, and underwear.

LIGHT

Include flashlights and candles—at least. Lamps and lanterns are also possibilities.

Don't forget matches and/or lighters and fuel and batteries. Lighters tend to lose their fuel so don't count on them alone for lighting purposes. Light Sticks are also good.

TOOLS

Pliers, saw, axe, file, sharpening stone, wire, cord, adjustable wrench, hammer, screwdrivers, duct tape, shovel, pocket knife and/or larger knife, can opener, bucket, nailbar or crowbar, and pick are all pretty basic.

SANITATION

Remember to include an emergency toilet, toilet paper, feminine hygiene products, disposable diapers, premoistened towelettes, plastic sacks, shovel, soap, towels (paper and cloth), and disinfectant. (See also the discussion in "Sanitation," chapter 13.)

PERSONAL ITEMS

Some suggested personal necessities are: toothbrush and toothpaste, hairbrush and/or comb, shaving gear, deodorant (indispensable), mirror, nail clippers, and personal medication. (Baking soda can double as toothpaste and deodorizer for room or person.)

FIRST AID

Take a first aid course and review the information often. Keep immunizations current—especially tetanus. Keep a first aid kit and a good manual with the kit. (See "First Aid", chapter 17, for more specific suggestions on first aid kits, manuals and supplies.)

VALUABLES

Cash, personal papers, licenses, treasured books, important photos, insurance policies, contracts, deeds, social security information, passports, birth certificates, checkbook, charge cards, wills and testaments, genealogical records, jewelry, and so on should be placed where they can be readily retrieved.

WATER

There should be at least seven gallons for each person, and preferably fourteen gallons. (Portability becomes a problem here.) Water should be stored in containers no larger than fifteen gallons (depending on the container, fifteen gallons will weigh upwards of 130 pounds) otherwise, it is very difficult or impossible to carry. There are sturdy, heavy-duty plastic carboys that would probably withstand a substantial earthquake and are ideal if there is someone around who can lift them. The heavy-duty, one-to-five-gallon plastic jugs or buckets with lids and insulated jugs are also good. In a pinch, five gallons of water could keep a person going for a period of two weeks of shelter living; but if you were doing heavy work in southern Arizona in the summer, that amount would not cover it. A purifier and/or purification tablets should also be included (see Chapter 10). Large plastic bags contained in a burlap bag, pillowcase, or makeshift cloth sack will hold water in the absence of something better.

Other options are retort water packets and mylar-cardboard box containers.

SPECIALTY ITEMS

Other important items are weapons, communication gear, "survival kit," and war protection equipment. *Weapons* are discussed in chapter 14 of this book. Weapons that may be included in this kit are a matter of individual choice. Most civil defense instructions proscribe firearms in this type of situation, but I think that is somewhat unrealistic—especially in some areas of the country and in some forseeable circumstances.

Communications gear should include a portable radio (with spare batteries), compass, map of your area, an emergency signaling device (mirror, whistle, flare gun, etc.), and possibly even some CB or other equipment.

A *"survival kit"* for procuring small animals might include a razor blade, wire, monofilament line, picture-hanging wire, fishing hooks, aluminum foil, a few finishing nails, and a knife.

War protection equipment consists of rain suit or improvised suit to keep nuclear fallout or chemical agents from touching your body or from being carried into your shelter (see "anti-contamination suit" in chapter 4), monitoring equipment for nuclear fallout (i.e., a dosimeter and a survey meter); and, depending on what provision you have made, materials for an expedient fallout shelter including a ventilation device.

MISCELLANEOUS

Pencils and notebook; polyethylene sheeting; sewing kit (including needles, thread, safety pins and scissor); aluminum foil; fire extinguisher; survival manual; recreational equipment such as games, musical instruments, song books, other books, paints, glue, paper, scissors (see Chapter 16 for additional ideas); small piece of hose (for siphoning); some insect screen; and dust masks. Dust masks are included because many of the potential emergencies, such as a volcano, can produce a lot of dust. A large bandana could also serve as a dust mask and impart other utility as well.

One final very important item is at least one-half tank of gas in your car.

CONTAINERS

Containment of items in this kit could probably be best done in a soft, waterproof duffle bag or in good backpacks that can be easily carried; but other containers such as trunks, suitcases, new metal garbage cans, could also do the job. Avoid cheap plastic garbage cans; they crack and break with time and in cold. If you use garbage cans, don't overload and make them too heavy to easily lift. Some of these items listed may not lend themselves to sitting in storage—some of the tools and valuables for instance. Keep a list of these things. Don't be misled into thinking you could gather "everything" up at a moment's notice. You can't do it! If you don't believe it—try it!

Preparations to enjoy backpacking and camping can also, at the same time, be important emergency preparations. As part of fourteen-day emergency kit gear, backpacks can be packed with their ordinary backpacking trip containment. Each family member could carry an appropriate-sized kit that would enable his survival if separated from the family group.

As is also mentioned in "software" (chapter 11), in a pinch if you have to leave home and pack things in a hurry—lay out three or four layers of sheets and/or blankets. Place things in the middle, fold the opposite ends together and tie them in square knots. Wrap a rope or strap around the knots and tie firmly. Such a bundle can be carried "Santa Claus" style, or two such bundles may be tied to the ends of a shovel handle or other sturdy bar or board and carried "yoke" fashion. A good pad (perhaps another blanket) would be necessary to cushion such a yoke from the shoulders.

Don't be overwhelmed by the thought of putting together a 14-Day Emergency Kit. Most of it is already around the house. Just start gathering it up. Begin simply and go from there. It's just fine if you have to start out with a piece of plastic, a roll of paper towels, a jug of water, a shovel, a sewing kit, a sack of hardtack, a can of tuna fish, a coat, valuables, and a pencil and notepad. Then with time, keep building it up as you can.

As part of emergency departure preparations all responsible household members should know how to turn off the utilities (gas, electricity, and water). Some will probably require tools to turn them off. If you ever do leave home in the emergency mode it's a good idea to lock the doors and windows too.

In addition, many who have thought considerably about the possibilities for needing emergency kits see the need for a very basic lightweight emergency kit made up of carefully selected items. This kit would then be always carried in the car or wherever the owner goes away from home, including work and other daily pursuits. It makes sense.

There are two main ideas on emergency kits. The first is that if an emergency threatens your home you need to, most importantly, escape with your hide. The second is that if you leave home in a hurry, you still need to be able to function. Most reasons for leaving home in a hurry seem to indicate a threat to the home. Hence, the kit should be transportable by car but may need to be quickly rearranged to less bulky proportions.

14-DAY EMERGENCY KIT CHECKLIST

Food
- [ ] Food
- [ ] Utensils (pots, plates, spoons, etc.)
- [ ] Cleanup gear (dishpan, towels, soap)

Shelter
- [ ] Tent(s)
- [ ] Poncho
- [ ] Tarp, plastic, space blanket

Heat
- [ ] Stove or heater
- [ ] Matches, lighter

- [ ] Fuel

Sleeping Gear
- [ ] Night clothes
- [ ] Bag and/or blankets
- [ ] Ground pad

Clothing
- [ ] Complete change of clothing
- [ ] Suitable outerwear (coat, poncho, hat, gloves)
- [ ] Sturdy footwear

Light
- ☐ Candles
- ☐ Flashlight, Lightsticks
- ☐ Lamps or lanterns
- ☐ Fuel, batteries
- ☐ Lighters, matches

Tools
- ☐ Pliers, wrenches, and screwdrivers
- ☐ Knife, axe, saw, and hammer
- ☐ Shovel, pick, and bar
- ☐ Cord, tape
- ☐ Bucket, can opener
- ☐ File, sharpening stone, and sharpening steel

Sanitation
- ☐ Toilet paper
- ☐ Toilet (see Chapter 13 for ideas)
- ☐ Disinfectant
- ☐ Soap
- ☐ Diapers
- ☐ Hygiene items

Personal Items
- ☐ Toothbrush, toothpaste
- ☐ Hair care equipment
- ☐ Shaving gear
- ☐ Nail clippers
- ☐ Personal medications

First Aid
- ☐ Take a first aid class and put your notes in the kit

- ☐ Kit
- ☐ Manual
- ☐ Immunizations

Valuables
- ☐ Personal papers & records
- ☐ Cash, jewelry, credit cards
- ☐ Books, photos
- ☐ Deeds, contracts, policies, certificates

Water
- ☐ Contained water
- ☐ Water purification device
- ☐ Container (bucket, pan, plastic sacks)
- ☐ Solar still

Specialty Items
- ☐ Weapons
- ☐ Communication gear
- ☐ Survival kit
- ☐ War protection gear

Miscellaneous
- ☐ Radio
- ☐ Survival manual
- ☐ Pencil and paper
- ☐ Fire extinguisher
- ☐ Sewing kit
- ☐ Dust masks
- ☐ Fuel for auto
- ☐ Games
- ☐ Insect Screen

SELECTED REFERENCES AND NOTES

Christiansen, John, Reed Blake, and Ralph Garrett. *Disaster Preparedness*. Bountiful, Utah: Horizon Publishers and Distributors, 1984.

Crockett, Barry G., and Lynette B. Crockett. *72-Hour Family Emergency Preparedness Checklist*. Orem, Utah: published by the authors, 1983.

Essentials of Home Production and Storage. Salt Lake City, Utah: The Church of Jesus Christ of Latter-day Saints, 1978.

SOURCES

Nitro-Pak Survival Foods and Supplies, 325 West 600 South, Heber City, UT 84032, (801) 654-0099, (800) 866-4876.

Major Surplus and Survival, 435 West Alondra, Gardena, CA 90248, (800) 441-8855, (310) 324-8855.

Emergency Essentials, 165 South Mountain Way Dr., Orem, UT 84058, (801) 222-9596, (800) 999-1863.

Country Harvest, Inc., 325 West 600 South, Heber City, UT 84032, (801) 654-5400.

CHAPTER 6

Nutrition

There is an old expression that "an army travels on its stomach." It is true that some of the great masters of the flesh have survived many days of fasting; but the fact is that everyone must have regular nutritional intake or ill health and eventually death will result.

Estimates predicated on representative studies indicate that many Americans do not receive adequate diets during normal times, and it follows that this could become much more pronounced during the high stress and less abundant supply of a prolonged crisis.

NORMAL NUTRITIONAL REQUIREMENTS

A list of average nutritional needs stated as Recommended Dietary Allowances (RDAs) is published by the National Academy of Sciences-National Research Council. These allowances take into account that size, sex, current amount of stress, metabolism, and so forth produce wide variations in nutritional needs. A partial U.S. RDA vitamin chart is presented in Table 6.1. Another commonly used nutritional standard is the U.S. Recommended Daily Allowances (U.S. RDA). These are protein, vitamin, and mineral standards set by the Food and Drug Administration (FDA) for food labeling and are derived from the National Research Council's RDA values, mostly representing the highest values from among the various sex-age categories. Basically. both of these standards are extentions of the simple fact that the body needs water, protein, carbohydrate, fat, vitamins and minerals.

A brief review of basic nutritional requirements can help in formulating emergency storage as well as improving current eating habits. At any point a wide variety of well-selected foods eaten in moderation generally provides a good diet.

Water

Water is necessary to essentially all body functions. The preservation of a usable source of water is a foremost survival necessity. Salts (especially sodium chloride—table salt) are also necessary to preserve proper water balance in the body.

Protein

Protein is necessary for growth and repair of bones, muscles, organs, genes, blood and so on. It consists of amino acids in various combinations. There are about twenty-two common amino acids used by the body. The body itself manufactures all but nine of them, which must be taken in from other sources. (One other amino acid, arginine is also necessary for growth in children). Of the nine, five are common to many foods and four are not. The four that are less common—lysine, isoleucine, methionine, and tryptophan—must be acquired by eating foods which contain them in sufficient quantities.

TABLE 6.1

VITAMINS

Vitamin	U.S. RDA	Sources	Functions	Symptoms of Deficiency
Fat Soluble				
Vitamin A	5000 International Units (IU)	Egg yolks, liver, butter, margarine, whole milk, yellow vegetables and fruits, dark green leafy vegetables (including alfalfa), sweet potatoes; some sprouts contain small amounts	Eyes, skin, mucus membranes, and bones all need Vit. A for proper function; also helps resist infection	Eye deterioration, dry skin & nails, pimples and other skin eruptions, improper growth in children
Vitamin D	400 IU	Fortified milk, butter, margarine, oily fish, exposure to sunlight	Needed by the body to utilize calcium and phosphorus for bone growth and tooth growth	Defective bone growth in children (rickets); hardening of bones in adults
Vitamin E	30 IU	Whole grains, cereal, nuts, seeds, legumes, vegetable oils	Helps body utilize some nutrients, aids in proper cell function and circulation	Muscle cramps, possible decrease in some glandular function
Vitamin K		Green leafy vegetables; also manufactured by bacteria in the stomach	Needed for production of platelets necessary for normal blood clotting	Hemorrhaging—inability of blood to clot in air; rare
Water Soluble				
Vitamin B_1 (thiamine)	1.5 mg	Whole grains, milk, liver, pork, peanuts, brewers yeast, legumes, nuts, asparagus, collards, okra	Needed for metabolism of carbohydrate and protein and for function of nerve and muscle tissue	Causes improper growth in children; also beriberi, numbness, loss of appetite, nausea, abnormal heart function, depression
Vitamin B_2 (riboflavin)	1.7 mg	Milk, eggs, liver, kidney, heart, turnip greens, yogurt, almonds, beef, spinach, peas, yeast	Needed for metabolism of carbohydrates, amino acids and fats; needed for a wide variety of body functions	Nerve problems, low energy, low immune response, general malaise, skin disorders, bloodshot eyes, hair loss
Vitamin B_3 (niacin)	20 mg	Lean meat, poultry, fish, liver, kidney, whole grains, peanuts, yeast, potatoes, summer squash, peas	Needed for tissue repair and for metabolism of sugars and tryptophan	Pellagra—headache, depression, sores on skin and mouth, insomnia, inability to concentrate
Folic Acid	0.4 mg	Liver, kidney, yeast, oranges, fish, legumes, nuts, asparagus, broccoli, some leafy greens	Needed for protein metabolism, red blood cell formation, cell transport	Anemia, fatigue, nervousness, constipation, deficiency in pregnant women can cause spina bifida in babies

Vitamin	U.S.RDA	Sources	Functions	Symptoms of Deficiency
Vitamin B$_6$	2 mg	Liver, fish, poultry, whole grains, legumes, potatoes, broccoli, oranges, bananas, milk, beef	Needed for red blood cell production, proper metabolism, soft tissues (skin, digestive system, nervous system)	Nervous disorders such as depression, irritability; also nausea, anemia
Vitamin B$_{12}$ (biotin)	0.006 mg	Milk, liver, fish, poultry, eggs, cheese, yeast	Needed for proper cell functions and red blood cell production	Weakness, nerve degeneration, increased susceptability to pernicious anemia
Pantothenic Acid	10 mg	Liver, eggs, yeast, many legumes, whole grains, fish, beef, potatoes, many other foods	Needed for proper digestion, proper growth, red blood cell production, production of some hormones	Headache, nausea, fatigue, improper growth in children
Vitamin C	60 mg	Citrus, tomatoes, berries, potatoes, broccoli, raw cabbage, many other fresh fruits and vegetables, ascorbic acid pills; Vit. C is one of the vitamins most easily destroyed by cooking	Vital to healing, growth, and immune systems of the body	Tendency to bruise or hemorrhage easily, scurvy, degeneration of gums, weakness

The body's ability to manufacture a given protein compound can be limited by the available amount of any one of its amino acid constituents, just as a recipe can be limited by the quantity of one of its components. Even if all other amino acids are present in abundance, the absence or scarcity of one of the components will limit the amount of finished protein that can be produced. Hence, it is important to provide combinations of foods in your diet and storage which can provide all necessary amino acid building blocks.

Protein foods can be divided into two categories according to their amino acid content; namely complete proteins and partially complete proteins.

Complete proteins are those foods which have balanced essential amino acid content. Some examples are milk, meats, eggs, poultry, and fish. *Partially complete protein* are those which are deficient to some substantial degree in one or more of the essential amino acids. Grains, legumes, nuts, and seeds all fit into this category.

Combining foods to achieve a proper amino acid balance in a diet can greatly improve the protein quality. A notable example of mixing food to provide better protein in the survival diet (or any other) is that of mixing grains and legumes. The Chinese mix soybeans with rice, and the Mexicans eat beans with their corn tortillas. A practical ratio for this mixing is about 2 parts grain to 1 part legumes (by weight). This could be reasonably stretched to a higher grain value, such as 6 parts grain to one part legumes or even greater for healthy individuals.

The leaves of the prolific leafy green comfrey plant, long used as an herb, is also an easily grown, low-grade source of less common essential amino acids (generally higher in amino acids in spring, lower in fall). (See also "Comfrey", Chapter 18.) Wheat can also be mixed with amaranth to make a complete protein.

Table 6.2 shows some of the food combinations which can provide whole proteins by complementing each other.

TABLE 6.2
COMPLEMENTARY PROTEINS

Whole grains combined with dairy products, legumes, nuts, seeds, or eggs at any one meal can supply complete protein. Below are some suggested examples of each category. Meats and fish, although not listed below, are excellent complements to grains or other partially complete protein foods.

Whole Grains	Dairy Products	Legumes*
Barley	Cheese—cheddar,	Black beans
Buckwheat	cottage, or processed	Blackeyed peas
Corn	Milk—fresh or	Chickpeas
Millet	reconstituted dry	Kidney beans
Oats	Yogurt	Lentils
Rice		Lima beans
Rye		Mung beans
Wheat		Navy beans
		Peanuts
		Pinto beans
		Soybeans

Nuts	Seeds	Eggs
Almonds	Pumpkin	Chicken
Brazil nuts	Squash	Duck
Cashews	Sunflower	Goose
Chestnuts		Other
Hazel nuts		
Pecans		
Pinenuts		
Pistachio nuts		
Walnuts		

*Sprouted beans such as mung and soy may also be used.

Carbohydrates

Carbohydrates provide energy and facilitate body processes. They can be simple, such as honey and sugar, or complex, as is the case in most naturally occurring foods. Fruits, vegetables, grains, legumes, nuts, and seeds are all good sources of complex carbohydrates. The B vitamins are necessary to utilize carbohydrates. One pleasant form of carbohydrate in a survival situation might be some sweet breads, cookies, or cakes. They could produce an emotional as well as physical lift.

Fat

Fat is often thought of as being undesirable in the diet. It is, however, a very concentrated form of energy, and some fat is essential for proper nutrition. The liquid oils are usually substantially higher in the essential fatty acids (EFA) than the solid forms. Shortening, liquid vegetable oils, peanut butter, margarine, butter, nuts, seeds, mayonnaise, and salad dressing are all high in fats and oils. A survival diet should include at least one ounce per person per day of fat or cooking oil which, in the form of vegetable shortening or vegetable oil, would add about 250 to 270 calories to the diet. This totals about twenty-three pounds for a one year supply for one person. (Liquid vegetable oil weighs a little less than eight pounds per gallon.)

Vitamins

Vitamins are essential to good health and proper body function. Table 6.1 lists common vitamins and food sources that can provide them. In a survival diet vitamins A, C, and D are most likely to be deficient; such deficiencies can cause serious health problems (see Table 6.4). Vitamins A and C can be acquired in small quantities by eating sprouts (see "Sprouting," Chapter 7). Vitamin D is not necessary for adults but is necessary for growing children and can be acquired by exposure to sunlight. In some emergencies these sources may not be readily available; for this reason it is wise to store some multiple vitamin pills or at very least some vitamin C tablets. A ration of 365 multiple vitamins per family member can be relatively inexpensive and gives a sense of nutritional security. Even one vitamin-and-mineral supplement for every two or three or more days could help insure dietary adequacy. Vitamin pills lose their potency over a period of time and cannot be stored indefinitely. Keep them stored in a cool place, and check the labels frequently so that they can be used up and replaced as needed. Finally, it is also important to understand that many vitamins taken in overabundance are potentially toxic. Moderation is especialy important in concentrated vitamin intake.

Minerals

Some minerals are also necessary for good health. Minerals can be divided into two groups: those which are needed in relatively large quantities (sometimes called macrominerals), and those which are needed in only small quantities or trace amounts (sometimes called microminerals).

The major macrominerals are calcium, chlorine, magnesium, phosphorous, sulfur, potassium, and sodium. Of these, sulfur is given no RDA and is abundantly present in most diets, as are chlorine, magnesium, and phosphorus.

The trace or microminerals that are currently considered essential are chromium, cobalt, copper, fluorine, iodine, iron, manganese, molybdenum, nickel, selenium, silicon, vanadium, and zinc. Of these, cobalt, manganese, molybdenum, nickel, silicon, vanadium, and selenium are probably sufficient in almost any reasonable diet. Table 6.3 lists various minerals and some sources that can provide them. There are other trace minerals, some of which are needed by animals, that may be useful to the human body; but their need has not been scientifically established.

The minerals most likely to be deficient in a survival diet are iron, calcium, and possibly magnesium. Fluoride for children may also be deficient. For this reason, supplements of these minerals may also be desired storage items. Some common greens that are high in calcium are beet, turnip, mustard, dandelion, and kale.

Iron intake may be increased by cooking acidic foods such as tomatoes in iron pots or by taking small quantities of iron solutions. One way to make an iron solution is to place iron nails (do not use galvanized nails) in a container of vinegar and let it sit for two to four weeks until small pieces of iron rise to the surface. One teaspoon of this solution will contain around thirty to sixty milligrams of iron (two to four times the U.S. RDA).

TABLE 6.3

MINERALS

Mineral

Major

	U.S. RDA	Sources	Functions	Symptoms of Deficiency
Calcium	1 g (1000 mg)	Milk, cheese, cottage cheese, leafy green vegetables, clams, eggs, some nuts and grains, alfalfa	Needed for bones, nerves, muscles	Weakening of bones and teeth, nervousness, muscle spasms
Phosphorous	1 g	Milk, cheese, eggs, dried peas, meat, fish, poultry, whole grains	Works with calcium	Rare; most American diets contain too much phosphorous
Magnesium	400 mg	Whole grains, nuts, dark green vegetables, meat, beans	Similar to calcium, activates enzymes	Rare; most Americans receive more than adequate supplies of magnesium
Potassium		Meat, fish, poultry, grains, fruits, esp. bananas, vegetables, legumes	Needed for nerve transmissions, muscle action, fluid balance	Rare; weakness, nervousness; diarrhea can cause deficiency
Sodium		Table salt, milk, meat, fish, poultry, eggs, cheese	Help maintain proper fluid balance	When salt is low in blood, nausea, fatigue, heat stroke can occur
Chloride		Table salt	Helps maintain proper fluid balance	See Sodium above

Trace

	U.S. RDA	Sources	Functions	Symptoms of Deficiency
Chromium		Meat, whole grains, brewers yeast	Aids in sugar metabolism	Improper sugar metabolism
Copper	2 mg	Liver, meat, seafoods, legumes, whole grains	Needed for proper utilization of iron, tissue repair and growth, nervous system	Weakness, degeneration of several body functions; especially needed by infants
Fluorine		Naturally fluoridated water, fish, tea, spinach, kale, parsley	Can aid in reduction of tooth cavities	Soft teeth
Iodine	0.150 mg	Iodized salt, seafood	Necessary for proper growth and metabolism	Goiter (thyroid enlargement); improper thyroid function
Iron	18 mg	Liver, meat, egg yolk, fish, prunes, molasses, whole grains, legumes, leafy greens, vegetables	Necessary for proper blood function	Weakness, fatigue

Trace	U.S. RDA	Sources	Functions	Symptoms of Deficiency
		(Vitamin C helps absorption of iron)		
Manganese		Legumes, whole grains, nuts, tea		Rare in reasonable diet
Molybdenum		Organ meats, whole grains, legumes, leafy vegetables		Rare in reasonable diet
Selenium		Seafoods, meat, whole grains		Rare in reasonable diet
Zinc		Meat, seafoods, whole grains, legumes	Needed for proper metabolism, and cell growth	Poor healing, impaired sexual function, poor appetite

Calories

In addition to the foregoing dietary considerations the total caloric intake and its sources are also important. Table 6.4 shows the nutitional content of some possible survival diets (grain and beans only). This illustrates the need for a substantial weight of food to provide the needed calories. A survival diet should consist of at least 2600 calories per person on the average. A thirty pound two-year-old needs about 1400 calories per day; a sixty pound nine-year-old about 2500 calories per day and one-hundred fifty pound eighteen-year-old male would normally need about 3400 calories. Teenage and adult females normally require slightly fewer calories than males. A man involved in heavy labor in cold weather could require 5000 or 6000 calories per day. Metabolisms, weather, efforts, and other factors have great effect on the need for caloric intake.

Some additional ideas on foods which would provide adequate nutrition appear in Chapter 7 ("Food Storage") and there are some suggestions on feeding babies in Chapter 16 ("Miscellaneous").

KEEPING NOURISHED DURING TIMES OF STRESS

During the extreme upsets and stresses present in most survival situations there is a tendency to not take the time and effort to eat properly. It becomes easy to eat candy and pleasant things that give quick energy, and it becomes more difficult to eat solid meals. This is one of the great reasons to have some food on hand that is part of the normal diet and to not depend totally on grain or other storage items that may be seldom used in normal times. People have become seriously undernourished during times of crisis when there was plenty of food available. But the food was not normal fare and was not appealing. Eating sugars, extreme stresses, and lack of rest can also put a special drain on body reserves of B-vitamins and calcium, which are vital to the nervous system.

A final point is that there is also a tendency to rely on eating to cure all body ills and to do without proper rest in times of difficulty. This may be necessary for a short time to

TABLE 6.4

APPROXIMATE NUTRITIVE VALUES IN A SURVIVAL DIET OF GRAIN AND BEANS

	1 lb. wheat + 4 oz. soy beans[2]	1 lb. wheat + 4 oz. kidney beans	1 lb. yellow corn + 4 oz. soybeans	1 lb. yellow corn + 4 oz. kidney beans
Protein	101	89	81	69
Fat	27.6	9.6	38.6	20.6
Total energy (calories)	1950	1930	2059	2039
Vitamin A (RE)	0[3]	0[3]	461[4]	461[4]
Vitamin C	0	0	0	0
Vitamin D	0	0	0	0
Thiamine B_1 (mg)	3.7	3.1	2.9	2.3
Riboflavin B_2 (mg)	.9	.8	.9	.7
Niacin B_3 (mg)	22	22	11[5]	11[5]
Calcium (mg)	425	305	329	209
Iron	23	24	19	19.6

[1]The information in this table is approximate and is adapted from values found in U.S. Department of Agriculture and the Science and Education Administration. *The Value of Foods*. 3d ed. Oxford, Oxon, England: Oxford University Press, 1979.

[2]Lentils, dried peas, and other legumes also provide high nutritive values and could be well used in combination with the foods mentioned in the chart.

[3]Wheat only.

[4]Corn only - This is less than half the RDA for an adult male. Other sources of vitamin A should be available, either in foods or in multiple vitamin tablets.

[5]The niacin in corn is bound in a form that is not readily usable by the body. By soaking the corn in lime water (about 1% lime) several hours or overnight the niacin is liberated and becomes available for use by the body. The practice of soaking and/or boiling corn in lime water is common in Mexico, where significant quantities of corn are used. This prevents pellagra—a combination of muscular, mental, and skin disorders caused by lack of niacin.

meet the immediate demands of the problem; but in a prolonged crisis staying healthy, including proper rest when time is available, can mean the difference in surviving and not surviving.

SELECTED REFERENCES

Briggs, George M., and Doris H. Calloway. *Nutrition and Physical Fitness*. 11th ed. New York: Holt, Rinehart, and Winston, 1984.

Fisher, Patty, and Arnold Bender. *The Value of Foods*. 3d ed. Oxford, Oxon, England: Oxford University Press, 1979.

Guthrie, Helen Andrews. *Introductory Nutrition*. 3d ed. St. Louis, Missouri: C.V. Mosby Company, 1975.

Kearny, Cresson. *Nuclear War Survival Skills*. Cave Junction, Oregon: Oregon Institute of Science and Medicine, 1987.

Ruff, Howard J. *Famine and Survival in America*. Salt Lake City, Utah: Publishers Press, 1974.

U.S. Department of Agriculture and the Science and Education Administration. *Nutritive Value of Foods*. Home and Garden Bulletin No. 72. [Washington, D.C.: U.S. Government Printing Office, 1981].

The USDA's new and enlightened Food Guide Pyramid is helpful. A representation is here included:

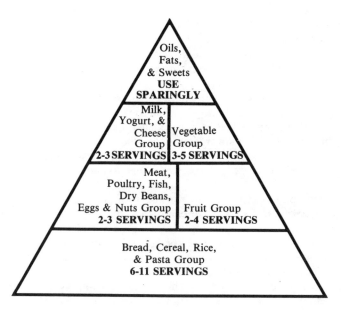

SOURCE: U.S. Department of Agriculture, U.S. Department of Health and Human Services.

CHAPTER 7

Food Storage

Many church and civic leaders recommend storage of at least a one-year supply of food and other essentials. If this is not a familiar concept, it could border on what you might consider impossible. BUT IT IS NOT!

One important concept to grasp in any emergency or survival situation is that you will not necessarily starve to death if you are out of frozen pizzas and fruit pies. The human body is able to adapt to and can thrive on diets consisting of very basic foodstuffs.

We obtain most of our food by short-term purchase, but there are alternatives (including storage, growing and gathering). A long-term storage based on what would keep a person alive and nourished is not terribly expensive and can be easily accumulated by purchasing a little toward it on each grocery shopping trip. This kind of storage program is a bulwark against economic as well as physical perils and leads to greater use of basic foods, especially if you store what you will use and use what you store.

There are many good reasons for a storage program. Strikes, wars, social disturbances, earthquakes, or bad weather can disrupt our distribution system at any time. Personal emergencies such as unemployment, illness, and accidents could arise without warning. Buying ahead can also be a hedge against inflation, can cut down on trips to the grocery store and can effect savings by making purchases carefully and in large quantities. When the kinds of calamitous events mentioned above occur, the time for preparation is past. Whether the need is that of your own household or other family members or friends, peace of mind is a product of proper preparation.

WHAT TO STORE

So what will maintain life and nourishment and yet store easily and inexpensively? Basics, dehydrated foods, canned foods, SAP foods, and retort foods can help.

Basics

The ancients spoke of lands flowing with milk and honey, bread as the staff of life, and men being worth their salt. These observations may show a bit of literary and historical license; but grains, milk, salt, and honey or sugar are all easily stored and not expensive in the amounts recommended. Legumes and oils must be added to these, as well. Vitamin and mineral supplements can be added to the list as an inexpensive and convenient way to insure proper nutrition. As previously mentioned, vitamins C and D are especially important supplements to diets of storage foods. Another fact that may be surprising is that these basic foods can be prepared in a staggering variety of ways to provide palatable nutrition. (See "Recipes" and references, this chapter.)

Wheat can be considered the backbone of an inexpensive storage program. It is highly rated by nutritionists as one of the most biologically valuable foods. It stores well for many years; provides protein, carbohydrates, and oils; and lends itself to a great variety of preparations. Wheat is low in the amino acid lysine, but this may be compensated for by adding lysine-rich foods such as milk to the diet. The oils in wheat can be stored in the unbroken kernels for many years without becoming unusable.

Other grains that are very high in nutrition are rice, oats, barley, corn, and rye. Whole rice can go rancid in a relatively short time, but polished rice will keep for many years. Dried peas and beans are also highly nutritious and most varieties store well.

Unless you have a cow, the only practical way to store milk is in the powdered or condensed form. Many storage charts say powdered milk can only be stored for short periods; but good quality nonfat dried milk, if kept cool and dry, will store with little change for a few years (some say as many as fifteen to twenty years). Most people find properly prepared noninstant powdered milk very palatable, differing in taste only slightly from fresh skim milk; and if nothing else were available it would be a welcome addition to the diet. If you prefer not to drink it on a regular basis but still want to rotate it to keep your supply fresh, you may wish to use it for cooking. Hot chocolate, puddings, etc. made with powdered milk taste little different from those made with fresh milk.

Spray-dried milk has more biologically available lysine than roller-dried milk and is therefore more desirable. For storage, extra grade is better than standard grade because of lower moisture content. The milk should also be fortified with Vitamins A and D.

Honey and/or sugar is needed to supply energy. Sugar can be stored indefinitely if it is kept dry and clean. In most markets it is considerably less expensive than honey. Honey stores for long periods of time if not contaminated. If it crystallizes, it can be easily reliquified by heating the container in a water bath. Many claims of virtue have been made for honey. Its chemical composition, however, is very similar to sugar. Diabetics can not eat honey or sugar without medication or difficulties will arise. Honey's principal advantages seem to lie in the facts that it contains very small quantities of some other nutrients such as Vitamin C, iron, riboflavin, and nicotinic acid; that it tastes sweeter than sugar and hence, may be a more effective sweetener; and that, according to many people, it just plain tastes good. Taste is probably honey's foremost asset. Honey should not be fed to infants less than one year old because of the possibility of getting botulism poisoning from spores contained in it.

Fat or oil is a concentrated form of energy and a certain amount is a dietary necessity. It is recommended that a survival diet include at least one ounce of fat per day and more if possible. One ounce per day is little less than 24 pounds for a year. Oils can be stored in the form of liquid vegetable oil (with a normal storage life of two years or more); shortening (with a storage life—unopened—of 15 years); dairy products, such as cheese or evaporated milk; nut products, such as nuts and nut butter; seeds; salad dressings; and meat, such as bacon. Vegetable oil has been prized by the starving.

Salt (sodium chloride) is a vital component of our biological processes that is regularly lost through several avenues of body function. It must be replaced. Also, when a strictly vegetarian diet is maintained, there is a special need to replenish the sodium content of the body.

It is estimated that 300 pounds of wheat, rice, corn, oats, or other cereal grains, 60 pounds of dried legumes, 60 pounds of sugar or honey, 75 pounds of powdered milk, 20 pounds of fats or oil, and 5 pounds of salt will provide about 2300 calories daily for one year and could fill the needs of an average adult. (There are differing estimates on nutrition from basic foods printed in many places. These figures are from *Essentials of*

Home Production and Storage, The Church of Jesus Christ of Latter-day Saints, p. 10—full reference at the end of this chapter.) Recent updates recommend less milk and more grain.

A storage plan should provide at least these basic ingredients in some form or other or suitable substitutes. Grains, legumes, sugar or honey, salt, oil, milk (nonfat dry), and adjuncts would keep a body running through a lean or emergency period. In addition, include about one pound of yeast and one pound of baking powder per person.

When deciding how much to store, it is important to allow for differences in requirements for different size, age, sex, and other factors effecting the caloric requirements. For example, a two-year-old may require about half of the 2300 calories mentioned in the sample above. A 20-year old male would normally require a third to a half more than the amount listed (about 3400 calories per day). An adult female would require about the 2300 calories.

Teenagers and pregnant and nursing women have an excess need of calcium and vitamin A. This should be allowed for by extra milk rations. The calcium, in wheat and many other grains is bound by phylic acid and is not readily useable by the body, so the powdered milk is an important part of the survival diet. Fifty pounds of powdered milk will provide about 800 mg per day of calcium for one year. A growing teenager or a pregnant or nursing woman should have about one and one-half this much or about 1200 mg per day.

Jack A. Spigarelli (*Crisis Preparedness Handbook*, p. 86) recommends 2600 calories per day for at least a year starting with what he calls the "7-PLUS" plan with the option of altering the plan according to individual needs. (His "7-PLUS" plan consists of: 1) 8 pounds salt - half iodized for eating and half non-iodized for other culinary uses; 2)75 pounds nonfat dry milk; 3)36 pounds of oil - 2 gal. liquid and 21 pounds shortening; 4)65 pounds sugar; 5)325 pounds of grain such as wheat, rice, and corn; 6) 50 pounds legumes such as beans, peas, and lentils; and 7) 365 multi-vitamin and mineral daily supplements; "PLUS" leavening agents - yeast and baking powder, and seasonings and flavorings.) His book is, unfortunately, out of print.

This is a reasonable storage program which allows for proper nutrition and a little waste—which surely would occur.

The vitamin content of wheat (especially vitamins C and A) and seeds and other grains may be enhanced considerably by sprouting. This is a simple process and is covered in the recipe section of this chapter along with other ideas for use of these foods.

It is important to evaluate your own storage program and think about what it can do for you and what would be required to use it. For example, if beans and grain are mixed together and cooked into a mush, an adult ration would be nearly a gallon per day with oil and salt added to insure adequate intake. A stark prospect for the frozen pizza people. Stocking a variety of grains and legumes, becoming familiar with recipes and processes, and keeping an open mind will all help. But it is certain that most people would be struck with considerable digestive adjustments if grains and legumes suddenly became the total diet.

It is also of interest that there are vast farm reserves of wheat, corn, soybeans, and grain sorghums in the U.S.A.

Other commodities that may be effective storage items if used and rotated regularly are:

Fruit and vegetable juices	Canned meats and stews
Canned fruits and vegetables	Dried fruits and vegetables, including raisins

Canned soups
Crackers
Desserts (gelatins, puddings, etc.)
Vitamin pills
Spices, flavorings, and other adjuncts (ginger, cinnamon, clover, sesame seeds, onion and garlic salt, pepper, cayenne, onion flakes, bouillon cubes, nutmeg, vanilla flavoring, blackstrap molasses-high in calcium, almond flavoring, maple flavoring, food coloring, dried cheese, toppings, beverage mixes, cornstarch, corn syrup, etc.)
Dried eggs (an excellent addition to the storage—can be stored for two to four years)

Dehydrated Foods

In addition to the foods mentioned above there is extensive availability of dehydrated and freeze-dried foods. Some of these foods maintain a high percentage of their natural nutrition and some will store well for long periods of time. Others will not. When properly prepared, these foods are palatable, and some are even delicious. They also take up minimal storage space and weigh less than foods prepared for storage by other methods.

The disadvantage of these foods is that a steady diet of them, as with grains and legumes, without an adjustment period is likely to be difficult. They are also more costly than foods prepared for storage in some other ways, and most dehydrated foods do not maintain their original nutritional values "forever", as some would purport. They have a tendency to deteriorate with time, and to lose their nutritional potency, sometimes even becoming unfit to eat. Dark color, rancid odors, and hydration (wet in the can) are some of the clues of deterioration. Vegetables and legumes seem to have a good shelf life. Items high in oil content go rancid after relatively short periods of time. Nonfat dry milk stores very well.

Freeze-dried foods are more expensive than air-dried foods with, in many cases, no particular advantage but variety. Meat, for example, can only be freeze-dried.

If you have already purchased a "unit" or "units" of dehydrated food, take the time to try living on some of the food for a couple of days to make certain that what you have would do the job for you if it became necessary to use it. Many of these "units" fall short on calories, quality nutrition, and palatability. They may be the last word in storage convenience, but they are expensive and certainly not always the best answer to food storage. My own opinion is that the dehydrated foods are most effective when used selectively to complement other food storage.

Canned Foods

In spite of the bad name given foods canned in metal containers (canned goods), they remain a "best buy" and, according to scientific study, retain high nutritional content in storage. They should be rotated regularly but are pleasant and easy to use and cost-effective to buy.

SAP Foods

Sterile aseptically packaged (SAP) foods are sterilized rapidly at high temperatures and packed in supersterile containers. Milk, juices, margarine, and some other foods are available packaged this way. Storage shelf life is a few months for most of these products and they must be refrigerated or used soon after opening.

Retort Foods

Another packaging method in current vogue is that of cooking foods in flexible foil and plastic pouches in retort ovens (specialized pressure cookers). Foods packaged this way are used widely by the U.S. government—mostly the military, and are called in military jargon "meals ready to eat" (MREs). (See also Chapter 5.) The pouches of food need no refridgeration in water or on a warm surface. They have a shelf life approximately equal to canned goods, are reasonably priced, widely available, and convenient to use. Entrees vary greatly in strength of flavors, seasonings, and so on. Sample the flavors before buying more than one or two. (See Sources at the end of the chapter.)

HOW TO STORE

It is important to take the necessary steps to protect stored food against the ravages of the elements and pests. Usually a few simple steps can bring some "healthy" dividends.

Grain

Some care needs to be taken when storing grains at home. Clean, insect-free, low-moisture, high-quality product from a reputable dealer is recommended. Wheat should be at least 11 percent protein and less than 10 percent moisture. The grain should be stored in a cool, dry place. Molds and other undesirables must have oxygen and moisture to grow.

It is nearly impossible to purchase insect-free grains and cereals unless they have been specially prepared for storage. My son once asked his grandfather who put the weevils in the wheat; the answer: "You don't have to put them there; they just happen." A little care in storing grains will prevent frustrating losses to spoilage and insect infestation.

There are four fairly effective and safe procedures for treating grains at home: freezing, fumigating, heating, and treatment with diatomaceous earth. Another method is that of adding bay leaves or chewing tobacco; this may be only partially effective. The leaves do, however, act as a repellent to many intruders.

Freezing   Grain may be frozen in a deep freeze for three to four days at about 0 degrees F. in appropriate-sized batches. This will usually destroy the "animal life" present. The effectiveness of this method may be measured by spreading the grain or a representative sample of it on a flat surface for a few hours at room temperature. If there are live insects present they will probably become visible through close inspection.

Fumigations   Fumigation is normally done with dry ice (solid carbon dioxide) or nitrogen gas. Dry-ice fumigation is accomplished by spreading a layer of crushed dry ice on a layer of about three inches of grain in the storage container, then adding the remainder of the grain to the container. Proportion estimates vary from about one-half pound to one pound of dry ice per hundred pounds of grain. Use at least one-half pound. Care should be taken to keep the grain dry. The lid should then be placed loosely on the container for at least one-half hour or until all the dry ice has sublimed (become gaseous) and replaced the air in the container. There is nothing to lose by leaving the lid loosely on the container for three to five hours. The lid should then be placed tightly on the container and could be sealed with sticky tape if necessary to make a good seal.

The only significant danger involved in this process is in tightening the lid too soon and building up excessive pressure in the container. This could conceivably cause the container to bulge or perhaps even break. If the container begins to bulge, simply remove the lid and relieve the pressure. The carbon dioxide should remain in the container for some time.

Nitrogen may also be used to replace the air in the container if it is available. Both of these processes should control all adult and larvae infestations in the grain but may not destroy all of the pupae or eggs. They can protect the grain for several years, however.

Heating   Grain may also be heated in shallow layers to about 145 degrees F. for at least one-half hour to kill insects, and then cooled and returned to a clean container. This process may inhibit the ability of the grain to sprout, however.

Diatomaceous Earth, Expedient Storage, Drying Agents   Diatomaceous earth can be used to protect especially small kernel grains from weevils and other insects. It is a very fine, hard, inert material that is added to and mixed with the grain in storage at about the rate of one quart per hundred pounds of grain. The powdery material suffocates or otherwise harms the intruders. Essentially no taste difference in culinary productions results from its use, and it is not nutritionally harmful. (In fact, diatomaceous earth is an ingredient in many toothpastes.) Perma-Guard D-10 is one diatomaceous earth product that is packed for this purpose.

Expediently grain can be stored for short periods of time in a plastic sack buried in earth or in a grass- or dust-lined hole also covered with earth.

Some people place drying agents such as silica gel or calcium chloride in storage containers to keep moisture from the grain. These agents may be contained in a cloth sack or other porous container. Silica gel is more expensive but is easily regenerated after it has become damp by heating it in a moderate oven (250-300 degrees F.) for about ten minutes. In damp climates a drying agent is probably recommendable, but they need to be regenerated or replaced periodically.

Canned Goods

In many of the imaginable disasters such as earthquakes or explosions, storage containers are very susceptible to breakage. Plastic and glass bottles suffer much less damage when properly stored. A recommendable procedure is that bottled goods such as home-canned fruits and vegetables be packed in cardboard boxes lined with plastic sheeting (visquene). Old newspapers or pieces of cardboard can then be placed between the bottles to prevent breakage. The plastic would contain any breakage that did occur. This process is simple and affords considerable protection from messes. A few items at a time can be placed on open shelves for convenient use. It is also advisable to place retainer strips of wood or wire at the edges of storage shelves to prevent storage items from being tipped off the shelves during a possible earthquake. Heavy containers should be stored close to the floor to avoid damage or injury during quakes or other ''earth-shaking'' difficulties.

STORAGE LIFE

It is not easy to predict how long goods will store. During long-term storage the chemistry of almost any food gradually changes. There are many variables, including the food itself, processing, packaging, storage conditions, contained organisms, and standards of the consumer. Charts giving shelf-lives of various preserved foods vary so widely that there is often a hundred-percent difference in two charts. It is my considered opinion that most of the storage lives given are usually very minimal for peak preservation of color, texture, and nutrition. If the food is stored longer and has not spoiled, it will still provide nutrition and will be edible. It is not reasonable to think that after six months or eighteen months (or whatever length of time) a food is suddenly unsuitable for human consumption. The process of breakdown is gradual.

Vitamins may be altered or destroyed by this ongoing chemical change occurring in foods. Vitamins C and A and thiamine are changed by the presence of oxygen, especially when accompanied by elevated temperature. Riboflavin is very sensitive to light and can be destroyed by exposure. Minerals are essentially unaffected during storage.

Some people that do home canning say they can tell what year their peaches have been canned by their color. Chemical reactions in many foods cause a darkening with length of storage.

Canned foods with any bulging should be discarded. They may or may not be harmful but it is difficult to tell the difference. Cans with dents in the side or end seams are suspect and should not be purchased. If these are found in storage, check for leakage and discard leaky cans. Rusty cans may also be rusted through; if so, they should be discarded. Home-canned fruits and fruit jams and jellies are acidic and likely to be subject only to fermentation or mold. These forms of contamination are easily detected, and spoiled items should not be eaten.

Home-canned meats and vegetables are subject to botulism, food poisoning which is not detectable by smell and taste and which is deadly. These toxins can normally be destroyed by boiling for at least 15 minutes. Boil all home-canned meats and vegetables in a covered container for at least 15 minutes before they are eaten.

Under ideal storage conditions (cool temperature and little light), highly acidic and highly pigmented foods such as citrus juices, black and red cherries, prunes, plums, and berries generally have a shorter maximum shelf life (one to two years) than fruits such as peaches, pears, applesauce, and apricots (two to three years). Canned vegetables' maximum usable shelf life is in the neighborhood of three to four years. Poor storage conditions can greatly shorten these times. In fact, many storage-life charts show a maximum storage life of six months to one year for all canned foods.

Heat and oxygen readily make fats go rancid, although some fats may be used even when rancid if the flavor can be disguised. Preservatives such as BHT and BHA offer some protection from this process and extend the storage life of fats and oils. Canned vegetable shortening can be stored for several years and is an excellent source of oils in a storage program.

Proper rotation of foods insures that the food will not be wasted and that if an emergency occurs your taste buds will be able to endure eating the food. Studies have shown that people under stress are very reluctant to eat foods they are not accustomed to eating.

To insure normal taste, all canned foods should be consumed within two years or less. We have developed a general rule of thumb about storage at our home that we feel is well founded: All canned meats and vegetables should be used within a year. Usually we can vary consumption or purchase in order to have about a year's supply on hand at a given time. Our home-canned fruit is on a rotation schedule of a little less than two years: for instance, we would eat the fruit canned in the fall of 1980 during the fall of 1981 and winter and spring of 1982. And we try to hold items such as wheat, powdered milk, sugar, and honey in adequate supply at all times. (In all frankness, for us it doesn't turn out to be quite as organized as this sounds!)

STORAGE CONDITIONS

All foods that have been dehydrated should be stored in moisture-proof, airtight containers. (When foods are dried at home by primitive methods to the extent necessary to slow the growth of mold and other microorganisms they sometimes end up being a bit hard, but that is better than having spoilage.) All grains, legumes, flour, sugar, salt, baking mixes, and other commercially or home-dried foods should be kept off floors and away from areas where condensation can occur.

Dry food staples, canned foods, and most other foods and food products store best where it is cool (sixty degrees F. or lower), dark, and dry. Warm temperatures greatly accelerate the breakdown and deterioration of food. Fats usually melt at around ninety-five degrees F. Also, most insects are inactive below forty-eight degrees F. Packaging should provide protection from air and moisture. Canned foods should be protected from freezing.

Protection from rodents is usually afforded by storing food contained in less durable containers or loose foodstuffs in metal containers. Some plastic containers are sturdy enough to withstand pests, but some are not and can be chewed through by rats, mice, gophers, and so on. New garbage cans, steel barrels, or storage cans are more effective storage containers. If food is stored in plastic containers, make certain the plastic is approved for food storage. Some plastics are not safe for the purpose. Plastic also has a tendency to break when it is cold.

Additional food storage ideas are found in Chapter 8.

USING WHAT YOU STORE

Stored food may sometimes need some help to make it palatable and nutritious. The regular fare such as canned goods should be easy to use, but a few recipes are included to help with the palatability of the basics. Additional steps that render especially grains and legumes more digestable or nutritious than they are in the uncooked whole kernal are sprouting, growing grain grasses, and making gluten. Grinding seeds is discussed in "Nuts and Seeds" in chapter 8.

Recipes

The scope of this book is not sufficient to include many recipes. It is, however, important to have some ideas that make a diet of grain, powdered milk, oil, and honey or sugar seem survivable. For that reason, recipes and ideas for use of some of the basic storage items follow. Some recipes for substitute items, tips on making bread, and a few other basic goodies are also included.

If you are going to store basic foods, and plan ever to use them, then it is essential to sit down and plan menus for at least a week to see what you will be "suffering through" when you eat it. (Then prepare and eat some of it.) There are some books listed at the end of the chapter that are tremendously helpful in this endeavor, and I recommend them highly. Some of the authors make you feel like a chef at the Hilton when you have only grain, beans, and sugar.

Bread   Breads of various descriptions can (and for some, do) occupy a considerable portion of the nutritional intake. Using whole grain and grinding it fresh just before the bread is made can enhance the nutrition of the loaves considerably. Some nutrients in whole grain begin oxidizing and degrading as soon as the grain is cracked or ground. Even young children can learn to eat and enjoy whole wheat bread made from grain ground at home, although it is usually somewhat coarser than "store-bought" bread.

More than one grain can be used to further enhance nutrition or to give variety. Soy beans are good for this; so are corn, oats, rye, seeds, and so on. However, wheat should be used for at least two-thirds of the flour to give a good texture to the finished product. It is the only grain with a high enough gluten content to produce a good texture. Here are some tips for successful breadmaking:

1. *Take care of the yeast.* A water temperature of 110° F. is about right to activate dry yeast. If the water is too hot it will kill the yeast; if too cool it will not activate it. Yeast in sweetened water should bubble within five to ten minutes.
   Yeast is now available that may be stored in the unopened package for long periods of time without refrigeration.

2. *Take care of the dough.* (At least that's what Sylvia Porter says.) Mix the dough thoroughly to develop the gluten. When possible, cook some cut-up potatoes in water and use the water in the bread. It does help. The addition of 100 or 200 miligrams of powdered vitamin C per batch can also improve the texture. Mix yeast water with about half of the flour and let the yeast work before adding the oil and salt. Add these with the rest of the flour. (Use any standard bread recipe or the one included here.) Keep the temperature of the rising bread around 80 to 85° F. Do not allow it to rise to more than twice its bulk. If you need to, punch it down and let it rise another time.

3. *Treat the loaves right.* Fill the pans about two-thirds full but no more. Use a convenient size of pan. The standard 8½'' x 4½'' x 2¾'' pans are about right. Shape the dough into a nice rectangle evenly distributed in the pan. A good way to do this is to make a tight roll a little longer than the pan and then press into the corners. Lightly oil or grease the top of each loaf.

4. *Bake it right.* When the dough has risen in the pan to about double its original size, it should be ready to bake. When it is, a finger mark poked into the side of the loaf will spring partially back out, leaving only a small indentation. Properly raised loaves will continue to rise a bit in the oven. This is called oven spring. If the bread falls, it is overrisen. When the bread is done it should be brown to golden-brown (depending on the grain used). The smell of yeast should pretty well be gone and the loaf should sound hollow when you tap it with your finger. Shake the hot loaves out of the pans and allow them to cool on the counter top or table.

Here are a few tips on *trouble-shooting* bread problems:

1. Poor texture:
   Too fine—low in yeast
              —improper volume in pan
              —rising time too short
   Too coarse—rising time too long
                —low-protein flour

—oven not hot enough
—improper volume in pan
"Side-break"—oven too hot
—dough overmixed
—rising time too short
2. Poor Crust:
Dark crust—oven too hot or baking time too long
—too much sugar or milk
—insufficient rising time
Blistered crust—too much liquid in dough
—incorrect fermentation
—too much oil or grease on top of loaf
—oven not evenly heated—too much heat from the top
Low volume—low in yeast (may be too little or improper handling)
—oven too hot—no oven spring
—poor flow—probably old or low in protein

Whole Wheat Bread Instructions are for 4 loaves, measurements for 5 loaves in parentheses.

5 C. (6 1/4) warm milk	2 T. (2 1/2) yeast
2/3 C. (3/4) oil	4 (5) eggs
1/2 C. (2/3) honey	14 C. (17) flour
2 T. (2 1/2) salt	

Optional ingredient: 1/3 C. (1/2) whey or powdered milk.
Optional ingredient: 1/3 C. (1/2) wheat germ or cracked wheat (if desired).

Dissolve yeast in warm milk. Add honey, whey, wheat germ, eggs & 7 cups (8) flour. Mix & let rise till double. Punch down. Add oil & salt & 7 cups (9) flour. Mix well—6 minutes in mixer or kneaded by hand until it stays together well and is spongy. When mixing by hand cut back on oil & use the rest on kneading surface to keep bread dough from sticking to it.

Form into 4 (5) loaves. Place in greased pans. Let rise to just over the top of pans. Bake 35 minutes at 325°-350°.

Optional: May substitute 1/2 cup soy flour for 1/2 cup wheat flour.

Indian Fry Bread

6 C. flour	1 t. salt
2 t. baking powder	

Mix. Add enough warm water to make a stiff dough when thoroughly mixed. Form into 2-3" patties about 1/2" thick and fry in hot oil until golden brown on both sides. Or brown them on a hot griddle.

Whole Wheat "Graham" Crackers Mix 1/2 cup honey, 1 cup brown sugar, 1 teaspoon vanilla, 2 beaten eggs, and 1 cup cooking oil together vigorously so that the oil is homogenously blended into the other ingredients.

Add a mixture of 1/2 cup water, 1/2 cup powdered milk and 1 tablespoon lemon juice or vinegar to the above mixture. Then add 6 cups wheat flour, 1 teaspoon salt, and 1 teaspoon soda and mix thoroughly. Roll out to 1/8" thickness on greased and floured baking sheets, prick the dough with a fork, etch bite-size sections, and bake at 400°F for 12 to 15 minutes.

Rice Pudding Mix 1 cup rice, 1/2 cup sugar or honey, 2/3 cup raisins, 1/2 teaspoon nutmeg, 1 teaspoon vanilla, and 1 quart milk (fresh or reconstituted) in a casserole and bake for 1 hour.

When cooked rice is used as part of a meal, put some of the cooked rice, honey, raisins, and milk in a glass or bowl and eat it as a dessert.

Rennet From Thistle   Soak the thistle in very little water until the plant is saturated. Smash and pound it, let it soak, and repeat until the water looks like tea and has a brown color. Use one or two teaspoons of this solution per quart of milk to speed up the curdling of the milk.

Candy   A fondant can be made by combining 1/2 cup honey with 1 1/2 cups noninstant dry milk. Mix it thoroughly and use as desired: cover it with chocolate, roll it in nuts, or just eat it plain.

Egg Substitute   Mix 1 teaspoon unflavored gelatin with 5 tablespoons hot water as a substitute for each egg called for in a baking recipe.

Vinegar   A favorite material for making vinegar is apple juice. Other fruit juices also work. Place the strained juice in a clean crock, jug, jar, or watertight wooden container, leaving about a fourth of the volume of the container unfilled to allow for expansion during the fermentation. Cover it with a piece of cloth to allow air contact but to keep out insects, dust, and light. Tie the cover tightly over the opening and store the container in a cool dark place. After three or four months check the mixture to see if its acidity is suitable for your purposes. If it is too weak, leave it stored for another month or two. A scum will form on the top of the liquid. This is commonly called the "mother" or "mother of vinegar". When the vinegar is at the strength desired, the "mother" should be removed to retard the fermentation process. This may be done by straining the mixture through a cloth, but the "mother" should be saved for adding to subsequent batches of vinegar; its addition will greatly speed up the process. Store the strained vinegar in capped jars or jugs.

Pickling with homemade vinegar is difficult because of the uncertainty of its acidity.

Some people use yeast in their vinegar and say it improves the flavor. It can be used as follows:

Allow the juice from any of many fruits including apples, grapes, or pears to stand at room temperature in a very clean wood, glass, or earthenware container for a day or two and then decant the clear liquid away from the pulp into a clean container similar to the first. Yeast may be added at this point. If you add the yeast, do so at the rate of one-fourth cake per gallon of fruit juice. Cover with a loosely woven cloth such as cheesecloth thick enough to keep out light and let stand for three or four more days at room temperature. Stir it once or twice a day.

Add one part "mother of vinegar" from an old batch to four or five parts juice in a clean container and let stand four or five months covered as above with a loosely woven cloth. Do not stir.

Filter it through a damp cloth, bottle it, and age it for a few months. The bottles should be corked or capped.

Sourdough Starter or Wild Yeast   Mix 2 cups flour, 2 cups warm water or potato water, and 2 1/2 teaspoons sugar or honey in a glass or crockery container. Allow the mixture to stand at room temperature for 5 days, stirring or otherwise agitating it several times a day. Use it as a substitute for yeast. After some is used, replace the used volume with an equal-parts mixture of flour and water or potato water and allow it to stand out for 24 hours, at which time it will be ready to use again. Reserve at least 1/2 cup for starting the next batch. Refrigerate the unused portion between uses and shake it occasionally.

Everlasting Yeast   An everlasting yeast can be made by mixing 2 teaspoons dry yeast, 2 tablespoons sugar, 1 teaspoon salt, and 2 cups flour with a quart of warm potato water. Allow the mixture to rise in a warm spot for a few minutes until it is used in baking. Reserve a small amount and refrigerate it in a glass jar until it is needed again, then add all ingredients but the yeast and allow it to rise again. The process can be repeated indefinitely.

Sourdough Bread   Combine 2/3 cup sourdough starter (see above) with 1 1/3 cups warm water, 1 teaspoon salt, 1 1/2 tablespoons dry milk, 3 teaspoons honey, and 2 1/3 cups flour. Mix, then knead vigorously for about 10 minutes. Cover the dough with a damp towel and allow it to rise for about 5 hours or until it doubles in bulk. Divide it into two small loaves, place these in well-greased and floured pans, and allow them to rise again for 2 to 3 hours. Bake them at 350° F. for about 1 hour.

Wheat Cereal   Most recipes for cracked wheat cereal recommend about 1/4 cup wheat per serving and 2 cups water and 1/2 teaspoon salt per cup of cracked wheat. At our house we use about half that much salt and twice the water: about 1/4 cup cracked wheat, 1 cup water, and about 1/16 teaspoon salt for each serving. How the wheat is cooked (open kettle or with a lid; slow or fast) makes a considerable difference in the amount of water needed. We use low heat. Cracked wheat can be cooked in 20 or 30 minutes.

A crock-pot type of cooker is an excellent device for cooking whole wheat. Place ingredients (one part wheat to four or more parts water plus salt to taste) in pot just before bedtime and cook on low heat until breakfast. Or, just before bedtime, place whole wheat, water, and salt in a pot and boil for 5 or 10 minutes, turn off the heat, put a good lid on the pot, and leave it until breakfast. Warm it up again just before eating, or eat it cold. The hot wheat and water can also be held in a vacuum bottle (thermos bottle) overnight; this cooks the wheat. Serve it with honey, sugar, jam, or fruit for sweetening and milk or cream.

Soybeans in Cooked Cereal   Soybean meal may be soaked in hot water for 2 or 3 hours before being cooked with grain meal to make cereal. After the bean meal is soaked it should be boiled for a few minutes before adding the grain meal. By adding 1 part soaked bean meal and dry grain meal to about 3 parts salted water, the mixture will produce a good "mush" in 15 to 20 minutes by slow cooking and frequent stirring. (A ratio of 2 parts wheat to 1 part beans or even up to 5 or 6 parts wheat to 1 part beans provides good protein.)

Other Soybean Uses   Soybeans are a low-carbohydrate, high-protein food that can be used in many ways. The high nutritive value in soybeans makes them an excellent food to store and to use. However, since they are very high in oil content they should be rotated often in storage to prevent rancidity.

Soybeans may be made into milk, ground and used as a supplement to wheat or other grain flours, sprouted, used as a source of oil, cooked as a bean dish, and used in other ways.

The flavor of soybeans is quite strong and the flour is lacking in gluten. For this reason no more than 2 or 3 tablespoons of soy flour per cup of wheat flour should be substituted in baking. Otherwise, the lack of gluten causes the baked foods to fall apart and the flavor can be altered.

Dry soybeans may be cooked by soaking them in cold water for a few hours, discarding the water, and then cooking them in salted water until tender. Partially cooked beans may be used in baking, sauteing, or in other methods of preparation.

One of the very useful things that can be done with soybeans is to make milk from them. The best way to make it is with a blender. First, soak the beans a few hours in cold water and then discard the water. Put about 2/3 cup of beans and 2 1/2 cups of water in the blender and run it at high speed for a few seconds or until the beans are ground well. Pour the puree into a pan, add an additional 1 1/2 cups of water, and simmer it over low heat for 12 to 15 minutes. Strain this mixture through a cloth, squeezing out the milk and refrigerating it. Save the pulp (sometimes called okara) for use as a meat extender or for other cooking purposes.

In the event your blender is out of commission at the time of need, place 1 cup soy flour in a pan with 5 cups water and simmer it for 15 minutes over low heat, then cool it and strain it through a cloth. Milk made this way may be somewhat bitter, depending on the beans.

Soybean milk may be used as a drink, or as an ingredient in cooking, or in the preparation of other foods such as cheese or cottage cheese. To make soy cheese, allow the milk to sour in a warm area (add lemon juice or vinegar to speed up the process) strain it through a cloth, rinse the curds, and then season them to suit your taste.

Pinole   Mix 1 heaping tablespoon brown sugar with 1 cup cornmeal and spread the mixture on a cooking tray or pan bottom. Heat it in the oven or on a cooking surface until the sugar begins to melt. Cool the mixture. Break it into pieces, and eat it as it is.

A mush of fresh berries may be substituted for the sugar. In this case, mix the smashed berries thoroughly with the cornmeal and dry the mixture in the oven, in the sun, or over a fire.

Hardbread   Mix together 4 cups whole wheat flour, 6 tablespoons cooking oil or melted shortening or margarine, 3 tablespoons sugar, 4 tablespoons honey, 4 tablespoons molasses, 1 heaping tablespoon powdered milk, 1/4 teaspoon salt, 1/2 teaspoon baking powder, and 1 cup water. Bake at 300° F. for 1 hour, then turn the heat down low, open the oven door a crack, and let the bread dry. This may take a few hours.

Some Other Sources of Sweet   In the absence of refined sugar other things can be used to "sweeten" life up a bit and to provide calories in the diet.

Cut up sugar beets and boil them down to a thick liquid. Use it as sweetening. The same thing can be done with carrots and corn cobs. (Strain the pulp and cobs from the liquid before it gets too thick.)

Extract the sap from maple trees in the early spring and boil it down to a syrup.

A sweet syrup can be acquired from pumpkin and some kinds of squash. Break them open or cut a hole in the top and clean them out as you would for a jack-o-lantern. Leave them out (protected from pests if possible) in the cold for a few nights until a syrupy liquid will become apparent on the surface of the inside. Pour or spoon off the liquid and boil it down to a sweet syrup.

Ripe and overripe fruit can be boiled, strained to remove the pulp, and boiled down further to produce a sweet, fruit flavored syrup.

Powdered Milk Whipped Topping   Whip 1 cup non-instant powdered milk and 1 cup cold water until they thicken. Add sugar, lemon juice, and flavorings such as vanilla to suit your taste.

Corn Bread

1 C. corn meal or flour (grind your own in your mill)
1 C. sifted whole wheat flour
1/4-1/2 C. honey
1/2 tsp. salt

4 tsp. baking powder
1-2 eggs
1 C. milk
1/4 C. soft shortening

Sift dry ingredients in a bowl. Add eggs, milk and shortening. Beat with egg beater 1 minute or until smooth. Do not overbeat. Pour into greased cake pan. 8 x 8''. Bake 425°, 20-25 minutes.

Wheatcakes (Pancakes)

Flour - 5 heaping T. per person (about 3/4 cup per person)
Baking Powder - 1 tsp. per 2 people (1/2 t./person)
Salt - 1/8 tsp. per person
Eggs - 1 per person   Oil & Milk
Optional: Sugar - 1 T. per 2 people
          Cracked Wheat - 1 T. per 2 people
Mix ingredients, cook and serve.

Oatmeal Cake

Pour 1 1/2 cup boiling water over 1 cup quick oats. Add 1 cube margarine. Let stand until it melts. Beat 2 eggs, 1 cup brown sugar, 1 cup white sugar. Add this to 1st mixture then add:

1 1/2 C. Flour
1 tsp. soda
1 tsp. nutmeg

1 tsp. cinnamon
1/8 tsp. salt
1 tsp. vanilla

Bake 35 minutes at 350°. Ice with **penuche** icing.

Penuche Icing

1/4 C. margarine
1/2 C. brown sugar

3 tsp. milk

Melt in saucepan. Let it come to boil. Remove from heat, add sifted powdered sugar until icing thickens. Double for layer cake.

Pickled Beans (Dill)

4 lbs. green beans (small, tender)
1 1/4 tsp. red cayenne pepper
4 cloves garlic (optional)
7 large heads dill

5 C. vinegar
5 C. H$_2$O
1/2 C. salt
(pack beans tightly)

Wash & stem beans. Pat dry & pack in clean, hot jars. To each jar add 1/8 teaspoon red pepper, 1/2 garlic clove, 1 head dill. Measure vinegar, water & salt into saucepan and bring to a boil. Pour boiling hot brine over beans filling to 1/2'' from top of jar. Place a clean lid on each jar and screw band on firmly. Process in boiling water bath for ten minutes plus one additional minute for each 1,000 feet above sea level. Yield: 7 pints.

Dill Pickles

fresh cucumbers, washed
dill weed, fresh or dried
garlic cloves, peeled

salt
vinegar

To make the brine bring 1/2 cup salt, 2 quarts water, 1 quart vinegar to a boil. Pack cucumbers firmly into one quart or 2 quart jars. Add 1 sprig (with stem) dill to each quart and 1 clove garlic, if desired. Add brine to within 1/2 inch of top of each jar. Wipe top of jar. Place clean lid on and screw band on firmly. Place on rack in canner. Cover with water and bring to boil. Process in boiling water for fifteen minutes plus one additional minute for each 1,000 feet above sea level. Remove and let cool. Check seal by tapping lid-high pitched sound, also lid will be slightly concave.

Care and Uses of Milk

Milk should be strained through a clean cloth or paper filter immediately after being drawn to remove any foreign particulates; it should then be cooled. If you desire to pasteurize the milk, do so before the cooling by heating the milk to 142°-145°F. and holding it at that temperature for at least 30 minutes or by heating it to 160°-161°F. for fifteen seconds. Pasteurization kills many harmful germs, prolongs the milk's storability, and is a recommendable procedure.

When refrigeration is not available, milk (and other things) may be cooled by placing them in a secure container in a shaded area of a cool stream. After several hours of storage, the cream may be skimmed off the top of the milk for use as whipping or table cream or for churning into butter. The skimmed or unskimmed milk may be used for drinking or for making cheese, cottage cheese, or yogurt.

Everyone knows how to drink milk but there are, of course, other forms it can take that are nutritious and delicious.

Cottage Cheese Cottage cheese is traditionally made from raw milk after the cream has been skimmed from it. It may also be made from pasteurized milk with the aid of some form of rennet to coagulate the milk. Rennin, an enzyme in rennet that coagulates milk, is found in the gastric juice and stomach lining of some animals and also in plants (see "Nettle" in "Edible Naturally Occurring Plants", chapter 8).

Leave raw milk in a warm room for a day or two until the milk clabbers and the curds and whey separate. Small amounts of rennet, cultured buttermilk, or yogurt placed in the warm milk will cause pasteurized milk to curdle and speed up the process in raw milk. Cut the curd into pieces no larger than 1 1/2 inches to 2 inches, heat them in a water bath to about 115° F., and hold them at this temperature for 20 to 30 minutes with occasional stirring. Strain the mixture through a coarse cloth or cheesecloth over a colander. Tie the corners of the cloth together and hang the cloth long enough to remove the whey from the curds. Retain the liquid whey, which is rich in B vitamins and minerals, for use in cooking or beverages or as animal feed. The curd may also, if desired, be rinsed with cool water.

To add zest to the cheese, add cream and/or season it with salt or herbs, such as dill or chives, to suit the taste. You can expect about 1 1/2 pounds cottage cheese from 1 gallon milk.

Yogurt Yogurt is an age-old food that was and still is a staple in many areas of the world. It has been carried in saddlebags and packs the world over. Although caution should be used to keep it as fresh as possible, it is not nearly as fragile as the expiration date in the grocer's showcase might indicate.

There are yogurt recipes in cookbooks everywhere. There are also fancy yogurt-makers and other devices for the cause. Our family has enjoyed, what seems in retrospect, like tons of yogurt made by a simple and inexpensive method. Two recipes are included, one using regular milk and another using powdered milk. There is evidence that when large quantitities of yogurt are consumed, especially by children, it is more healthy if it contains significant amounts of butterfat, which makes whole milk an excellent base for making it.

Yogurt Recipe
1 C. yogurt for start
1 1/4 C. powdered milk
1 pt. milk
5 C. warm water
2 Cans milk

Beat in order given until completely mixed. Pour into four 1 qt. jars (or smaller, if desired). Place uncovered jars in water bath so that water comes to within 1/2 inch of yogurt line on jars. Heat until water is 115°F. Turn off heat but leave pan on burner and covered; best if wrapped in a towel. Leave for about 8 hours. Yogurt is done if a slight imprint is left on top when touched lightly with finger. Remove, cover & chill immediately.

Yogurt Recipe with Powdered Milk Only   Add 1 cup yogurt start to 10 cups reconstituted warm milk. Proceed as in the other yogurt recipe.

Cheese   Add 1/4 cup cultured buttermilk to 1 quart milk in a clean jar and stir or shake it thoroughly. Allow the mixture to stand at room temperature for 24 hours. Add 1 cup of the above buttermilk mixture to 2 gallons fresh milk and heat this to 87° F. in a stainless or enameled steel pot (do not use aluminum). Allow the ingredients to stand for 1 or 2 hours. Add 1/4 rennet tablet dissolved in 1/3 cup water, and allow the mixture to stand for an additional hour or until a good curd has formed. Before the curd is properly formed the mixture is somewhat elastic.

Cut the curd into small pieces no larger than about 1/2'' to 3/4'' and heat the mixture over low heat with constant stirring until it reaches a temperature of about 105° F. and the whey has separated from the curds. Remove it from the heat and allow it to stand until the curd settles and has somewhat hardened in the bottom of the pan—about 1 hour. Pour off the whey and use it for cooking or animal feed. Add about 1 teaspoon salt—adjust the amount to suit your taste—and mix it into the curd with a spatula or large spoon.

Place at least a double layer of cheesecloth in a colander and pour the cut, salted curd into it. Pick up the corners of the cloth and twist them together, then squeeze the curd to press out the whey. Hang the ball of curd up for 1/2 to 1 hour to drain. Then place the cheese ball in an appropriate-sized container (bowl, dish, plate, pan, mold, etc.) and weight the top of it to press the curd together into the desired shape for one day.

For about the first week of the aging process turn the cheese several times a day. When a hardened "rind" has formed on its surface, replace the old cheesecloth and cover the entire outside with a thin layer of melted paraffin. Age it further in a cool place for about two or three months before using. If the cheese swells, it is a sign of undesirable organisms working on the cheese and it should be discarded. This can usually be avoided by taking greater care in "ripening" the milk.

Sprouting, Grain Grasses, and Gluten

The following instructions can alter the useable nutrition content of grains.

Seed Sprouts   Most grains or seeds need to be cooked in order for the human body to be able to fully use the nutrients in them. Sprouting seeds is a useful way of generating nutritional entities and breaking down the nutrients to forms more easily asimilated by the body without cooking them. Vitamin C is a primary dietary necessity that can be acquired by eating sprouted seeds. The U.S. Recommended Daily Allowance (U.S. RDA) for Vitamin C is 60 miligrams. About a quarter of this amount or 15 miligrams can prevent scurvy, however, and this can be supplied with approximately 35 grams (1 1/4 ounces) of wheat or beans sprouted until the sprouts are about 1 to 1 1/2 times as long as the seeds. This can be done in about 48 hours at normal room temperature. In some areas of the world sprouted seeds have been a long-time Vitamin C source, especially in winter.

Sprouting seeds may also increase the availability of riboflavin, niacin, and folic acid, as well as some minerals in them. If sprouted until they are green, sprouts also produce small amounts of Vitamin A. The Vitamin A content is maximized in wheat and alfalfa sprouts by sprouting them for about three days in the dark and then in the light for the last few hours. Sprouting in the dark generally increases the B-Vitamin content, and sprouting in the light increases the Vitamin C content. Sprouting generally reduces the caloric content of seeds and does not produce additional protein from the grain. Sprouts are also generally rendered more digestible and hence more nutritious by light cooking. To avoid excessive loss of Vitamin C, sprouts should be boiled or sauteed over low heat no longer than about 2 minutes.

The first step in any of endless methods of sprouting is to remove all foreign material and broken seeds from among the seeds to be sprouted. The next step is to soak the seeds for about 12 hours in water, then drain and rinse them. From here the numerous methods diverge, only two of which are detailed here. (Note: If water is at a premium, the water that ''clean'' wheat has been soaked in is very good to drink, or it may be used in cooking. It may take a while to adjust to the flavor, but it is good and good *for* you.)

Method One: Place the drained seeds in a jar and cover the opening with a nylon sock or other coarse cloth, securing it with a string or rubber band. Turn the jar upside down but slightly ajar to allow air circulation and to drain. Most sprout "recipes" say to sprout seeds in the dark. We sprout them on the kitchen counter top which has indirect and artificial light most of the time during sprouting. Keeping the above-mentioned nutritional factors in mind, take your pick.

Rinse the seeds with cool water at least 3 times a day and leave the jar upside down with the nylon sock cover after each rinse to drain off excess water. When the sprouts of wheat or beans are about as long as the seeds, they are ready to eat. Alfalfa sprouts are good at about 1 1/2 inches long.

Method Two: Clean, soak, and drain the seeds as outlined above. Place them in a layer 1" or less deep in a plastic sack lying flat or in a jar lying on its side. If a plastic sack is used, moisten some pieces of paper, roll them or crumple them, and place them in the sack on top of the seeds to keep the plastic from collapsing on the top of the seeds and allow air circulation. Place a wet paper towel in the mouth of the sack or jar, leaving only a small opening of 1 or 2 square inches to allow air to circulate. Leave the seeds alone until they are sprouted. Sprouting is usually sufficient in two or three days at room temperature. Colder temperatures significantly retard the sprouting process.

Grain Grasses   Wheat, barley, corn, oats, sunflower, and rye grass can be grown indoors and can add nutrition to the diet. Place grain that has been soaked for a few hours (or overnight) on a layer of soil 1'' to 2'' deep. This can be done in any flat container, including a piece of plastic with a board frame around it. The kernels can be very close—shoulder to shoulder if desired. Sprinkle them with a fine layer of soil (about 1/4'') and dampen the soil. Keep it damp (not wet) until the grass is cut. Between one and two weeks after planting, the grass will be six to eight inches high and may be cut. Cut the grass just above the soil (scissors work well for cutting), and allow it to grow again as before. Replant the container with fresh seed and fertile soil after one or two growths of grass, since the grass generally becomes bitter and decreases in chlorophyll content after the first planting. Grind the grass and use in "green" drinks (with juice), cereals, soups, or salads.

Gluten   The protein part of the wheat may be separated from the rest and used in a variety of interesting and nutritious ways. Wheat gluten has more than twice the amount of essential amino acids per pound than beef.

Make gluten by mixing 1 cup of cold water for about every 3 1/2 cups of wheat flour. This mixture will form a ball that should be kneaded, stretched, pounded, and generally worked over for several minutes (about 10).

Next, completely cover and soak the dough in cold water for 45 minutes to 1 hour. Then put the dough-ball in a small amount of hot water and work it so that the starch is released into the water. The water will become milky. Change the water several times. When the wash water is clear the starch is sufficiently removed. All wash water may be saved to use as a source of starch for soups, gravies, drinks, and desserts, or as compost for the garden.

This process should produce 3/4 to 1 cup of gluten for every 3 1/2 cups of flour. Substitute this gluten for meat. Flavor it and cook it in varieties of ways to suit you: Roll it thin, make small balls, cook it in a lump. Simmer it, bake it, toast, it, roast it. Use it in soups as dumplings, as sausage, as meatballs, etc. Use the old imagination and have some fun!

A simpler alternative method is to soak 1 cup flour in 3/4 cup water for at least 1 hour. Then drain it, wash out the starch, and use it as gluten.

Weights and Measures

3 teaspoons = 1 tablespoon	1 lb. rice = 2 cups
2 tablespoons = 1 liquid ounce	1 lb. pitted dates = 2 cups
4 tablespoons = 1/4 cup	1 lb. cheese = 5 cups grated
5 1/3 tablespoons = 1/3 cup	1 square bitter chocolate = 1 ounce
16 Tablespoons = 1 cup	1 package cream cheese = 3 oz. or 6 Tbsp.
1 cup = 8 ounces	1 cup chopped nutmeats = 1/4 pound
2 cups = 1 pint	16 marshmallows = 1/4 pound
4 cups = 1 quart	1 cup egg whites = 8 to 12 egg whites
8 quarts = 1 peck	1/2 pint heavy cream = 1 cup whipped
4 pecks = 1 bushel	No. 1 can = 1 1/2 to 2 cups
2 Tablespoons fat = 1 ounce	No. 2 can = 2 1/4 to 2 1/2 cups
1/2 lb. butter or fat = 1 cup	No. 3 can = 4 cups
1 lb. water or milk = 1 pint	No. 10 can = 12-13 cups
1 lb. granulated sugar = 2 cups	8 ounce carton cottage cheese = 1 cup
1 lb. confectioners sugar = 3 1/2 cups	1/4 pound (1 stick) butter = 1/2 cup
1 lb. flour = about 4 cups	1/2 pint half and half = 8 2-Tbsp. serving
	(1 cup)

Equivalents

1 Tbsp. flour = 1 Tbsp. cornstarch
1 C. cake flour = 7/8 cups all purpose flour
1 C. honey = 1 to 1 1/4 C. sugar plus 1/4 C. liquid
1 oz. chocolate = 4 Tbsp. cocoa plus 1/2 Tbsp. shortening
1 C. butter = 7/8 to 1 C. hydrogenated fat plus 1/2 tsp. salt
1 C. 20% cream = 3 Tbsp. butter plus 7/8 C. milk
1 quart ice cream = 6 medium servings
1 C. sweetened condensed milk = 1 1/3 cups
1 tall can evaporated milk = 1 2/3 cups
1 small can evaporated milk = 2/3 cups
1 C. corn syrup = 1 C. sugar plus 1/4 C. liquid
(as replacement for 1/2 of sugar in recipe)
1 Tbsp. flour for thickening = 1/2 Tbsp. cornstarch plus 2 Tbsp. tapioca (quick)
1 Tbsp. herbs = 1 tsp. dried
1 C. fresh milk = 1 C. reconstituted nonfat dry milk plus 2 Tbsp. butter

SELECTED REFERENCES AND NOTES

Dickey, Esther. *Passport to Survival.* Salt Lake City, Utah: Bookcraft, Inc., 1969. This is a classic on the use of basic foods.
———. *Skills for Survival: How Families Can Prepare.* Bountiful, Utah: Horizon Publishers and Distributors, 1978.
Emery, Carla. *Old-Fashioned Recipe Book: An Encyclopedia of Country Living.* New York: Bantam Books, Inc., 1977.
Essentials of Home Production and Storage. Salt Lake City, Utah: The Church of Jesus Christ of Latter-day Saints, 1978.
Hornberger, Sally, Kay Martineau, and Marlynn Phipps. *Recipes for Self-Sufficient Living.* Salt Lake City, Utah: Deseret Book Company, 1984.
Moulton, LeArta. *The Gluten Book.* Provo, Utah: The Gluten Company, Inc., 1974.
Nelson, Louise E. *Project: Readiness.* Bountiful, Utah: Horizon Publishers and Distributors, 1974.
Salisbury, Barbara. *Encyclopedia of Family Preparedness, Just In Case.* Salt Lake City, Utah: Bookcraft, Inc., 1975.
Stevens, James Talmage. *Making the Best of the Basics: Family Preparedness Handbook.* Salt Lake City, Utah: Peton Corporation, 1977.
U.S. Department of Agriculture from the Superintendant of Documents. *Soybeans in Family Meals.* Home and Garden Bulletin No. 208. Washington D.C.: U.S. Government Printing Office, June, 1974.

Sources for Food Storage Products

Country Harvest, Inc., 325 West 600 South, Heber City, UT 84032. (Good source of food products and other items. They also have an attractive home distributor program.) (801) 654-5400.

Out N Back, 1797 South State, Orem, UT 84058.

PermaPak
40 East 2430 South, Salt Lake City, UT 84155
(Dehydrated and freeze-dried meals and foods)

Oregon Freeze-Dry Foods, Inc.
P.O. Box 1048, Albany, OR 97321

Nitro-Pak Survival Foods and Supplies
325 West 600 South, Heber City, UT 84032, (801) 654-0099, (800) 866-4876.
(Good source for dehydrated foods, also retort foods, survival seed pak, and
other supplies.)

Major Surplus and Survival
435 West Alondra, Gardena, CA 90248, (800) 441-8855, (310) 324-8855.

Emergency Essentials
165 South Mountain Way Dr., Orem, UT 84058, (801) 222-9596, (800) 999-1863.

Back to Basics
11660 South State, Sandy, UT 84070
(Food preparation equipment. They make one of the best and least expensive—hand
grinders.)

Crystal Springs Packing Co.
P.O. Box 2924, Petaluma, CA 94953
(707) 778-0567 (Color-coded silica gel that detects moisture levels.)

See also the Sources at the end of Chapter 8 for grain mills and other equipment.

Diatomaceous Earth

Perma Guard Co.
P.O. Box 29453, Phoenix, AZ 85034

Diatect Corp.
410 East 48th Street, Holland, MI 49423

CHAPTER 8

Growing, Gathering and Preserving Plant Food

At almost any given time one of the best ways to provide food is to grow or gather it. The first thing to come to mind when speaking of growing is usually a vegetable garden or fruit orchard. Animals may also be raised (even in relatively short periods of time) for use as food. The object in the next two chapters is to give enough details to arm the reader with sufficient knowledge to help provide for himself successfully in a crisis situation (or maybe even some other time). Headings in this plant chapter are: "Gardening," "Some Methods of Storing and Preserving Food," "Edible Naturally-Occuring Plants," and "Grains and Seeds."

GARDENING

An acquaintance of mine once said that if we ever had to survive on our gardens most of us would probably starve. The following pages should help avoid this drastic prognostication.

No matter what conditions a garden is planted under, it will undoubtedly be most useful if it is well planned. Even a small garden planted with foods you like and will use can successfully augment your food procurement efforts at any given time. Gardens can also be grown hydroponicly in nutrient solutions. Some of the important considerations for a successful garden plan are garden site, soil, fertilizer, control of pests, tools, where and how to plant, and the care and harvesting of garden plants.

Garden Site

Ideally a garden should be planted close to home, be fenced to keep out predators, have full sunlight on it, be made in fertile soil that is well drained, and be close to a usable irrigation source if needed. It is also desirable to have the garden where it is open to air currents to help eliminate early or late-year frost damage. For instance, a low-lying area may incur frost damage while a hillside where air currents are passing may not.

Direct sunlight is necessary for most crops, but many leafy vegetables can be grown mostly if not entirely in indirect sunlight.

Soil

Gardening soil should be properly cultivated and fertilized. All soils should be periodically tilled—whether this means spading, plowing, or power tilling—to reduce weeds and to mix crop residues and other organic and fertilizing materials into the soil. In the fall a cover crop such as wheat or rye grass, may be planted or the garden may be mulched to reduce erosion.

In dry areas, soils should be tilled in the fall and left rough to allow winter moisture to enter the soil.

Heavy clay soils can sometimes be made more workable by fall tilling, to allow the maximum soil surface to be exposed to freezing and thawing through the winter. Sandy soils can be mulched or cover-cropped and tilled in the spring. Soil should be worked as early in spring as possible, but not until the moisture content is low enough to prevent lumping.

Fertilizing

Fertilizing is a perpetual task in gardening that can be accomplished by using available organic materials and manure, earthworms, and commercial concentrates (which may or may not always be available). Compost and mulch can also be sources of plant nutrient and soil conditioner.

In some areas of the world, nothing organic is thrown away; every single scrap is composted and returned to the soil. Culinary or kitchen scraps and waste, hay or other grasses, leaves, manure, sawdust, sod, straw, spoiled crops or crop residues, and so on may be piled and allowed to decompose. The object is to get it to break down or decay to forms usable as fertilizer. Composting materials should be kept moist in a pile or bin, but a bin should not be airtight or watertight. If an auxiliary source of nitrogen—such as commercial fertilizer—is available, it may aid the decaying process when added periodically. If your soil is known to need lime, lime may be added to the compost. Periodically turning the compost with a pitchfork, especially in warmer weather, will also aid decomposition by allowing oxygen into the pile. The compost may be used for mulch or worked into the soil.

Commercial fertilizers may be used according to the manufacturer's directions. Most areas have agricultural extension services nearby that will test or otherwise tell you what your soil needs to be productive. Excessive and sometimes even moderate use of commercial fertilizers is hard on earthworms. Many chemical pesticides also kill earthworms.

Earthworms break up, aerate, and fertilize. They burrow several feet into the ground and produce approximately their own weight in fertilizer every day. They do well when nurtured with organic materials. There are several varieties; some are more prolific and active than others. Indeed, here is a creature to be encouraged in your garden.

Manure is a very good material to introduce minerals and nitrogen into the soil. Stir it into the soil for best results. A fall covering of manure may replace a cover crop. Animal urine and manure combined in bedding from animal pens and corrals make it a good source of plant nutrient. Some animal manure such as from chickens is very strong and should be applied sparingly.

The garden is not just a place to grow radishes; it is a biological environment. Encouraging the proper ecoculture will increase garden fertility and usability.

Pests

Pests invade every garden. Viruses, bacteria, nematodes, fungi, rodents and other small animals, insects, and last but not least, weeds all can strike blows at your garden. In simple terms, here are some ''nontechnological''defenses against some of these pests.

First and foremost, control all pests as early as possible. The underground animals can be especially frustrating. If you are planning to plant where these are prevalent it is a favorable investment to obtain poisons and traps.

Some reptiles, birds, and insects make good garden "friends". Toads, lizards, praying mantises, snakes, ladybugs, chickens, ducks, and lacewings all prey on small garden-gulping pests. Mantis castings (unhatched egg cases), for example, may be collected from thickets and shrubs where they have been left and moved to or near the garden or into a hothouse. The whole stem or branch they are on should be moved and placed where they will have vegetation to go to upon hatching in order to avoid their enemies. Other beneficial insects and reptiles may be imported as they are found. Ducks and chickens are good bug-eaters and can do a great deal toward debugging a garden: but if given access to the garden while it is too young they may also be disastrous to the plants.

Many worms and insects can be physically removed from crops and destroyed by hand, or they can be crushed to make a solution to spray on the garden. We have some friends who do this and use an old blender to make the ''soup''. Grasshopper spores that grow inside grasshoppers and eventually consume them can be applied by this method.

Here are some other natural pest remedies reported to be of benefit:
• tobacco and garlic boiled together until fairly thick, and then the mixture can be diluted and applied to the garden.
• soapy water sprayed on the garden.
• tea made from nettle leaves and sprayed on the garden as protection from aphids.
• wood ashes placed in a ring around plants such as celery to help them resist worm infestation.

Further, an entomologist has found urine to be effective as a fungicide, helping to reduce orchard mildew and scab. If kept in a bucket and stirred occasionally, urine loses its smell rapidly. Use a 1-to-2 (one part urine to two parts water) or a 1-to-4 solution while plants are growing and up to full strength in spring and fall. These solutions will approximate concentrations of urea used commercially for the same purpose.

Some plants interspersed in the garden may also help. It has been said that:
• chives deter aphids.
• garlic deters aphids.

- mint deters cabbage moth and ants.
- nasturtiums deter aphids, beetles, and bugs.
- oregano, horseradish, and pot marigolds deter many pests.
- rosemary deters beetles, cabbage moths, and carrot flies.
- tansy deters worms, ants, beetles, bugs, and some other pests.
- thyme deters cabbage and other worms.

The flowers of some chrysanthemum species (e.g., coccineum, cinerariifolium) contain pyrethrum, which can control aphids and soft-bodied insects. Crush the dried flowers and apply as a dust or in a water solution as a spray. This can also be used to control some pests on persons or animals.

Marigolds are not only beautiful but produce chemicals that help keep nematodes away from your plants. The plant also has an odor that is said to repel aphids, beetles, and other pests. (This point has been argued, but we found it to work—we simply plant a border of marigolds around some sections of the garden.)

It has always been perplexing to me why weeds can grow so much faster than anything else. Elbow grease (work) is one way to eliminate them in a small garden. Weeds can be pulled, or chopping them off under the surface of the earth with a hoe or shovel when they are about three to four inches high will sometimes kill them. Another method that is tremendously helpful to the extent it can be implemented is mulching. Grass clippings, spoiled hay (cheap from farmers), straw, and newspaper, all make good mulch material. Just place a thick layer everywhere except where your crops come out of the ground. Weed thoroughly before mulching. This can nearly eliminate the need for subsequent weeding as well as helping the soil to retain moisture and guarding it against frost. In addition, it composts right there in place to increase soil fertility and workability.

Black plastic can also be used to cover all but where the garden plants are.

Tools

As stated elsewhere, a shovel is an absolutely indispensable survival tool. Spading, ditching, cultivating, and harvesting can all be done with a shovel. Some other hand tools that are very helpful are a bow rake, common hoe, pitchfork, digging fork, stakes and cord for laying out rows, grubbing hoe, large wheel cultivator, some buckets, hose (for siphoning if nothing else), and wheelbarrow. A file for sharpening implements is also very useful. A word of advice is in order here: buy good quality tools. A rake that comes apart the first time it is used is unnerving, as is a shovel that bends when it hits its first rock.

If you have a rototiller or small tractor—vive la machine! These are tremendously helpful in gardening but not absolutely necessary for a small garden. Even if you have a machine, you still need hand tools; and you also need fuel and spare parts.

Where and How to Plant

There are some general rules that may help your garden. First, a garden can be planted in many unused or underused spaces besides the "garden". Look around and find some of them; flower beds, small strips of ground, and so on. They can be attractive if properly planted. In fact, plants should normally be rotated to different areas of the garden or yard from year to year to prevent buildup of specific crop pests in that area and to allow soil nutrition to stabilize.

Companion planting can be helpful in promoting healthy crops. This is based on the fact that plants actually practice chemical attraction and warfare. Chemical entities produced by various plant species may be harmful to soil microbes, insects, or nematodes and thus repulse them. Other chemical entities (alleopathic chemicals or chemicals that produce "mutual harm") are harmful to other plants and thus repulse them. Hence, some plants may ward off certain "pest" infestations for each other and planting "friendly" garden plants together and "enemies" apart is certainly a beneficial practice.

In the Table 8.1 several garden plants are listed by companionability. Some plants are detrimental to each other and are called *detrimental neighbors*. Some plants are mutually beneficial when planted together and are called *beneficial companions*. There are also plants which provide protection from enemies or otherwise assist the growth of some garden vegetables, and these plants are labeled *beneficial neighbors* in the chart. Several of the *beneficial neighbors* and their uses in gardening are mentioned above in "Pests".

If your garden is planted on a hillside, plant the rows across the incline to prevent erosion and assure good irrigation. When arranging your crops, remember that tall plants grown too near short ones may interfere with the growth of the smaller by providing too much shade. Also, permanent garden species (perennials, such as asparagus and rhubarb) should be planted where they will be out of the way of regular cultivation.

Many plants can be started indoors, then replanted outside when frost danger is past. Some can be grown completely indoors. Follow seed planting instructions. Bear in mind, too, that late maturing crops may be planted in areas from which early crops have already been harvested. For example, early peas could be followed by carrots or cabbage. In some areas it may be possible to have succeeding crops planted all year.

A reserve supply of seed is a storage item of the wise. Many of the seeds available today are hybrids, and when the plants from these seeds are used for seed the succeeding generations do not reproduce in their true form and are thus decreasingly useful. I prefer nonhybrid seeds when they are able to provide reasonable results. Nonhybrid seeds packaged in a storage container designed to retain their viability for longer than normal periods of time are avilable from some seed companies. Another reason for storing seed against a year of difficulty is that some plants take a full year to produce seed. For example, if a carrot is left in the ground through one winter it will produce seed the following summer which may be too late to plant until the following spring.

Most agricultural information booklets give the following seed storage-life approximations: one to two years for corn, leeks, onions, rhubarb, salisfy, parsnips, and all hybrids; three to five years for beans, brussels sprouts, carrots, asparagus, cabbage, cauliflower, celery, kale, lettuce, okra, peas, peppers, radishes, turnips, and watermelon; five or more years for beets, cucumbers, eggplant, muskmelon, pumpkins, squash, and tomatoes. Seeds should be stored in a cool, dry place in a sturdy container. Sad experience has taught that rodents and insects can make quick work of seeds. Store them in a plastic or metal bucket with a lid.

At planting time, seeds may be sown in the garden area they are assigned to in successive plantings in order to provide their fruits over a longer period of time. For instance, if you want to plant forty feet of something, plant ten feet each week for four weeks or every other week for eight weeks depending on time to maturity and when the first frost is expected in the fall. Gauge the time carefully to avoid getting a crop almost

TABLE 8.1
COMPANION PLANTING

Plant	Beneficial Companions	Detrimental Neighbors	Beneficial Neighbors
Asparagus	Basil, parsley, tomatoes		Pot Marigolds
Beans	Cabbage family, carrots, celery, chard, corn, cucumbers, eggplant, peas, potatoes, radishes, strawberries	Onions, garlic (can inhibit growth)	Marigolds, nasturtiums, rosemary
Beets	Cabbage family, lettuce, onions, bush beans	Pole beans (can inhibit growth)	
Cabbage family (including broccoli, brussels sprouts, cabbage, cauliflower, kale, kohlrabi)	Beets, celery, chard, cucumbers, lettuce, onions, potatoes, spinach	Tomato (can inhibit growth of kohlrabi)	Chamomile, garlic, dill, mint (improve growth); catnip, rosemary, sage, mint, nasturtiums, tansy, thyme (deter enemies)
Carrots	Beans, lettuce, onions, peas, peppers, radishes, tomatoes	Dill (inhibits growth)	Chives (aid growth); rosemary, sage (deter some pests)
Celery	Beans, cabbage family, tomatoes	Carrots, parsnips	Chives, garlic, nasturtiums (deter some pests)
Chard	Beans, onions, cabbage family		
Corn	Beans, cucumbers, melons, peas, potatoes, pumkins, squash	Tomatoes (worms also attack corn)	Odorless marigold, white geraniums (deter some pests, such as Japanese beetles)
Cucumbers	Beans, cabbage family, corn, peas, radishes, tomatoes	Sage	Marigolds, nasturtiums; oregano, tansy (deter some pests)
Eggplant	Beans, peppers		Marigolds
Lettuce	Beets, cabbage family, carrots, onions, radishes, strawberries		Chives, garlic
Melons	Corn, pumpkins, radishes, squash		Marigolds, nasturtiums, oregano
Onions	Beets, cabbage family, carrots, chard, lettuce, peppers, strawberries, tomatoes		Chamomile (improves growth)
Parsley	Asparagus, corn, tomatoes		

TABLE 8.1 (CONTINUED)
COMPANION PLANTING

Plant	Beneficial Companions	Detrimental Neighbors	Beneficial Neighbors
Peas	Beans, carrots, corn, cucumbers, radishes, turnips	Garlic, onions (can inhibit growth)	Chives, mint
Peppers	Carrots, eggplant, onions, tomatoes		
Potatoes	Beans, cabbage family, corn, eggplant, peas	Tomatoes (blight also attacks potatoes)	Horseradish, marigolds
Pumpkins	Corn, melons, squash		Marigolds, nasturtiums, oregano
Radishes	Beans, carrots, cucumbers, lettuce, melons, peas	Hyssop	Nasturtium
Spinach	Cabbage family, strawberries		
Squash	Corn, melons, pumpkins		Marigolds, nasturtiums, oregano
Strawberries	Beans, lettuce, onions, spinach, thyme	Cabbage	Thyme
Tomatoes	Asparagus, carrots, celery, cucumbers, onions, parsley, peppers	Corn (same worm attacks both); potatoes (same blight attacks both); dill, kohlrabi	Basil, marigolds, pot marigolds, chives, mint
Turnips	Peas		

to maturity only to have it spoiled by frost. Some crops, such as winter squash, should be harvested all or nearly all together in the fall as late as possible and properly stored.

Some plants are so hardy they can be planted as soon as the soil can be worked. Others can hardly survive any frost at all. Seed requirement for space, soil temperature, etc. are found in Table 8.2. You may want to keep a farmers almanac around or consult your local agriculture agent for average temperatures and times of frosts in your area if you don't already know.

Delicate plants may be started indoors in pans, boxes, pots, egg cartons, and similar containers before outdoor weather is sufficiently warm to place them outdoors. They should be planted in sandy soil and moss or just an easily drained soil.

The soil should be kept moist. This may be aided by covering the soil around the plants with a plastic film. The plants should be placed where they will receive at least six hours of direct sunlight. Fluorescent lights placed less than a foot from the plants, can also serve as a light source for them.

Once the plants are germinated and growing well, they should be moved to a larger facility such as a hothouse or cold frame (a glass- or plastic-covered garden area enclosed and able to receive maximum sunlight or use an auxiliary heat source). When weather is sufficiently warm to accommodate the plants in a normal garden setting,

TABLE 8.2
SEED REQUIREMENTS FOR TIME, SPACE, AND TEMPERATURE

Plant	Planting depth[1] (inches)	Germination Temperature Min.	Germination Temperature Optimum	Distance between seeds of plants (inches)	Distance between rows (inches)	Days to Harvest[1]	Hardiness of Plant[2]
Asparagus	seed 1½ crown 8 (perennial)	50	75	24	36-48	perennial (3 yrs.)	A
Beans: dry	1-1½	55	75	3	18-24	90-100	C
snap bush	1-1½	55	75	2-3	18-30	60-65	C
snap pole	1-1½	55	75	4-5	36-48	65	C
lima	1-1½	60	75	4	18-24	65-85	D
soy	1-1½			2	18-24	90	D
Beets	¾-1¼	40	85	1-2	14-18	55-70	B
Broccoli	3-4 for sprouts ½-1 for seed	40	85	24	30-36	55-75	A
Brussel sprouts	1-2			12-18	24-30	80-100	B
Cabbage	1 for seed 4 for sprouts	40	85	12	24	60-100	A
Cantaloupe	1-1½	60	95	24-48	48-60	70-90	D
Cauliflower	¾ for seed 4 for sprouts	40	80	18	30-36	55-65	B
Celery	½ for seed 3 for sprouts	40	70	6	18-24	100-125	C
Chard (swiss)	1	40	80	6	18-24	60	B
Chives	½			8	10-16	80-90	A-B
Corn (sweet)	1-1½	50	85	12	30-36	65-90	C
Cucumbers	1-1½	60	75	24	48	50-70	C
Eggplant	½ for seed 4 for sprouts	60	85	18		24-30	D
Garlic	1			2-4	12-18	90 for sets	A
Kale	½-¾			12	18-24	55-70	A
Kohlrabi	½-¾	40	85	2	18	55-65	A
Leeks	½-1			2-4	12-18	80-100	A
Lettuce: leaf	¼-½	35	75	4-6	14-18	40-60	B
head	¼-½			12	14-18	60-80	B

TABLE 8.2 (CONTINUED)
SEED REQUIREMENTS FOR TIME, SPACE,
AND TEMPERATURE

Plant	Planting depth[1] (inches)	Germination Temperature Min.	Optimum	Distance between seeds of plants (inches)	Distance between rows (inches)	Days to Harvest[1]	Hardiness of Plant[2]
Melons: Water-melons other	1 for seed 4 for plants	75	90	18-24	36-48	75-100	D
(several varieties)	1-1½	90	110	24-48	48-60	80-100	D
Mustard	½			2-6	12-18	40-60	B
Okra	1			16-18	30-36	55-70	C
Onions sets	1-2	35	80	1-2	14-18	45-120	A
seed	½-1	35	80	1-2	14-18	45-120	A
Parsley	¼-½	40	80	12	14-18	75-85	B
Parsnips	½-¾	35	70	2-3	14-18	105-120	B
Peas	1-1½	40	75	1-3	18-24	60-70	A
Peppers	¼-½	60	85	18	24-30	60-80	D
Potatoes	4-6	40	80	9-12	30-36	90-125	B
Pumpkins	1-1½			24-36	72-120	80-120	C
Radishes	½-¾	40	85	½-1½	10-16	20-45	A
Rutabaga	½-¾			2-4	14-18	80-90	B
Spinach	½-¾	40	75	2-4	14-18	45-65	C
Squash: summer	1-1½	60	95	18	36-48	45-50	C
winter	1-1½	60	95	24	48	85-110	D
Tomatoes	½ for seed 4-6 for plants	50	75	24	36	45-75	D
Turnips	½-¾	40	85	1-2	14-18	55-60	A

Information in this table is largely taken from U.S. Department of Agriculture bulletins.
[1]Wide varieties in planting depths and days to maturity are usually dictated by different varieties.
[2]A—Plant as soon as possible
 B—Plant 1 to 2 weeks after A
 C—Plant about the time of last frost
 D—Plant 1 to 2 weeks after C when soil is warmer

harden the plants by reducing watering and lowering temperature over a period of one to two weeks, then move them to the garden.

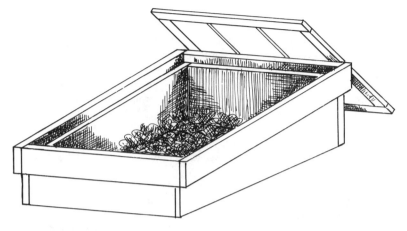

Figure 8.1   Cold Frame

Cold frames can be used to protect sensitive plants from cold and to harden plants into the outdoors.

Caring for and Harvesting Garden Plants

Here are some specific tips on planting and harvesting some garden plants. Some perennials are mentioned first, followed by other garden plants.

Perennials   The perennials should be planted where they will not interfere with cultivation and other work in the rest of the garden. Here are a few to think about:

Asparagus plants (crowns) are carefully grown from seed in a controlled environment and set out when one year old. Set the crowns with roots spread out and one to two inches under the soil surface. Do not cut them the first year. Cut the shoots while they are only a few inches in length for the best texture and flavor. Cut them at ground level or just below. Harvest them for only one to two months, then allow them to grow through the summer and fall. In fall, cut them to the ground and burn the tops.

Chives may be started from seed or from bulb division. The greens are decorative and tasty.

Comfrey is discussed under "Herbs" in this book.

Horseradish may be propagated by planting root cuttings from existing plants. Place the cuttings on an angle with their tops near the surface. Harvest a portion of the root crops as needed, leaving some root to reestablish itself. (See also "Herbs", Chapter 18)

Mint plants or roots may be transplanted from existing patches. Harvest the leaves or the entire plant as needed. (See also "Herbs")

Rhubarb may be started from seed but is usually propogated from roots. Plant them three or four feet apart. Harvest a few leaf stems the second year, but don't fully harvest until the third year. Seed stalks are generally removed as soon as they form. The leaf stems are edible, but the leaves are poison: do not eat them or feed them to animals. Hills should be divided and reset every few years to maintain plant vigor.

Other Garden Plants   *Chard* (swiss chard) may be planted in rows and thinned for use to about six to ten inches apart as the plants mature. The outer leaves may be harvested without injury to the plant for continual production through the whole season.

Lettuce, spinach, cornsalad, and *endive* may be planted in early spring or even late fall for early production of appetizing greens. Many greens are very hardy.

Celery should be planted indoors and set out under mild conditions. Celery germinates quite slowly. Plants should be kept upright.Before being harvested, celery is usually blanched by wrapping the portion of the stalks that is above the ground in paper or some other suitable material for a few days. In the fall when weather is cold, the unpicked celery may be kept for later use by banking it with dirt and covering the plants with straw to keep them from freezing, or it may be harvested and stored in a cellar. Banking early celery in warm weather causes it to rot and promotes worm infestations.

Beets are fairly hardy to grow and store. Plant them in rows and thin them as needed. Beet greens are excellent food. Harvest the beets after frosts begin, and store them at thirty-five to forty degrees Fahrenheit in a moist area.

Carrots are hardy. Successive plantings provide good carrots nearly the whole year. If a late planting is covered with straw or some other suitable insulative material to keep it from freezing, carrots may be left in the ground all winter. Some of the sweetest carrots I have ever eaten have come out of the ground in February when snow was on the ground. They may also be stored in cellars at thirty-five to forty degrees Fahrenheit.

Kohlrabi, rutabaga, and turnips should be harvested after early frosts when the soil is dry. If washed, they should be thoroughly dried before storing. Store them at thirty-five to forty degrees Farenheit in a moist area.

Potatoes are a very productive crop with good nutritional value and storability, but they are also a cool-weather crop and don't do as well in extreme heat. Also, heavy clay soil and poor watering may cause undersized and misshapen potatoes. Plant pieces of potato with at least one eye; and ideal size would be about one to one half ounce but can be much less. Moderate spring frosts do not usually permanently damage young potatoes; but extended periods of cold, wet weather may cause the seed to rot before it begins growing. Keep hilling up around the plants and don't over water.

Fresh potatoes may be harvested without damage to the rest of the plant (take a few from each plant), but the vines should be fully mature before the potatoes are harvested for storage. Keep unharvested tubers covered with dirt.

Potatoes continue to grow until the vines die. If practical, an early mild frost may be allowed to kill the vines before harvesting—some say this makes the potatoes more storable. Be careful not to skin the potatoes during harvesting. Also, do not leave potatoes lying in the field exposed to sunlight or wind; this renders them undesirable for use after a short time. For best results, cure the harvested potatoes for one or two weeks in a damp dark area at seventy degrees to 75 degrees Fahrenheit (not below fifty degrees or above eighty degrees) to allow the potatoes to "heal" skinned or cracked areas. Then store them at a temperature of forty degrees to fifty degrees in a dark, moist area. Any part of a potato that is frozen will spoil very rapidly when stored: keep them from freezing.

Sweet potatoes need long growing seasons and must be started indoors or in hot boxes in colder climates. They should be harvested before frost and allowed to dry thoroughly. They should then be cured by storing for a week or ten days at eighty-five degrees Fahrenheit and then stored at fifty to sixty degrees. Be careful not to bruise the roots when harvesting.

Turnips and rutabaga are widely used. Rutabagas do better in cooler climates than turnips. Turnips may be thinned by using young plants for their greens, which are very nutritious.

Radishes grow everywhere, although it is harder to keep them moist for growth when it is very hot. Use small successive plantings for continual harvest.

Cucumbers need fertile soil and warm weather, and are susceptible to pests and disease. Plant them in hot boxes or indoors for early crops. Indoors or out, keep the soil loose until the plants are established. Harvest them before the seeds are large and hard and while the rinds are still dark green. Harvest them as they ripen to assure continual production. Some varieties are small and especially well suited for pickling, but any variety may be pickled.

Muskmelons grow well under conditions similar to cucumbers and produce delicious fruits. Casabas and honeydews generally do well in areas where irrigation is necessary.

Pumpkins should be planted after frost danger is over, and they do well in areas of partial shade such as among corn, potatoes, or other crops. Harvest them before frost and store them at fifty to sixty degrees Fahrenheit.

Squash grows just about everywhere. Summer squashes such as zucchini, crookneck, straightneck, and cymling or bush scallop should be used while young and tender before their skins and seeds harden. Winter squashes such as Hubbard, Table Queen (acorn), Boston Marrow, and Delivious should be harvested only when they are fully mature—but before hard frosts—and stored in a cool dry place. If properly stored, some of these will keep until late winter and would be a welcome addition to a survival diet. Hubbards are especially storable.

Watermelons take a lot of space and a fairly long growing season but can be a wonderful treat.

Beans are some of the most productive and useful garden plants. Green beans (snap and lima) are probably most useful types to a home or survival gardener. Successive plantings may be made to insure a continuous crop. Keep young, tender beans harvested. Provide a good climbing surface for pole beans and an access to bush varieties. Lima beans are usually grown only where long warm summers prevail. Snap beans are more resistant to cold and grow well in most northern areas.

Soybeans have growing requirements similar to bush beans but need a longer growing season. Plant them about the same time as tomatoes. Harvest soybean pods when the seeds are fully grown but before the seed pods turn completely yellow. Boiling or steaming the pods for three to five minutes greatly facilitates the somewhat difficult process of removing the beans from them. The dried shells of beans allowed to mature or dry for seed or other uses are also difficult to remove. Grating them between two hard surfaces, such as concrete and a board or two boards, helps. A windy day facilitates winnowing to remove the chaff.

Peas are delicious and nutritious and at least one variety may be grown almost anywhere. English peas are a cool-weather crop and should be planted early. Sugar peas may be harvested when young and tender and cooked shell and all, similarly to bush beans. Black-eyed peas (cowpeas) need warm weather to grow, but they are easy to grow and very nutritious.

Broccoli is a very productive plant. Plant it indoors and set the plants out when heavy frost danger is past. Cut the sprouts carrying flower buds before the buds open. A long continuous harvest may be maintained—we are still able to harvest broccoli after the first fall frosts come. Broccoli may also be planted in many areas in the fall for an early spring crop.

Brussels sprouts are fairly hardy and may be planted early or "wintered over" in southern areas. As the heads appear, break away the lower leaves to make room for further growth. Leave the upper leaves. Harvest the whole plant for storage and remove the sprouts as desired for use.

Cabbage is another very useful garden plant. In fairly mild climates, cabbage plants can be grown or set out in the fall and will winter over for an early spring crop. In colder areas they may be grown indoors and set out early in the spring, or planted later for a fall crop for storage. Plant cabbage indoors a month or six weeks before time to set it out. Early cabbage can be planted between potato plants to conserve space; later cabbage can be planted where early crops such as peas and spinach grow and have been harvested. Harvest and store the heads before heavy frost.

Cabbage may be stored in a "pit" with the outer leaves and roots left on, or in damp sand or earth. For storage in a cellar or basement, remove the roots and outer leaves and wrap the heads with burlap, newspaper, or some other suitable material. The fact that cabbage emits quite a strong odor may influence where you store it. Storage temperature should be thirty-five to forty degrees Fahrenheit.

Cauliflower has requirements similar to cabbage. Tie the leaves together when the heads begin to form to keep it from yellowing.

Garlic is planted from cloves or sections of bulb and is fairly "fussy" about growing conditions. Harvest the mature bulbs; dry them; and store them in a cool, well-ventilated area. (See also "Herbs.")

Onions may be grown from seed or sets (small, dry onions grown the previous year). Onions do well with abundant moisture and moderate climate conditions. Different varieties may be grown for longer storage, for use as small "green" onions, and for short-term storage and use. Pull, dry, and store onions in a cool place in mesh bags or by some other method that allows for free air circulation.

Peppers are very sensitive to cold. They may be grown indoors and set out when the danger of frost is past. Start them six to eight weeks before setting them out in the garden.

Everybody grows *tomatoes*. They are very sensitive to cold. Plant varieties of your choice indoors about six weeks before time to place in the garden. Allow sufficient growing space for roots in containers.

When fall frosts begin, harvest all tomatoes, green and red, and store in a dark, cool place. (We place them on several layers of newspapers on the basement floor in a dark corner and have fresh tomatoes as they ripen until after Christmas.) They may also be stored in dry sand.

Corn takes substantial space to grow. Plant it in rows or in hills with several plants to the hill. Plant several short rows together rather than one or two long rows; otherwise, adequate pollination will not occur and the crop will be poor. Plant corn when the danger of frost is past. Soak the seed overnight before planting it. Succession planting will provide a continuous crop.

Corn may be dried and ground for flour or meal (cornbread and cornmeal mush are both delicious.) To dry it, pull the shucks back, tie several ears together, and hang them in a well-ventilated area to dry. Rub the dried ears to remove the corn kernels and store them in a vermin-proof container. Parching may increase storability. Nonhybrid corn may be dried and saved for seed, but hybrid corn will not produce successive crops of the same quality as the original.

Kale grows similarly to broccoli. Plant it indoors and set it out after frost danger or sow it in July or August for a fall harvest. The leaves and young flowers are edible.

Parsley should be grown in early spring in an area of partial shade. Keep it moist. Flowers on plants not used for seed should be cut back; letting them grow adversely affects affects the quality of the greens. Harvest parsley by cutting the stems about one inch above the ground. Do not use it as a garnish to be scraped off the plate with the refuse: eat it! It is very nutritious. It may also be dried and used as a seasoning.

In folk medicine, a tea of parsley roots and leaves is said to assist the function of several internal organs—including the liver, spleen, and gallbladder—and to soothe chills, colic, fever, and gas. A tea made of crushed parsley seed is said, in folk medicine, to help remove scalp vermin when applied locally.

Nasturtiums do well in full sunlight and flower until frost. The seed pods may be pickled in vinegar while still green. The seeds are peppery. The greens may be eaten like watercress, and the flowers are also edible.

SOME METHODS OF STORING AND PRESERVING FOOD

With the modern methods of packaging and preserving, we have almost any food at any season. Some of the in-home and primitive methods of preserving food also bring satisfactory results. Storage in a root cellar or pit, pickling, canning (bottling), drying, and sulfuring are all methods of preserving that can be used at home in (or out of) the emergency mode. Salting of fish and jerking of meat are presented in "Gathering, Raising, and Using Animals," Chapter 9.

Storing in a Root Cellar or Pit

An underground or partly underground cellar is a good place to store many root crops and fruits. These buildings or areas should provide protection from temperature extremes, ventilation, and usually a moist environment. They have been made of earth, wood, stone, concrete, metal and various combinations of these materials. A crawl space or cool corner of a basement may also be used for the storage area.

There are hundreds of ideas for many "corners" for root cellars, if you want to build one during normal times. A good root cellar may also double as a fallout shelter if it is properly made. Many of our forefathers built root cellars by digging into a bank or hill, building a small house of logs or stone, and—after proper bracing—covering all but the doorway with earth and/or sod (figure 8.2).

A cone-shaped "pit" is a very convenient storage device (figure 8.3). Simply pile some straw on the ground, stack fruit or vegetables on the straw (do not mix fruits and

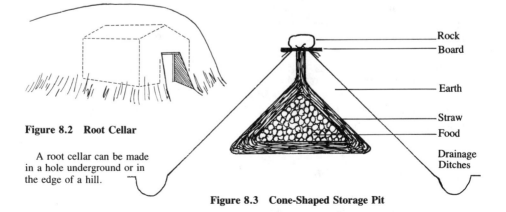

Figure 8.2   Root Cellar

A root cellar can be made in a hole underground or in the edge of a hill.

Rock
Board
Earth
Straw
Food
Drainage Ditches

Figure 8.3   Cone-Shaped Storage Pit

vegetables in the same pit), cover the produce with more straw, and then mound the pile over with a few inches of dirt, leaving a trench at the sides for water run off to be carried away from the pit and a "chimney" of straw through the top to provide ventilation. Compress the dirt by patting with the back of a shovel or a board to help prevent erosion and leakage. Place a weighted board or piece of plastic on top of the pit to help keep water from entering the straw "chimney". Keep the pits small and make several, because once they are opened in cold weather it is usually difficult to reclose them.

Another convenient storage pit is made by placing a trash can, barrel, or other container in the ground and then stacking food in the container in sacks or boxes. Put the bottom layer of food on blocks in the bottom of the container to allow air circulation around stored items. Cover the can with a ventilated lid and straw or other insulating material.

Fully mature fruits that may or may not be fully ripe generally require storage at near-freezing temperature (thirty-two degrees Farenheit) and moderate moisture. Many fruits readily absorb odors, making it somewhat undesirable to store them near vegetables with strong odors such as potatoes, cabbage, or onions.

Pickling

Some foods may be cured in a brine and/or acid solution. The brine should be made with noniodized salt and the pickling should be done in earthenware or wood containers. The acid is usually introduced as vinegar or citric acid. Metal containers have a tendency to corrode and contaminate the food. Stainless steel is quite resistant to this; but other metals, especially aluminum, are not.

When a high-salt brine sufficient to eliminate spoilage (fifteen percent salt) is used for preservation, the food is difficult to use. If less salt is used, acid must also be added to prevent fermentation. (Salt is added at the rate of approximately one-fifth cup per gallon of water for each percentage level of salt desired in the solution. A five percent solution, then, is one cup per gallon; a fifteen percent solution is three cups per gallon.)

Two simple, but favorite, recipes for dill pickles and dill beans are included in "Recipes" in chapter seven.

Home Canning

Rather than give complete canning instructions here, I refer the reader to the information available from agricultural extension offices, and two makers of glass jars and lids for home canning—Ball and Kerr—who have booklets available. This information is usually either free or available at minimal cost (see end of chapter for addresses).

Another idea for home canning that may be of interest to some is a system for preserving food in cans at home. According to their ads, they are the only ones who have a home system like this one. If you are interested, send a large stamped self-addressed envelope to Embarcadero Home Cannery, 2026 Livingston Street, Oakland, CA 94606 for a brochure.

There are also other hand canning units available.

Drying Fruits and Vegetables

Most fruits can be dried with very satisfactory results. Most vegetables change in color, taste, and consistency and become less attractive as "finger food," but they are still nutritious and satisfying when reconstituted for soups, stews, casseroles, and so on. There is a concentration of many nutrients during the drying process, but there are significant losses of Vitamins A and C and some other nutrients.

Use unblemished, fully ripe fruits and vegetables for drying to obtain maximum nutrients and minimum spoilage.

Fruits   Fruits may be dried with no preparation beyond coring or pitting and cutting in appropriately sized pieces. A very effective tool for coring especially apples and pears is a piece of copper tubing.

Lemon juice or a solution of 2000 to 3000 mg of ascorbic acid (vitamin C) per cup of water can be used as a dipping solution that will act as an antioxident for preserving color and perhaps some vitamins. The fruits may also be dipped into a lye solution for a few seconds to crack the skins and promote drying, but this is not necessary. The lye must be removed by washing before the fruit is dried. (A suitable lye solution can be made of one or two tablespoons of lye in one gallon of cold water.) As a general rule, the thinner the slices are the faster they dry. Blanching of tough skinned fruits will also speed up the drying process. The fruit should, however, be cut into uniform-size pieces to promote uniform drying. Fruit is dry when it feels dry on the outside but is still slightly moist on the inside. A temperature of one hundred forty to one hundred fifty degrees Farenheit is about ideal for drying but a hot summer day will do fine. Fruit will sometimes ferment if not dried rapidly; do not allow drying food to become wet with rain or dew. Bring it indoors at night. Use a screened device or area to keep flies and other pests away from the drying food. After fruits and vegetables have been dried, they should be stored in sterile, pest-proof, moisture-proof containers.

Here are some specific suggestions for fruits:

- Core and cut apples and pears into thin slices or rings. Some people core them, cut them in rings, and dry them on a stick.
- Cut apricots and peaches in half and remove the pits. They may be dried in halves or sliced. Some people prefer to remove the peach skins before drying. Agricultural extention directives generally advise not eating the nuts from peach or apricot pits although some people eat the apricot nuts. Pits may also be dried and used as fuel.
- Dry whole *berries,* or half or slice them if the fruit is large, such as strawberries.
- Pit cherries and allow the juice to drain from them before placing them on drying trays.
- For grapes, just remove the stems and dry them. They may be dipped into boiling water to crack the skins and speed drying, but this is not necessary. Where raisins are made commercially, the grapes are dried on a special paper in the sun. They are turned during the drying process by placing another paper on top of them, turning the fruit and papers upside down, and removing the top (formerly bottom) paper. They must not be allowed to become wet while drying.

Fruit may be pureed and dried on a cookie sheet, plastic, glass or similar surface to make leather. If a blender or other device is not available to puree the fruit, it may be cooked until it is soft and homogenous and then dried.

Vegetables   Vegetables should be prepared in the shape in which they will be used, but the pieces should be kept small enough to dry in reasonable time. Most vegetables should be blanched by steaming or boiling them in water before they are dried. They should be completely heated but generally not cooked. Steaming requires longer times, but boiling may remove more nutrients from the food. The blanching process speeds drying times; deactivates enzymes; and can help preserve flavor, color and some nutrients.

Drying temperature should not exceed 150 degrees to 155 degrees if a drying device is used. If vegetables are dried in the sun, they should spend no more than one or two days in the direct sunlight to avoid surface deterioration of the food. The drying may be finished in a warm shady area. The faster the food is dried, the better it looks, and generally, the better it is nutritionally.

Here are some specific ideas for drying vegetables:

- Cook *beets* and *sweet potatoes* until they are done, then skin, slice or dice, and dry them on a screen, cloth, stick, or thread.
- Cut *peppers, cabbage, brussels sprouts,* and *broccoli* into strips, blanch them, dry them.
- Cut *zucchini* and *summer squash* into slices or strips without pealing them; then, blanch and dry them.
- Cut *pumpkin* and *winter squash* into strips about three-quarters of an inch wide and blanch them. Peel them just before drying them.
- An ideal way to do *tomatoes* is to blanch them and run them through a food processor to remove seeds and skins. Strain the juice from the pulp with a cloth or other device, then dry the pulp on glass or plastic. It will form tomato flakes. The juice can be used in drinks or cooking.
- Cut *kale, chard, spinach,* and other greens into strips and blanch and dry them.
- *Beans* can be dried on trays or on a string or thread. Lima and soybeans may be shelled before blanching. Snap beans may be left whole or broken into shorter pieces before blanching.
- Shell and blanch *peas* or dry them whole in the sun until they are hard and then crush
  the hulls and winnow them to remove the chaff.
- *Corn* may be blanched, cut off the cob, and dried for use as a vegetable or may be dried on the cob and then rubbed off the cob and stored or used. Drying on the cob without blanching is a common method of preparing corn to be used as a grain and ground into cornmeal. It can also be parched.
- Slice *carrots* into the desired shape, blanch them, and dry them.
- If *onions* are to be used as seasoning in such dishes as soups, simply cut them into the desired shape and dry them. If they are to be cooked for other uses, then peel them, cut them into the desired shape, blanch them and dry them.

Trays for Drying   Drying trays are usually made with wire mesh as the drying surface to facilitate the free passage of air around what is being dried. Cheesecloth, cleaned window screening, wood slats, and other such materials will also do the job. The material is usually stretched over and attached to a wooden frame. It is important to prevent insects and moisture from getting to the drying food.

Where to Dry   We have dried food indoors, outdoors, in a homemade dryer, in the back seat and under the back window of the car (with newspapers underneath to catch the drips), and in the oven. Others use greenhouses, plastic covered boxes and so on.

Sweetners and flavorings are sometimes mixed with fruits to increase the palatability of fruit leather.

Reconstituting Dried Fruits and Vegetables   As a general rule it takes about one and one-half cups of water per cup of food to rehydrate home-dried foods. One good way to do it is to pour the required amount of boiling water over the food. Vegetables usually absorb all the water they will hold in two or three hours. They should then be cooked. If they dry up in the pot, add more water and soak them longer before cooking them the next time. Some manufacturers recommend soaking their dehydrated food products for up to 24 hours.

Fruits do not need to be cooked, but require a longer soaking period of several hours.

Eating substantial quantities of dried foods that have not been thoroughly rehydrated can cause digestive difficulties. Start with eating small quantities of dehydrated foods and then increase the amounts you eat when you are sure of your body's tolerance for it.

Drying Herbs

If herb *leaves* are to be used, it is usually best to harvest them just before blossoming. This is, in most cases, the time the leaves contain the greatest quantity of nutrients, and oils. Dirty leaves should be rinsed with cold water.

Many herbs lend themselves to being dried on the stem; others are most easily dried by stripping the leaves and drying them on a flat surface. Herbs that are easily dried on the stem—such as sage, oregano, basil, and mint—may be cut and the stems tied together and the bunches hung in the sun. As soon as most of the moisture has evaporated from them, hang them in a warm dry place out of direct sunlight and allow them to dry thoroughly. When they are dry, tie a paper sack around the bunch to prevent them from becoming dirty or store them in other containers.

Parsley, thyme, raspberry, comfrey, and other herb leaves can be removed from their stems and dried in a similar manner as above, but on a tray or other flat surface. When they are thoroughly dried, store them in a suitable container to keep them clean. The leaves retain oils and other nutrients longer if they are not crushed.

These drying methods may be used for other edible greens, too. The greens should be cooked before eating.

Some herb *flowers* are usable in cooking and remedies. They should generally be gathered when they are just starting to open into full bloom. If only the petals are usable, as in the case of roses, harvest only them. If the whole head is used, as with nasturtiums and camomile, dry the whole head.

Seeds should be gathered just before they are dry enough to start falling off the plant or before the seed pod is dry enough to break. Seed pods should then be allowed to finish drying and hulls rubbed off and winnowed. A whole plant can be hung upside down in a paper bag to dry. The seeds will fall into the bottom of the bag as they dry.

Some herb *roots* are useful for food, remedies, beverages,and candy. To avoid damage to the plants from which they are taken the roots should be gathered during dormant periods. The easiest time is probably fall. Take only a few tender small or medium-sized roots from each plant to allow the plant to live. Scrub the roots and slice them lengthwise to a reasonable drying thickness. Dry them in the sun and store them for later use. Sun-drying may be difficult in inclement weather; a low oven or fire will also do the job. Such roots are in high favor with many pests, so store them in a sturdy container.

Sulfuring

Many fruits can be preserved to some degree against spoilage, color change, and vitamin loss (especially Vitamins A and C) by sulfuring. The process does impart a different taste and add sulfur to the diet. Large-pit fruits—such as plums, peaches, nectarines, and apricots—and apples and pears are some of the fruits that are frequently sulfured.

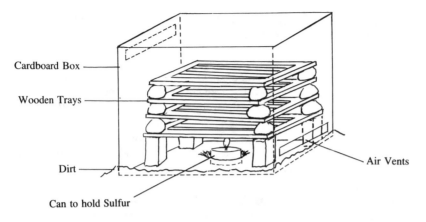

Cardboard Box

Wooden Trays

Dirt

Can to hold Sulfur

Air Vents

Figure 8.4   Sulfuring Box

Sulfuring should be done on slatted wooden trays outdoors by placing a cover over the stacked trays and burning the sulfur inside the cover in a suitable container, producing fumes which sulfur the fruits. Sulfur fumes will corrode most metals, so do not use screen or wire mesh. Allow one and a half to two inches between trays and keep trays about the same distance from the sides of the cover over the trays to allow free passage of the fumes over the fruit. Use only a single layer of fruit on the trays. Stack the trays on top of each other leaving a space about ten to twelve inches between the bottom tray and the ground.

Pure sulfur should be used in the preserving of food. Resublimed flowers of sulfur are of good enough quality to use. (In a situation where buying is impossible, sulfur can be found in deposits rich enough to burn in some geographic locations.) The sulfur should be placed in a clean container, such as an appropriately sized can, and then placed under the trays in such a way that no flames will reach the trays or fruit; the can will also become very hot and should not touch cardboard or wood. Light the sulfur and remove the match so that it does not remain in the container. Cover the stacked trays with a cardboard or wooden box that is only slightly larger than the trays (within a couple of inches). The box should have a slit about one-half inch by twelve inches at the bottom on one side and a slit of the same size at the top of the opposite side. The box should otherwise be pretty well airtight with the ground. Dirt may be pushed against the sides to insure a tight fit.

The amount of sulfur used for a given amount of fruit varies with the size of the fruit pieces (sliced, quartered, halved, etc.), the purity of the sulfur, the sulfuring system used, weather conditions, and the length of time used. Two cups (about ten ounces) of sulfur will normally burn more than two hours and sulfur about forty pounds of fruit on four or five trays. On a smaller scale use about two to two and one-half teaspoons of sulfur per pound of fruit. Let the sulfur burn completely. When it is completely burned, close the ventilation holes in the box until the time needed is completed. Sliced fruits will sulfur in about one to one and one-half hours. Quartered large fruits may take two hours or more. If subsequent drying of the fruit is to be done, the sulphuring time may be substantially reduced.

Other methods of sulfuring can also be used, such as putting less sulfur under a nonventilated tub or box with the food and leaving it all day and repeating the process two or three days in succession, or putting a tray or basket over coals of a fire and sprinkling sulfur over the coals.

EDIBLE NATURALLY OCCURRING PLANTS

Volumes have been written about edible plants. Extreme detail or knowledge of all edible species is not generally necessary, however, to be aware of much that can be gathered and used for food. It is usually better to stick with a few things you are familiar with than to try to find rare or unfamiliar species; this eliminates time-consuming searches and the danger of eating poisonous plants.

When food is gathered by groups of people in one area it is always important to practice conservation and not "bite that hand that feeds." Gathering wild edible plants is only a supplemental or short-term source of nourishment. Do not count on being able to survive very long on gathered plants alone unless you live in the Garden of Eden or are few in number on a gloriously provisioned tropical island. Also, most wild plants are low in the kind of energy that the human system can make use of, making it difficult to keep up body energy without supplemental sources of calories.

Knowing some general rules and becoming familiar with a few widely occurring specific plants can give a great boost to the quest for food. Drawings and photographs are helpful in distinguishing plants, but there is no substitute for firsthand experience in seeing and trying for yourself. Take the opportunity to become familiar with plants before the need arises; take a walk with a botanist or other knowledgable person and take some mental notes (and maybe even some notes on paper). Have an open attitude about wild edibles. Any time the diet is changed it takes a while for the taste buds to catch up, but—with a little patience—they do, to a great extent.

Some Rules of Edibility

The unreliability of many of the "hard and fast" rules makes it impractical on the level of this writing to include more than just a few general ones.

• Avoid fungi. About ninety-eight percent of all mushrooms and most puffballs and morels are edible, but the food value contained is usually not worth the chance taken, in my opinion. (The experts vary on this.)

• Avoid known poisons such as the cashew family, of which poison oak, poison ivy, and poison sumac are ubiquitous members. The leaves of plants from this family alternate on the stem, are segmented into three parts, and are pinnate (featherlike). Their berries (if any) are white and in autumn the leaves turn bright reddish orange.

• Beware of seeds or grasses having black or dark-colored growth in the seed heads. This is a sign of ergot growth, which is poison.

• Avoid plants with milky juice, unless you know them to be safe to eat.

• Eat unfamiliar plants in small quantities at first and observe their effect on you. Very few plants are lethal in small quantities. (This is not a reason to eat something you are unsure of.)

• Some poisons are rendered benign by cooking. If in doubt, cook it well.

• An unpleasant taste that causes a sensation of nausea or burning is a red-flag danger signal. This may be hard to distinguish because some safe foods are also quite bitter (juniper berries, for example).

• Foods that may be difficult for some people to eat alone may add nourishment to soups, cereals, and other mixtures; an example is alfalfa. Use this plant green; or dry the young leaves and then crush and add them to other dishes, or use them as a garnish or to make a tea.

• Avoid cucumber-like or melon-like plants unless you know for certain that they are edible.

• Many wild plants may be preserved by drying (see "Drying Herbs" in this chapter).

• Berries are a wholesome treat that can be found in many places. A general rule is that white berries are almost always poisonous; red ones are sometimes.

Trees produce some important food sources in fruit, nuts, bark, and sap. Nuts are very nourishing and most may be eaten raw. (Acorns, however, should be cut up and boiled with wood ashes to leach out the tannic acid and then dried or baked.) Nuts should be gathered when mature. Gently shaking a tree so that the nuts fall onto a tarp or piece of plastic is a good method and facilitates the gathering process. After gathering, the nuts should be hulled as soon as possible for preservation of good taste. Some nut hulls almost fall off; others, such as those of black walnuts, are toughies and can make hulling a grueling process. (One rather grandiose method I have heard of for cracking hard-hulled customers is to run over them with a car. Build a shallow wooden trough just the width of the tires, place a layer of nuts in it, and drive over them. The sides of the trough prevent the nuts from squirting. This process may also shell the nuts at the same time.) After hulling, nuts may be placed in water to separate the culls. The bad ones float to the top. (If you do not believe it, check the process out by opening a few.)

Ideally, nuts should be cured by drying for a few weeks after they are hulled and culled. Simply spread them out on a surface in a dry, well-ventilated area. Most nuts will rattle in the shells when dry. Once they are, store them in a cardboard box, paper bag, or perforated plastic sack in a cool, dry place. Moisture causes mold and mildew. Keep nuts as cool as possible. The more oil in the nuts the more likely they are to go rancid in a short time.

The inner, light-colored bark from many trees is edible. The thin green outer bark may also be used, but brown bark should generally be avoided. Young bark formed in the spring is most palatable. Bark is eaten raw, cooked, or dried and ground into flour. Aspen, cottonwood, poplars, birch, willow, and many pines have palatable inner bark. Teas may be made from the tender branch ends and needles of pines.

The sap of many trees can be eaten. The most notable example is the maple (Acer),

Figure 8.5 Collecting Sap From Maple Trees

Sap can be collected from maple trees in the early spring by inserting hollow tubes into holes made through the bark into the wood.

which produces a sugary sap. Tap the trees early in the spring by making a small hole to receive a hollow tube one inch to one and one-half inches long. Hang a bucket on the shaft and let the sap collect, then boil the sap down to the thickness or strength desired. Depending on the use, that usually occurs at about one-fifth to one-twentieth of the original volume. Thick syrup requires even greater reduction in volume.

Grass seeds and many flower seeds are edible and nutritious. Gather them by beating the grass heads or flowers into a piece of cloth, plastic, or maybe your shirt. Beating the heads or flowers with a bowl or ladle will help the seeds to fall in a given direction. Also, some types of seeds may be gathered by stripping them with your hand.

Seeds may be rubbed in the hands to break the chaff away. Some others, such as sunflower seeds, may need to be crushed or rubbed between two rocks (a small rock on a large one). Winnow them by blowing on small handfuls or by tossing them into the air in a cloth during a windy period. Some seeds, such as foxtail and fescue, need to be parched over low heat to loosen the hard outer hulls. Other seeds, such as arrograss, must be parched before being eaten to remove the poisonous hydrogen cyanide from them.

Wheat grass, rye grass, blazing star, bluegrass, amaranth, goldenrod, Chenopodium (lamb's quarter, pigweed, and so on), maple, and many other seeds are edible. Many seeds are more palatable and most are more nutritious when they are ground or at least cracked and cooked into a stew or porridge. Instructions on makeshift grinding are found under "Grains and Seeds" in this chapter.

Roots, tubers, and bulbs of many plants are edible and have high food value. They may be roasted, steamed, boiled, or dried and ground for flour. They are starchy and frequently need some kind of cooking to be digestible. Flower bulbs such as onion, Indian potato, tulip, daffodil, and lily may be eaten. Water lily seed pods, flowers, stems, tubers, and root stalks of various species are edible and are considered an important food source in some areas of the world. Some bulbs, such as hyacinth, are poisonous so make sure you know what you are gathering and eating. Underground stems or roots of cattail, ferns, and burdock, among many others, are also edible. Some roots should be peeled before cooking and eating.

Digging roots is done well with a shovel, but another excellent device is a digging stick. In his book *Outdoor Survival Skills* (pp. 34-35) Larry Dean Olsen describes the ideal digging stick as a stave of green, barked hardwood about three feet long and one inch in diameter which has been hardened (but not charred) by heating over a fire and sharpened slightly at one end. Ideal it is. Push the stick into the ground near the plant and with a lifting and prying motion extricate the root. With practice, it works well and eliminates digging up the countryside.

Greens represent a source of nutrient that grows almost everywhere. Dandelion, amaranth, lamb's quarter, shepherd's purse, asparagus, bullrush shoots, cattail shoots, burdock, ferns, nettle, miner's lettuce, mint, purslane, and the docks (sorrel dock, yellow dock, and so on) occur widely and many of them are very mild and delicious when eaten fresh or cooked as potherbs.

Eating the Edibles

Some greens may be eaten raw in salads. (Go out and pick several kinds you know are edible and taste each of them. Make a salad with two or three of the milder types.) The generally preferred order of cooking wild foods is boiling, roasting, baking or frying in that order.

Potherbs must be cooked in water. These can include stems, pods, seeds, and some leaves and roots. Wild greens are often bitter. If so, they may be boiled gently for a few minutes and tasted. If they are still bitter, change the water and boil them gently again in the fresh water. Repeat the process again if necessary.

Tubers and roots may be boiled, roasted, baked, or fried. Many roots and bulbs can also be dried and ground and used as flour. Seeds and nuts may also be ground into flour.

Some Specifics

Some notable edibles are described in greater detail below.

Cattail (Typha latifolia)   Cattail is an outdoor pantry and variety store found in swamp and marsh areas. Most parts of the plant may be eaten at various stages of development. The dried leaves can be woven or braided into mats, baskets or cordage, or other useful items. The stalks can be used for arrows, and the dry fuzz from the head makes good insulation.

The leaves are long and slender, similar to large blades of grass. The stem is long and jointless with the familiar "hot-dog"-shaped fuzzball seed head at the top with a short spike extending beyond it, and the roots are rope-like.

In the spring and early summer the cattail forms quantities of yellow pollen above the seed spike. This "bloom" can be cooked and eaten except for the center core, which is quite tough. As the male blossoms mature, the fine yellow-collored pollen can easily be gathered and used as flour for breads or stews, or eaten raw. Earlier in the season, before the pollen forms on the spikes, the green heads may be boiled and eaten. Later, when the heads are dry, they may be carefully burned to the point where the tiny roasted seeds are kept. The seeds may be eaten, but they are not abundant in quantity. They may be softened by boiling.

The young shoots are very good to eat. They may be boiled or eaten raw in summer or winter. Shoots under water or just breaking the surface may be pale green, yellow, or even white. The centers of the larger shoots may also be eaten.

The underground stems are starchy and very useful as food. They grow a few inches below the top of the mud and can be pulled up. They are usually reddish brown to black in color. After gathering and rinsing, the outside skin can be quite easily removed and should be discarded. The center parts of the roots may be dried and used for flour or in other cooking, or the starch may be extracted from them in water. This may be done by smashing the skinned roots in a container of clean water, stirring the mash, and allowing the starch to settle. The settling may take several minutes, or the mixture may be left to settle overnight. The clear water is poured off and the starchy layer used in bread or other cooking.

Red Raspberry (Rubus strigosus and other species)   A very good tea can be made from the leaves of this common garden plant. The tea can be used to soothe digestive upsets and has been used effectively for this, even in infants. The tea has also been used in folk medicine to aid in removing canker sores, to strengthen the walls of the uterus to aid in childbirth, and to work as a diuretic to help bring relief from water retention.

Of course, the ripe berries are delicious and nutritious.

Pinyon Pine (Pinus monophylla, Pinus edulis)   An evergreen that grows in many western areas, the pinyon pine, bears cones which contain edible nuts. The ripened cones may be gathered, roasted to loosen the nuts from the cones, and then beaten to remove the nuts, which make excellent food.

Figure 8.6   Cattail

(Typha latifolia)

Figure 8.7   Pinyon Pine

(Pinus monophylla, Pinus edulis)

Figure 8.8   Spearmint

(Labiatae)

Figure 8.9   Maple

(Acer)

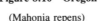

Figure 8.10   Oregon Grape

(Mahonia repens)

Figure 8.11   Biscuit Root

(Cymopterus bulbosus,
Cymopterus longipes, Lomatium)

Figure 8.12   Shepherds Purse

(Capsella bursa-pastoris)

Figure 8.13   Yellow Dock

(Rumes crispus)

Figure 8.14   Clover

(Trifolium spp.)

Mint   The smell and lanceolate leaves help identify the mint family. Catnip, horehound, hyssop, peppermint, sage, spearmint, thyme, and many others are familiar members. The leaves can be steeped for soothing, pleasant teas.

Maple (Acer)   The winged seeds of the Box Elder species of maple may be roasted and eaten. In addition, as already described, the sap may be drawn for use as a source of sugar.

Oregon Grape (Mahonia repens)   The deep purple berries of the commonly domesticated Oregon grape are edible but taste awful.

Biscuitroot   (Cymopterus   bulbosus,   Cymopterus   longpipes, Lomatium)   Biscuitroot has carrotlike leaves and grows in fairly dry areas in the west. Some poisonous species grow in wet areas and these areas should be avoided when gathering; stick to the hillsides instead. Find someone who knows the various species of this plant to teach you which are poisonous and which are not. If in doubt, leave it alone. The roots of this plant may be cooked and eaten or smashed and dried for later use. They found great favor with the Lewis and Clark expedition. The seeds of some species are also edible.

Shepherd's Purse (Capsella bursa-pastoris)   The common weed known as shepherd's purse has toothed leaves (somewhat resembling those of dandelions) and small winged seeds. The seeds can be ground into meal and the young leaves are usable as greens, cooked or raw.

Sourdock, Curly Dock, Yellow Dock (Rumes crispus)   The "dock" plants have brown seed clusters and large, dark-green, lanceolate leaves. The young leaves are good potherbs.

Clover (Trifolium spp.)   The seeds and greens of clover are edible. The leaves can be steeped for a tea.

Ferns (Pteridium aquilinum or bracken fern, and other species)   The young shoots of fern are called fiddleheads and make a good potherb or can be eaten raw. When mature, they may be toxic. The hairs covering fiddleheads can make them taste bitter and may be removed by rubbing the shoots in water.

Mallow, Cheesie Plant (Malva)   Mallow is found commonly as a weed almost everywhere. It has a large single root and is usually a few inches high. The leaves are roughly round and its flowers are small and light-colored. The small, flat, pumpkin-shaped fruits (or cheesies) that form are good to eat. The whole plant makes a very good tea—just wash and steep it for one-half hour.

Jerusalem Artichoke, Sunflower (Helianthus tuberosus)   The sunflower species known as Jerusalem artichoke has an edible potatolike tuber which is most harvestable late in the growing season. The plants are usually smaller and have more slender stems than the other common sunflower species.

Lamb's Quarter, Pigweed (Chenopodium album)   The common weeds called lamb's quarter and pigweed have edible stems and leaves that are delicious when young and tender, either raw or cooked. The seeds that form in clusters later in the year are also edible and may be ground for flour.

Purslane (Portulaca oleracea)   Purslane grows as a weed in many areas of North

Figure 8.15   Bracken Fern

(Pteridium aquilinum)

Figure 8.16   Mallow

(Malva)

Figure 8.17   Jerusalem Artichoke

(Helianthus tuberosus)

Figure 8.18   Lambs Quarter

(Chenopodium album)

Figure 8.19   Purslane

(Portulaca deracea)

Figure 8.20   Indian Potato

(Oroginia Linariafolia)

Figure 8.21   Arrowhead

(Sagittaria)

Figure 8.22   Burdock

(Arctium lappa, Arctium minus)

Figure 8.23   Dandelion

(Taraxacum)

America. Its stems are thick and juicy. It can be eaten raw or cooked and is very good when sauteed.

Indian Potato (Oroginia linariafolia)   The small Indian potato plants grow in the west and have slender leaves that have been said to resemble bird tracks. The bulbs are good raw or prepared as potatoes.

Arrowhead, Swamp Potato, Wapatoo (Sagittaria)   Wapatoo grows where it is wet. It has dark green leaves shaped like arrowheads and long rootstalks. In the fall it grows tubers which may be eaten like potatoes: baked, roasted, in soups, and so forth.

Burdock (Arctium lappa, Arctium minus)   Burdock has large leaves, pink to rose-colored flowers, and burrs that come off the plant very easily at the end of the growing season. The tender shoots and leaves may be cooked for greens. The roots (especially when young) may be peeled and cooked and eaten or roasted whole and then peeled and eaten.

Dandelions (Taraxacum)   Everybody knows about dandelions. The leaves make good greens, raw or cooked. (The young tender ones are not as bitter as the more mature ones). The roots also may be used raw or cooked. Watch out for chemical sprays on local dandelions.

Chickweed (Stelaria)   Chickweed is often found among sagebrush and in shady areas; it has white flowers. The roots creep and grow tubers at various locations along them. These tubers can be found most of the year and are good food, raw or cooked.

Bulrush, Tule (Scirpus)   Tules grow where it is wet and have seed clusters just below the top of their long, thick rootstalks. Its uses are similar to those of the cattail; tender shoots, pollen, rootstalks (peeled), and seeds (in season) may be used for food. A flour can be made from peeled rootstalks.

Watercress (Nasturtium officinale)   Watercress is tasty whether used raw in salads or as a seasoning. Its leaves have from three to nine segments and small white flowers. The white, threadlike roots are abundant. When gathering watercress beware of the poisonous water hemlock which also has small white flowers and finely cut leaves!

Wild Onion (Allium)   Wild onions look and smell like domestic onions. The bulbs are much smaller and not as round but are still very good. Beware of a plant that may look like an onion but does not smell like one; the poisonous death camas can be mistaken for an onion in looks, but not in smell.

Miner's Lettuce (Montia perfoliata)   Miner's lettuce is small and has round leaves that form a cup around the stem just below the pink or white flowers, as well as additional elongated leaves near the ground. Tender shoots and leaves can be eaten raw or cooked.

Mustard (Brassica)   Mustard is another plant that is common in many areas. It can have yellow or white flowers. The seeds and leaves (best when young) are edible.

Nettle (Urtica doica, ugracilis: Laportea canadensis)   Various varieties of stinging nettle are usable for food and fiber, though it is usually considered a weed. The hairlike spines produce a stinging sensation on the skin, but the greens may be eaten as a green or potherb. Steam or boil them. Small leaves at the top of mature plants are

Figure 8.24   Chickweed

(Stelaria)

Figure 8.25   Bulrush

(Scirpus)

Figure 8.26   Watercress

(Nasturtium officinale)

**Figure 8.27
Wild Onion**
(Allium)

Figure 8.28   Miner's Lettuce

(Montia perfoliata)

Figure 8.29   Mustard

(Brassica)

Figure 8.30   Nettle

(Urtica dioica, ugracilis;
Laportea canadensis)

Figure 8.31   Wild Asperagus

(Asparagus officinalis)

Figure 8.32   Sunflower

(Helianthus)

Figure 8.33   Wild Rose

(Rosa)

Figure 8.34   Prickly Pear Cactus

(Opuntia)

more palatable than larger bottom leaves. Young shoots coming up in the fall are tender ad palatable. They will often stay green after frosts begin. Nettle may also be dried and used for culinary purposes or animal fodder. As a bonus, a quart of salt added to a one- to two-quart infusion of nettle leaves may be used as rennet to curdle milk for making cheese.

Nettle is also one of the best and most common sources of strong plant fiber around. Its stems may be smashed, dried, and pulled apart for fiber. In addition, properties attributed to the nettle by folk medicine include asringent; antiasthmatic; diuretic; styptic; blood purifier; hair and scalp tonic; and antidote for hemlock, henbane, and nightshade poisoning.

Wear gloves when harvesting this plant. Stings from nettles are said to be relieved by applying the juice from the crushed stems of jewelweed (Impatiens).

Asparagus (Asparagus oficinalis)   A familiar vegetable, asparagus grows wild in many areas. The banks of waterways are likely places to look for it.

Sunflower (Helianthus)   The familiar yellow face of the sunflower is known widely. The seeds can be beaten from the heads, shelled, and eaten raw or roasted.

Rose (Rosa)   Roses, wild and domestic, grow almost everywhere. The ripened fruit of the rose is called a rose hip and is high in Vitamin C. They can be eaten raw, like apples, or cooked. The seeds are hard and may be ground before eating.

Prickly Pear Cactus (*Opuntia*)   This needly, flat, pear-shaped cactus is found in the southwest United States. The inside pulp may be eaten raw or cooked, and the seeds may be used as grain. (Barrel cactus and some other species may be similarly used. The barrel cactus can be opened and the pulp smashed or squeezed to produce a thirst-quenching liquid.)

GRAINS AND SEEDS

Grains and seeds are among the most usable of nature's foods, and in some cases they are abundant and easy to grow. In a situation where large farm machinery may not be available, harvesting grain would have to be done by hand or not at all. Grinding grain and seeds should also be considered, since many catastrophic occurences could put commercial grain mills out of commission and grains are normally more usable by the body when ground.

Harvesting Grains by Hand

Grain is normally harvested when it is fully ripe and about fourteen to fifteen percent moisture. Then it is cleaned and dried further. When grain is harvested by hand, it is necessary to cut it shortly before it is fully ripe to prevent losing the kernels from the heads during handling. When wheat is ready, its kernels will separate from the hulls if rubbed between the hands, and they should be hard if you bite into them. If the kernels are still milky the wheat is not ready. However, if you can shake the head a little and the grains fall out of the hulls, it is too ripe and will be difficult to harvest without a great deal of loss.

Oats should be harvested when the heads are yellow but some leaves are still green. You should be able to dent the kernels with a fingernail, but they should not be soft.

When grain is ready, it is cut, threshed, winnowed, and stored.

Cutting   Hand-harvesting is traditionally done by cutting the grain with a large scythe or cradle, binding it into bundles called sheaves, arranging several (usually ten to fifteen) sheaves so that they will stand together in shocks by leaning them against each other, and leaving them in the field to dry. The shocks are then gathered, threshed, and winnowed to remove the chaff.

Grain is still cut by hand in some areas of the world. A cradle is usually used for cutting; this is a scythe with tines against which the grain stalks gather as the implement is swung. The stalks of grain gathered in the stroke are then allowed to fall in piles which can then be easily gathered and tied into bundles or sheaves. A plain scythe or other cutting device may also be used to cut the grain; it just means more work gathering it into piles. Even a sickle-bar tractor mower could be used, but the tractor running over the cut grain "threshes" some of it onto the ground. A scythe or cradle should be kept very sharp (use a file and/or stone for sharpening) to keep the effort to a minimum. The strokes should not be exactly perpendicular to the grain but at an angle to it, swinging up at a slight angle from the ground level or down on the stalks toward ground level—up works better.

Sheaves are made by gathering bundles of cut grain about seven or eight inches in diameter or what you can almost enclose with both hands and tying it together with twine or straws of grain. The shocks are started by leaning two sheaves together with

Figure 8.35   Scythe With Cradle   **Figure 8.36   Sheaves and Shocks (groups of sheaves)**

the straws down and the heads up and then standing two or more at the sides of those two and so on until you have ten to fifteen sheaves stacked together. If twine is available, the entire shock should be tied together for stability. The grain should be allowed to stand in the shocks for one to two weeks until it is dry. If there is a way to prevent it from being rained on such as placing it under a roof, that is desirable. When it is dried it is ready to be threshed.

Threshing   A traditional way to thresh grain by hand is to flail it. A flail is made by segmenting a handle to meet your own desires. For example, a broom handle with a hole drilled in the end may be wired or tied with string or leather to a shorter stick with a similar hole in its end. A good size for the shorter stick is about an inch in diameter and about a foot long. Plastic pipe or other materials could also be used. Just string or cord may be substituted for the longer handle, but it is not as convenient. When the grain is ready, lay the sheaves on a hard, clean floor (of concrete, hard-packed earth, or something similar) and beat the daylights (or the grain) out of it. Remove the straws with a pitchfork or rake.

Another method of threshing is to beat the sheaves against a wall, allowing the loosened grain to fall onto a clean surface, into a container, or onto a cloth. The sheaves may also be beaten against the side of a barrel, allowing the freed kernels to fall inside.

The seeds of many grasses can be obtained without cutting by beating the seed heads into a bucket or onto a cloth when the seeds are fully ripe. Using a bowl to beat the seed heads can help direct the seeds toward the bucket or cloth.

Winnowing   The time-honored method of removing the chaff from the grain after it has been separated from the beaten sheaves of straw is to winnow the grain with the aid of the wind. While the wind is blowing, pour the grain slowly back and forth between two buckets or other containers, or throw it in a blanket or basket, allowing the wind to blow the lighter-than-grain chaff away. If facility is available to create an artificial wind—with a fan, for example—that is so much the better. Go ahead and remove most of the chaff, but you need not be too fussy. In most grains, a few hulls will not hurt a loaf of bread or other culinary delight—especially considering the kind of appetite you will have after this much work.

An appropriate-sized screen can also be used to help separate grain from chaff.

Storage   If for any reason the grain has become dampened or is still a bit green, it must be dried. This may be done on screens or in bags stored in a dry place on pallets. Turn the bags every few days to allow even evaporation of the moisture. It is customary

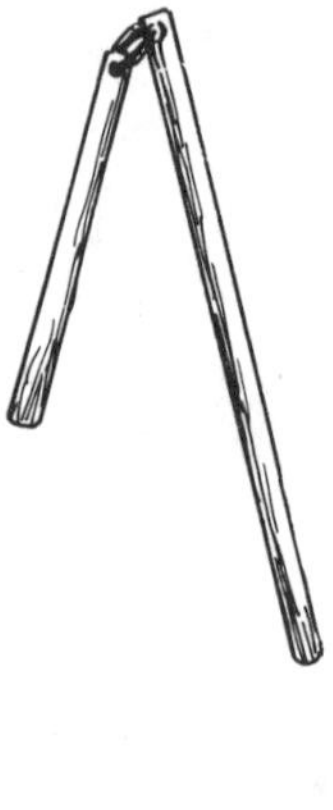

Figure 8.37   Flail

The small end of the flail can be about a foot or more long and about one inch in diameter and wired to a handle of choice.

to cure grain in this manner for at least four to six weeks before milling it for flour. Otherwise, it becomes somewhat gummy during grinding.

Grain should be stored where it is cool and dry and off concrete or dirt floors on slats of pallets. It should also be guarded against rodent and insect infestations. Rodents not only eat a lot, but they can also contaminate the grain. Sturdy containers of materials such as metal, glass, or plastic work well for storage. One favorite way is in convenient-sized sacks placed in metal drums. This way one sack at a time can be removed for use or to make moving any unwieldy drum easier. It is a good idea to put wooden slats between the bottom of a steel drum and a concrete floor to prevent condensation and rusting. Insects can also wreak havoc on a grain supply. See also "Food Storage".

An expedient storage facility is a hole two or three feet deep lined with grass or leaves. Place the grain inside, cover it with leaves, or grass or brush, and bury it with at least eight inches of earth. For short periods of time, the grain can also be buried in a tighly closed plastic sack.

Grinding Grain and Seeds

There has been a resurgence of popularity of homemade bread in recent years, and with it has come a wonderful variety of machines that will grind grains efficiently. It is wise to take advantage of this situation and invest in a grinder for two good reasons. First, since many nutritional entities break down or dissipate from the grain much more rapidly after it is ground, it is a good idea to grind it just before using. In addition, the use of grain solely in its whole state (which is for the most part the only practical way to store it) can cause some troublesome health problems, including sore mouths and loose bowels. This is especially true for the young, the feeble, and the unaccustomed. There are pioneer journals, however, which describe these early inhabitants of the American wilderness as living for weeks at a time on little more than a thin porridge made of wheat flour and water.

Power and Hand Grinders   Most of the grinding machines need electricity and cost in the range of $200 to $500. Some of them are or may be fitted to operate manually but almost without exception the conversions have problems that make them rather ineffective. A manual grinder of some type is an important item to have in your storage. Small, inexpensive hand grinders can be purchased for $50 to $75. Some will crack grain only, and if you think to opt for buying only one of these machines, don't. Spend more money and buy one that will also make flour—or buy both kinds. Probably the best option is to have (and use) a good electric mill and to also have an inexpensive hand mill, such as the Back to Basics handgrinder which sells for around $60. (See sources at the end of this chapter.) This mill will grind corn and soybeans, an important feature.

The "Country Living Grain Mill" is a top-of-the-line hand powered grinder that will produce a pound of fine flour in four or five minutes without causing undue perspiration. It sells for about $275 and will also covert to electric use. (See sources at the end of this chapter.) Another grinding mill, the Grain Country mill ($309), can be converted to bicycle power.

Grinding mills are available with stone or metal grinding surfaces. Both have advantages. The stone grinders run cooler and generally grind finer, but the metal plates grind faster and are much more versatile. The metal grinders will usually handle soy beans, peanuts and other hard or oily substances which would severly clog or ruin most stone grinding plates. In addition, a stone grinder usually produces a substantial quantity of flour when used to crack grains for cereal; the units using metal grinding surfaces usually do not. Two excellent grinders using metal grinding surfaces are the "Magic Mill III" and the "Kitchen Mill" (formerly the Magic Mill II). Both mills sell for around $300. The Jupiter sells for $250. (See the end of the chapter for sources.)

Expedient Grinding If a "day of reckoning" finds you with a barrel of wheat and no grinding device there are two expedient sets of devices that may be of assistance. The first set, the mano and metate, has been used for ages. The mano is a loaf-shaped, hand-held stone with a flat side. The metate is a bowl-shaped or flat rock. The grain or seeds are pounded and ground in small quantities on the metate with the mano. Various substitutes that may be made for these two items include a mortar and pestle, patch of cement and a rock, or a door knob and a bowl.

The second device that can be easily improvised was tested at the Oak Ridge National Labratory Civil Defense Project and found to be surprisingly efficient. To manufacture it, cut three pieces of pipe off squarely at about thirty inches. (The pipe works best if it is steel and about three-quarters inch in diameter.) In the middle and near the ends of each piece of pipe wrap a thin layer of string, cloth, or tape around the pipe so that the wrappings touch each other when the pipes are placed together in a bundle. Then tape, wire, or tie the pipes securely together at the places where they are wrapped. Next use a large fruit juice can with one end cut out (or some similar container) to hold a layer of grain about one inch deep. Place this on a hard surface such as concrete and, using the pipes as a ram, grind the grain by pounding it with about three-inch srokes. The finer flour may be filtered from the coarser pieces of grain by using cheesecloth, screen, or some other mesh. The coarser fraction may then be ground finer or used as is.

Barley, oats, and grain sorghums have hard fibrous hulls which should be removed. If these grains are ground dry, the hulls are brittle and fracture into small pieces. If the grain is dampened with about two-thirds of a tablespoon of water per pound of grain (about two percent) before grinding, the hulls toughen and stay in larger pieces. Sprinkle the water into the grain while stirring it, then allow it to stand for a few minutes. It will appear dry. After grinding, the hulls may be removed by stirring the meal into a shallow pan of water and floating them off.

SELECTED REFERENCES AND NOTES

Growing Plant Food

Acres, U.S.A. A paper published monthly by Acres, U.S.A., 10227 East 61st Street, Raytown, Missouri 64133.

Clemence, Richard, and Ruth Stout. *The Ruth Stout No Work Garden Book.* Emmaus, Pennsylvania: Rodale Press, Inc., 1971.

Country Wisdom Bulletins. Published by Garden Way Publishing, Charlotte, Vermont 05445. Tips on raising and storing.

Hills, Lawrence D. *Comfrey: Fodder, Food, and Remedy.* New York: Universe Books, 1976.

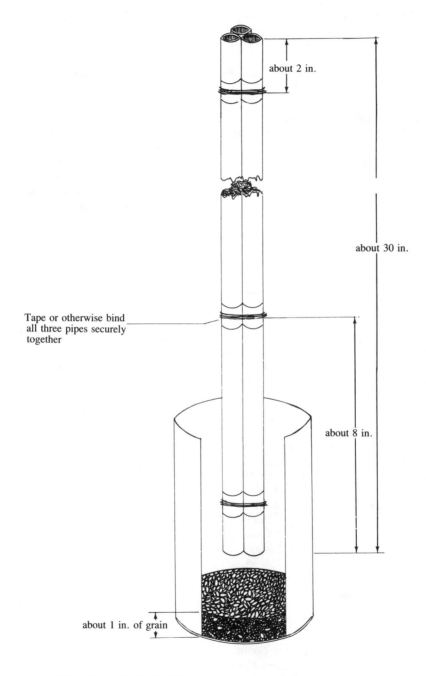

about 2 in.

about 30 in.

Tape or otherwise bind all three pipes securely together

about 8 in.

about 1 in. of grain

Figure 8.38 Make-shift Grain Mill

To use this grinding device the can must be resting on a hard smooth surface such as concrete.

Logsdon, Gene. *Small-Scale Grain Raising*. Emmaus, Pennsylvania: Rodale Press, Inc., 1977.

Mittleider, J.R. *More Food From Your Garden*. Santa Barbara, California: Woodbridge Press, 1975.

Natural Food and Farming Magazine. The official journal of Natural Food Associates, P.O. Box 210, Atlanta, TX 75551.

Philbrick, Helen, and John Philbrick. *The Bug Book: Harmless Insect Controls*. Charlotte, Vermont: Garden Way Publishing, 1974.

Raymond, Dick. *The Joy of Gardening*. Charlotte, Vermont: Garden Way Publishing, 1982.

Riotte, Louise. *Secrets of Companion Planting*. Charlotte, Vermont: Garden Way Publishing, 1975.

Rodale, Jerome Irving, editor-in-chief, and the staff of *Organic Gardening and farming Magazine*. *The Complete Book of Composting*. M.C. Goldman, comp. Emmaus, Pennsylvania: Rodale Press, Inc., 1971.

_____. *The Encyclopedia of Organic Gardening*. *Jerome Olds, comp. Emmaus, Pennsylvania: Rodale Press, Inc., 1971.*

Rogers, Marc. Growing and Saving Vegetable Seeds. Charlotte, Vermont: Garden Way Publishing, 1978.

Vuran, John. *The Manual of Practical Homesteading*. Emmaus, Pennsylvania: Rodale Press, Inc., 1977.

One very good resource of information on farming and gardening is your local State Agricultural Extension Office. There is usually one in each county. If you have a question or need information and can't find your local office, get in touch with the state office. The State offices, which are affiliated with state land-grant colleges or universities, are listed below.

Alabama: Auburn U., Auburn 36849
Alaska: U. of Alaska, Fairbanks 99701
Arizona: U. of Arizona, Tucson 85721
Arkansas: Box 391, Little Rock 72203
California: U. of CA, 2200 University Ave., Berkeley, 94720
Colorado: CO State U., Fort Collins 80523
Connecticut: U. of CT, Storrs 06268
Delaware: U. of Delaware, Newark 19711
District of Columbia: U. of Dist. of Col., 929 E. St., NW 20005
Florida: U. of FL, Gainesville 31611
Georgia: U. of GA, Athens 30602
Hawaii: U. of Hawaii, Honolulu 96822
Idaho: U. of ID, Moscow 83843
Illinois: U. of IL, 1301 W. Gregory Dr., Urbana 61801
Indiana: Purdue U., W. Lafayette 47907
Iowa: Iowa State U., Ames 50010
Kansas: Kansas State U., Manhattan 66506
Kentucky: U. of Kentucky, Lexington 40506
Louisiana: LA State U., Baton Rouge 70803
Maine: U. of Maine, Orono 04469
Maryland: U. of Maryland, College Park 20745
Massachusetts: U. of MA, Amherst 01003

Michigan: MI State U., East Lansing 48824
Minnesota: U. of Minn., St. Paul 55108
Mississippi: Miss. State U., Miss. State 39762—P.O. Box 5446
Missouri: U. of Missouri, Columbia 65211
Montana: Montana State U., Bozeman 59717
Nebraska: Div. of Institute of Ag. & Natural Resources, Ag. Hall, Lincoln, NE 68583
Nevada: U. of Nev., Reno 89557
New Hampshire: U. of NH, Durham 03824
New Jersey: Rutgers U., New Brunswick 08903
New Mexico: N.M. State U., Las Cruces 88003
New York: N.Y. Cornell U., Ithaca 14853
North Carolina: N.C. State U., Raleigh 27650
North Dakota: N.D. State U., Fargo 58105
Ohio: OH State U., 2120 Fyffe Rd., Columbus 43210
Oklahoma: OK State U., Stillwater (139 Ag. Hall) 74078
Oregon: Oregon State U., Corvallis 97331
Pennsylvania: Penn. State U., University Park 16802
Rhode Island: U. of RI, Kingston 02881
South Carolina: Clemson U., Clemson 29631
South Dakota: SD State U., Brookings 57007
Tennessee: U. of Tenn., Box 1071, Knoxville 37901
Texas: Texas A&M U., College Station, 77843
Utah: Utah State U., Logan 84322
Vermont: U. of Vermont, Burlington 05405
Virginia: VA. Poly Inst., Blacksburg 24061
Washington: Washington State U., Pullman 99164
West Virginia: W. Va. U., Morgantown 26506
Wisconsin: U. of Wisconsin, 432 N. Lake St., Madison 53706
Wyoming: U. of Wyoming, Box 3354, Univ. Station, Laramie 82071

Edible Naturally Occurring Plants

Angier, Bradford. *Feasting Free on Wild Edibles: A One-volume Edition of "Free for the Eating" and "More Free-for-the-Eating Wild Foods."* Harrisburg, Pennsylvania: Stackpole Books, 1972.

_____. *Field Guide to Edible Wild Plants.* Harrisburg, Pennsylvania: Stackpole Books, 1974.

_____. *Field Guide to Medicinal Wild Plants.* Harrisburg, Pennsylvania: Stackpole Books, 1978.

Benson, Ragnar. *Live Off the Land in the City and Country.* Boulder, Colorado: Paladin Press, 1981.

Hines, Frederick L. *Scout Handbook.* North Brunswick, New Jersey: Boy Scouts of America, 1972.

Olsen, Larry Dean. *Outdoor Survival Skills.* Provo, Utah: Brigham Young University Press, 1972.

U.S. Air Force Manual, 64-5. Boulder, Colorado: Paladin Press, 1979.

Brown, Tom, Jr., *Tom Brown's Field Guide To Wilderness Survival*. New York: Berkley Books, 1983.

Preserving Plant Food

Ball Blue Book. Muncie, Indiana: Ball Corporation, 1985.
   About $2.00. All about home canning. A real bargain.

Bubel, Mike, and Nancy Bubel. *Root Cellaring: The Simple, No-Processing Way to Store Fruits and Vegetables*. Emmaus, Pennsylvania: Rodale Press, Inc., 1979.

Densley, Barbara. *The ABC's of Home Food Dehydration*. Bountiful, Utah: Horizon Publishers and Distributors, 1975.

Garden Way Country Kitchen Catalogue.
Available from the Garden Way Catalogue Department, 90461-1300 Ethan Allen Avenue, Winooski, Vermont 05404.

Kerr Home Canning Book. Sand Springs, Oklahoma: Kerr Glass Manufacturing Company, 1985.
Similar to *Ball Blue Book* above. Send your name and address and $2.00 to Kerr Glass Manufacturing Co., 501 South Shatto, Los Angeles, CA 90020 (Attention consumer products division) to receive one by mail.

Rombauer, Irma S. *The Joy of Cooking*. Indianapolis, Indiana: The Bobbs-Merrill Co., Inc., 1943.

Stoner, Carol, ed., and the staff of *Organic Gardening and Farming Magazine*. *Stocking Up: How To Preserve the Foods You Grow, Naturally*. Emmaus, Pennsylvania: Rodale Press, Inc., 1974.

DeLong, Deanna. *How To Dry Foods*. Tucson, Arizona: H.P. Books, 1979.

U.S. Department of Agriculture. *Home Canning of Fruits and Vegetables*. Home and Garden Bulletin, No. 8. [Washington, D.C.: U.S. Government Printing Office, May 1983].

U.S. Department of Agriculture. *Storing Vegetables and Fruits in Basements, Cellars, Out-buildings, and Pits*. Home and Garden Bulletin, No. 119. [Washington, D.C.: U.S. Government Printing Office, January 1978].
Also available at many local county or state agricultural extension offices.

Some Unusual Supplies for Preserving

J.G. Durand International, Mellville, NJ 08332 (Luminarc glass storage jars)

Embarcadero Home Cannery, 2026 Livingston St., Oakland, CA 94606 (Home canning equipment for preserving in cans—send a self-addressed stamped envelope for catalog)

National Manufacturing, P.O. Box 700, Ventura, CA 93001 (Reuseable Canning Lids)

Tools for Hand-Harvesting Grain

Hand and Food Ltd., P.O. Box 611, Brattleboro, VT 05301.

Grinding Mills

Magic Mill Corp., 235 West 200 South, Salt Lake City, UT 84101 (Magic Mill III Grinder)

Back to Basics, 11660 South State Street, Sandy, UT 84070 (Back to Basics hand grinding mill and other mills and supplies)

K-Tec, 420 North Geneva Rd., Lindon, UT 84042, (800) 748-5400, (801) 785-3600.

Other Supplies

Organic Controls, Inc., P.O. Box 25382, Los Angeles, CA 90025 (Sources of Organic Controls, such as Preying Mantis)

Plants of the Southwest, 1812 Second St., Santa Fe, NM 87501 (Flowers, herbs, vegetables—many rare items)

Burpee Seed Co., 1009 Burpee Bldg. Warminster, PA 18974, or Riverside, CA 92502.

D.V. Burrell Seed Growers Co., P.O. Box 150, Rockyford, CO 81067

Gurney Seed & Nursery Company, 3112 Page St., Yankton, SD 57079

George W. Park Seed Co., Inc., 37 Cokesbury Rd., Greenwood, SC 29647

Stark Bros. Nurseries & Orchards Co., Box 12178, Louisiana, MO 63353.

U.S. Dept. of Agriculture Pamphlets available from the Superintendent of Documents, U.S. Government Printing Office, Washington, D.C. 20402

Nitro-Pak Survival Foods and Supplies, 325 West 600 South, Heber City, UT 84032, (801) 654-0099, (800) 866-4876. (Non-hybrid storage seeds packaged for storage, grinding mills.)

CHAPTER 9

Gathering, Raising, and Using Animals

Many Americans (and others) probably eat too much meat for good health. However, in times of hunger and/or cold, meat is an excellent nutritional source and can serve as a premium survival food. The animals can also provide skins and bones for other useful purposes. A point worth mentioning here is that many wild animals are low in fat and hence generally lower in calories than farm animals.

Some people criticize those who hunt animals for food or raise them and slaughter them themselves, but they think nothing of eating a hamburger or steak. This seems a little inconsistent. All life should be viewed with some measure of sanctity, and meat should indeed be used wisely. But if you are going to eat meat, you should be at least *willing* to take the life of the animal when it becomes necessary, and you should know how to do so humanely, using all the parts to the best advantage. Killing animals only for sport is, in my opinion, fundamentally wrong.

In a post nuclear war situation the glands of animals, which may be contaminated with a buildup of radioactive material, should not be eaten. If an animal is ill it should not be slaughtered and eaten. Even if the unhealthy animal is not contaminated with radioactive materials, the effects of radiation may have destroyed its disease-fighting ability and it may be toxic from bacterial infections.

This chapter provides some basic instructions in obtaining, raising (including some instruction on assisting some larger animals with birth) and caring for animals. "Trapping", "Hunting", "Fishing", "Raising Animals", and "Use of Animals" are the headings. "Use" contains some potentially helpful instruction on slaughtering, preserving meat, and tanning hides.

TRAPPING

Many animals can be taken by trapping. The easiest to take and most plentiful in most areas are rodents, birds, and other small animals. Pine hens, for example, are plentiful in some forested areas and can frequently be caught even without a trap. It is also possible to trap larger game. For best results, don't depend on one or two trapping devices; set several. Plain steel traps are good for many animals and birds if you happen to have some, but improvised traps can work very well.

To properly arrange a trap it is best to have an idea of what size of animal you are trying to trap and to construct your trap accordingly. Observe the animals in the area. The simplest and most easily made traps are deadfalls and snares. Unbaited traps should be arranged as inconspicuously as possible where there is heavy vegetation or some other natural channel. Baited traps may be baited with meat, entrails from previous catches, vegetables, roots, grains, or other appropriate attractants according to what animal is being sought. Carnivores such as bobcats or weasels like soft fresh viscera; scavengers such as racoons and opossums aren't so fussy about the meat or viscera being fresh, and vegetarians and some others like roasted roots as well as some of the other things mentioned.

Avoid leaving your scent on the trap by touching any of the bait—especially the meat. Cover up human scent on the trap parts by rubbing them with vegetation and/or smoking them by a fire for a few minutes. Construct the parts of traps in as inconspicuous a manner as possible; make them look natural. And last of all practice making them if you ever think you might have need to make one.

Trapping by use of the methods described here is prohibited almost everywhere in all but survival situations. It is also good to remember that traps are non-selective. They can catch domestic animals as well as wild ones.

Deadfalls

Deadfalls are made by raising an object heavy enough to hold the animal you are pursuing and propping it with a trigger mechanism that will be easily tripped by a visiting animal. Deadfalls may be set in trails, at the edge of thickets, at the base of cliffs, near burrows, and so on. Deadfall weights may also be replaced with cages or boxes to capture animals if this is for some reason more desirable. Figures 9.1 through 9.6 are some examples of deadfall traps. The heavy deadfall may be a rock, a log, or whatever is appropriate and available.

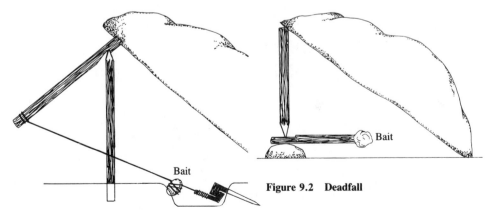

Bait

Bait

Figure 9.2   Deadfall

Figure 9.1   Deadfall With A Hooked Set

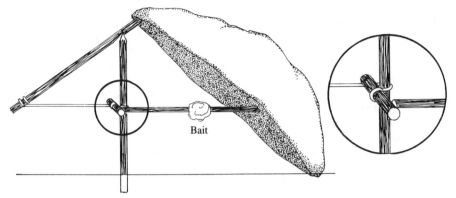

Figure 9.3   Deadfall

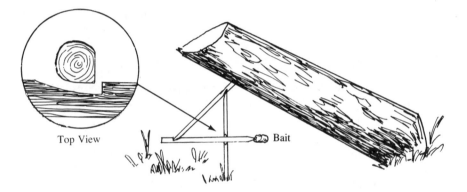

Figure 9.4   Log Deadfall With A Figure 4 Set

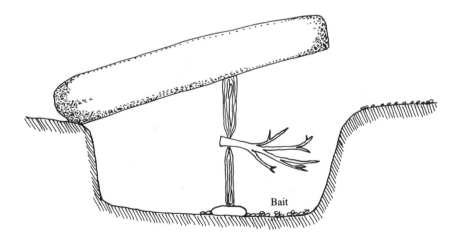

Figure 9.5   Bird Trap Deadfall

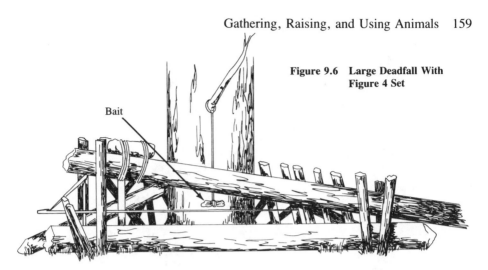

Bait

Figure 9.6   Large Deadfall With
Figure 4 Set

Snares

Snares are, at simplest, some cordage (rope, string, wire, or whatever) arranged with a slipping loop at one end to catch the animal you seek and with the other end secured and/or set in spring to elevate the captured animal. Sturdy fish line, string, or other cordage can be used for snares. Flexible wire and small cable, such as that used to hang pictures, are also very good. Set snares or traps for animals where their trails narrow at natural barriers such as trees or brush, in the edges of thickets, at the edge of bodies of water, and near bedding grounds and burrows.

When a snare is used for larger game it is important to stay near the set. After the larger animal is captured it must be killed quickly or it may pull loose. A snare should be set with enough spring to keep at least one set of the animal's legs off the ground and preferably both sets to prevent it from breaking the trap. Figures 9.7 through 9.9 are examples of snares. The loop of a snare should be tied with a slipknot. Adjustments should be made in the size of a snare according to the size of the animal you are after. A small cottontail could probably be caught in a snare two and one-half inches in diameter

Figure 9.7   Weighted Snare

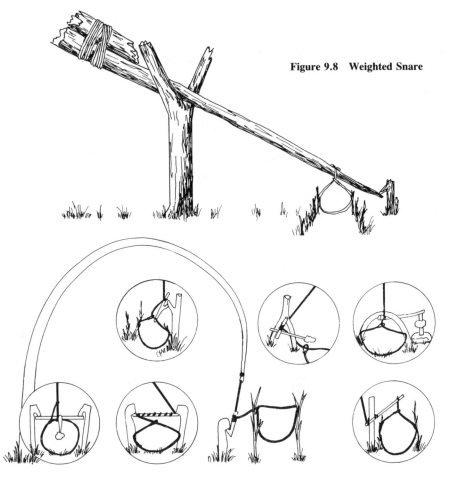

Figure 9.8   Weighted Snare

Figure 9.9   Spring Snare

Use a sapling for the spring. A variety of sets are shown. The set can be made on a trail or run, or it can be baited, or both.

with the bottom one and one-half inches off the ground while a larger rabbit might be better caught with a four-inch loop two and a half inches off the ground.

When trapping, don't overlook squirrels, rabbits, beaver, racoons, pheasants, quail, robins, blackbirds, pigeons, muskrats, and other small animals, many of which can be found in or near many cities and towns. These may not be usual in the average American diet, but they are edible and could mean survival when the going is tough. Ducks and geese may also sometimes be caught with improvised or commercial steel traps at feeding sites or source areas of sand if such areas are not plentiful (they eat sand and small rocks to help digest food).

Other Traps

Animals and birds that frequent streams and ponds may be snared with a weighted snare arranged so that when they approach the water, they will be pulled under by the

Figure 9.10 Weighted Water Snare

Some animals will head for the water when caught in a snare. If the snare is weighted the weight will hold the animal under water and drown it. A small stick at the end of the cordage will float and signal the location of the dead animal. The trap can also be baited. Check this type and all other traps frequently.

weight and drowned. A stick tied to the loose end of the snare will identify the location of the game (figure 9.10).

Another trap of interest for taking large birds, such as geese, is to dig an inclining trench about eighteen inches wide and about fifteen feet long. The trench should incline to a depth of about two feet. Bait the trap by making a thin trail of grain or other suitable bait down the incline. Secrete yourself near the trap, and when a bird eats its way to the bottom, hurry to the edge of the trap and hit the bird with a heavy stick, wring its neck or otherwise kill it. When the bird is at the bottom of the trench and it tries to fly, its wings can not extend and it can't fly away (or it will at least be delayed) allowing it to be captured. The dimensions may be varied according to the size of bird to be caught.

HUNTING

In a survival situation hunting is a method of providing that which may be difficult or impossible to acquire any other way, namely food and leather. It is, however, unrealistic to think that in a major disaster there would be sufficient large game to

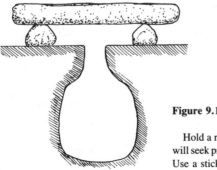

Figure 9.11 Rodent Trap

Hold a rodent roundup and drive them toward this trap. They will seek protection under the flat, round rock and fall in the hole. Use a stick to kill and retrieve.

provide long-term support. Still, some knowledge of hunting could make the difference between failing to survive and hanging on.

Game Animals

Man's abilities have enabled him to hunt virtually every creature. This dominion, however, does not necessarily mean meeting animals on their own terms. Man is a superior hunter because he can invent and produce devices to reduce an animal's advantages in his own habitat. The weapon used should be adequate for the job. A stick may suffice in some situations; a large, high-powered rifle may seem inadequate for others. An obvious first step in properly approaching a hunt is to know how to use your weapon of choice safely. Generally speaking, a high-powered rifle (30-06, .308, .270, .300 Winchester mag., etc.), used properly, is adequate to bring down any American game animal. But even then there can be difficulty with some large animals. You should become acquainted with the game animals in your locality and prepare yourself accordingly.

Finding game animals is, in most locations, not always an easy task. Most animals try to avoid people. They use sight, sound, and smell to detect danger. Their sense of smell is considerably better than ours. Many mammals, including the large game animals, are color blind. Blaze orange, for example, will show up as white to a deer or an elk. They can, however, be alerted by contrasts. Some hunters say that deer and some other animals will not take flight unless alerted by at least two of their senses.

Many animals feed mostly at night in open weather, and they will approach watering places early in the morning or late in the evening. By looking for tracks at such places to determine the likelihood of seeing game and then waiting quietly at a reasonable distance, you may be able to obtain an opportunity to take an animal. Remember the ability of animals to smell you, and observe and use wind patterns to your advantage. There is a rule of thumb in mountainous areas that air currents flow down the canyons and slopes in the morning until the heat of the day warms the lower air and causes it to rise up the canyons. Then later in the night, as the lower areas cool, the air again flows down. This is generally true even when it is cloudy. Sometimes in very cold weather the air will start its downward movement as early as just after sundown.

Heavily used game trails and bedding grounds are other source areas to watch for game. Animals usually bed in areas out of the wind and with low visibility; in hot weather, this includes shaded areas. When more than one person is hunting, one or more members of the party can go through a thicket or grove of trees and drive a large game animal into the open, where an armed comrade stationed at a strategic vantage point can make the kill.

In wooded areas, clearings are used as feeding grounds by animals and they sometimes stay in the edge of the clearings during most of the day. If you are hunting these areas, try to stay out of the clearings. If you can see from the opposite side of the canyon and it is not too far to shoot, watch from there with binoculars (if available) until you get a shot. Sometimes a startled animal will cross a canyon in wooded areas and stop in a clearing on the opposite side to see what is after it, thus providing a shot.

Another way to avoid being seen while stalking is to avoid skylines. When crossing from one side of a ridge to the other you become very visible to both sides. Cross where cover is available, or else maintain a low profile. Remember that noises such as crunching or rolling rocks or breaking sticks are very alerting to game. Following even a faint trail may be preferable to making a lot of noise by breaking your own trail.

If a visible animal you are stalking is alerted by your noise or movement, stop and remain perfectly still. After a short time the animal will probably continue its activities unless further disturbed. This is frequently true even if the animal can see you. Sometimes an animal will not take flight if you act uninterested in it and take a casual course not directly toward it.

A feeding animal will usually feed for regular periods of time—about ten to twenty seconds each—and then raise its head and look around. When it is grazing, its eyes are focused on what it is eating. During this time you can approach closer. If you do not alert one of the animal's other senses (smell or sound) you can move amazingly close. Just cautiously move toward the animal for about ten seconds as soon as it puts its head down to graze and then stop. Wait until it looks up and then goes back to grazing before moving again.

Another general rule that can be useful is that most big game animals do not often look uphill unless alerted by something. There are occasions when it may be helpful to stalk an animal from higher ground.

It is best in most cases to get as close to game as possible before shooting and to make deliberate, unhurried shots. One reason I feel that a bolt-action rifle makes a good hunting piece is that the user has a greater tendency to make sure of that first shot rather than counting on the second or third—or twenty-third—shot. Semiautomatic rifles are best for fighting wars, but they can burn up a lot of ammunition in a hurry. Some writers disagree on this point, and they have some good rationale for their conclusions. For further discussion on this subject, see ''Weapons.''

Aim at vital parts of the animal, such as the head, heart, or lungs, from the surest position possible. If a tree or rock is available as a ''dead rest'' for your shot, use it to steady your arms while taking aim. It is usually easier to make an accurate shot from a kneeling or prone position than offhand. Remember too, that overcoming emotional intensity is a major problem when hunting. The added stress of severe need, created by being in a survival situation, could add to the ''buck fever'' jitters. Be calm! Take the best control of yourself you can. You can only hit or miss.

It may be wise to take a shot from a place where there is cover and/or a good rest to take aim from, even though it may be a little farther from the target, rather than taking the chance to try to get closer. The decision to do this becomes easier when you know

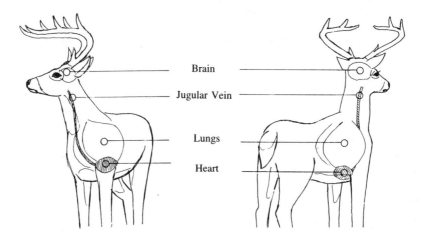

Figure 9.12   Vulnerable Places in a Deer

your capabilities. A shot of two hundred to three hundred yards, even at large American game animals, is not too far for a high-powered .30 caliber cartridge if the shot is well placed. It is, however, more advisable to limit shots to one-hundred yards or less. Even at these ranges it is necessary to act with skill and composure. At one-hundred yards or even fifty yards there is a lot of ''daylight'' around even the largest game animal (see figure 9.12).

When an animal is hit it will often go down only to get up and run on. When you have connected with a shot, keep your eye on the animal and wait. If it gets up again and you still have a good clear shot, shoot again. If the animal is seriously wounded and has run out of sight, wait for a few minutes before pursuing it. Then follow the animal or its trail of blood. It will probably lie down after traveling a short distance, and the wound will establish itself according to its severity. Usually a wounded animal will stiffen up and be much less agile after lying down for a few minutes. Use caution when approaching a wounded animal you are not sure is dead. They are fighting for their lives and can sometimes be vicious. For instructions on the handling of downed game see ''Use of Animals'' later in this chapter.

Following a nuclear war, animals could conceivably acquire "snow blindness" due to high ultraviolet light exposure and may thus become largely nocturnal and/or largely blind. There is a high probability that large game would become rather scarce.

The organs such as liver, kidneys, and thyroid glands of an animal that has been exposed to radioactive fallout or had opportunity to ingest radioactive feed or water, should not be eaten. If an animal shows signs of being sick, after exposure to radiation, it should not be eaten.

Edible Nongame Animals

Another type of hunting that may be useful, although not as exciting as big game hunting, is hunting for insects and other small creatures. Some of them are very nourishing and not really distasteful. Grubs, larvae, earthworms, and other immature insects are a boost to soups, stews, or porridge. The easiest and least conspicuous way to use them is to smash them and add them to the recipe. Large insects, such as grasshoppers from the field and hellgrammites from streambeds, may be cooked as above or else roasted. Well roasted ants have a sweet taste. All insects—especially the larger, mature ones—should be well cooked to kill parasites.

Frogs, lizards, snakes, and most other amphibians and reptiles are edible and may often be caught by hand. Some amphibians such as the pickerel frog have toxic glands and should not be eaten. For this reason it is best to be sure of the species. All such animals should be skinned before cooking. They may be roasted or boiled. Be careful that poisonous snakes do not bite themselves while you are killing them. Also be careful that no venom enters an opening in your skin while you are killing or cleaning the snake.

The exercise of common sense and your own instincts is one of your greatest assets in hunting, as in other survival activities. We are usually better equipped to do things than we think we are.

Predatory Animals

When handling a predatory animal after killing it, caution should be used to avoid contracting disease. Bubonic plague and other dangerous diseases are sometimes carried by predatory animals or their ''inhabitants'', such as fleas. According to the experts, the fleas will leave a dead animal within minutes after it dies. This is good reason to allow this kind of animal to cool for a few minutes before taking care of it. You may also want to wear gloves while skinning it.

Rabid Animals

If an animal is unnaturally agressive or tries to attack you, give it a wide berth—it may be rabid. In a survival situation, if you encounter an animal that you have good reason to believe is rabid, the animal should be killed and its carcass destroyed. It could be burned, using sufficient fuel to consume the flesh of the animal, and the remains buried.

FISHING

Fishing is a worthwhile survival activity and can provide substantial food supplies in some areas and at certain times of year.

Obtaining Fish and Other Aquatic Life

One of the lightest, smallest, and least expensive additions you can make to your survival gear is a fishing kit. This may be comprised of some monofilament line and an assortment of appropriate sizes of snelled hooks; or it may be much more elaborate and include flies, lures, leaders, poles, reels, and so on; or it may be something in between.

Everyone should be able to afford at least some hooks and line. A variety of hook sizes is advisable. This allows you to observe what the available fish are feeding on, catch it, and bait your hook with it. If bad goes to worse, hooks and line can be made from native materials such as bones, sticks, and naturally growing fiber (see Figure 9.13 for some ideas).

When fishing, experiment with the variables: time of day, kinds of bait, depth of water, and conditions of water. Many sea fish approach the shore when the tide comes in and may swim along parallel to the shore. Even in winter, fishing can be productive (through the ice if necessary). In cold weather, fish in inland waters move more slowly and feed mostly on the bottom.

Another method of acquiring fish is to catch them in nets. One person can drive the fish through a small channel while the other nets them. It may be necessary to get into the water and stir up fish from stream bottoms and edges. Muddying the water blinds the fish and it may be easier to catch them.

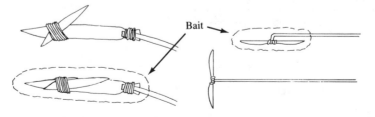

Figure 9.13   Expedient Fishhooks

Pieces of bone are good material for making these fishhooks. Wood or metal can also be used.

Trapping fish is an excellent way of acquiring them and may in some areas prove superior to hooks and lines. Figures 9.14 and 9.15 give some ideas for fish traps. Most of them are best made by pushing willows or other easily accessible materials near the waterway into the streambed. If several fish are caught in traps, it may be advantageous to keep them alive in a willow or other type of pen in the water and use them as needed. They can feed from water passing through the enclosure. This way they may be kept fresh until they are removed and eaten.

Other edible water life can be found in fresh water. Mussels, crayfish, and other smaller animals abound in many streams. Mussels leave a small trail to where they dig into the mud, at which point they leave a lump. A healthy mussel should close its shell tightly if touched. If it does not, it should be discarded. Crayfish walk the stream bottoms in the evenings. They can also be flushed from their hiding places under banks, rocks, or in holes. Mussels and crayfish should be cooked alive or the crayfish may be decapitated immediately before cooking. In fact, all fish and other water animals should be cooked soon after being removed from the water to avoid spoilage, or dried or salted for storage.

Sea animals such as clams, oysters, lobsters, crabs, and mussels will clean themselves of salt and dirt if left for a few hours in clean fresh water.

Cleaning and Preserving Fish

Fish should be bled and eviscerated immediately after being caught. Removing the gills is usually sufficient to bleed a fish. Some fish with coarse scales or heavy skin also need to be scaled or skinned.

Bad fish can be deadly to eat. Unusually unpleasant odors, slimy gills, sunken eyes, and flabby flesh are some of the danger signals.

Fish can be dried, smoked, or salted. Dry fish by first removing the head and cutting along the back down to the bone. Remove the spine and as many bones as convenient and dry the flesh in the sun or over a fire. Smoking it with nonresinous wood smoke will also help preserve it and add flavor. Moistening the fish with salt water occasionally while smoking or drying will also help preserve the meat. Larger fish will need to be cut into thinner strips to dry well. The whole fish (bones and all) may also be dried and then pounded into a meal for use. This conserves additional nutrients in the bones. The dried fish may be ground to a powder and used as flour or in soups, stews, and chowders. Cook dried fish before using it.

Figure 9.14   Fish Basket Trap

Figure 9.15   Fish Funnel

This trap can be used in conjunction with a fish funnel (Figure 9.15) or alone in a waterway—preferably at a narrow point. The trap must be anchored with some cordage.

Many variations of this trap can be made. Drive the fish toward the trap and capture them there. The enclosure should be a small, shallow area.

Fish may be salted by cutting the fish in two along the backbone, removing the backbone, cutting the flesh to the desired length, and packing it in a barrel or an earthenware crock with alternating layers of noniodized or rock salt and fish. The fish is usually layered skin-side to skin-side and meat-side to meat-side between thin layers of salt. The salt will pull the moisture from the fish and make a brine of its own. After one to two months (depending on the size of the pieces) the fish should be ''cured.'' Soak the pieces for a few hours in fresh water before cooking them.

Fish may also be kippered by soaking the pieces in a brine, of one and a half cups of noniodized salt to a gallon of water, which has been heated to put the salt in solution and then cooled before soaking the fish. After one to three hours, remove the fish from the brine and soak it in clean water two or three times for twelve to twenty-four hours. Then dry it at a moderate temperature and smoke it at a fairly high temperature.

RAISING ANIMALS

Obtaining animals to raise and use in a survival situation may be a valuable aid to sustenance. A farmer with more animals than he can adequately care for in the situation may be glad to trade some breeding stock for something of value to him. You may even be able to trade labor for animals, or work in some sort of cooperative arrangement. In most cases, the best source of information on the care and use of animals in a given area is those who are raising them. Open dialogue and diplomacy can be of great value in obtaining and successfully raising farm animals (and much else).

With any culture of animals, disease is a perennial villain. Whole flocks or herds have been wiped out by unchecked diseases. It is not within our scope here to give a comprehensive protection and healing guide for every animal discussed. You may want to include a veterinary manual in your library. However, the most important ingredients in disease prevention are adequate diet and cleanliness. Keep the animal's quarters cleaned up and you will be a long way down the road to disease prevention. Water and feeding facilities should be kept clean and arranged to avoid waste. Diseased animals should be isolated and if they die they should be buried or burned to prevent spread of the disease. Their quarters should then be cleaned with a good disinfectant.

The species included here are selected because they are prevalent, they are useful, and they can be raised (when necessary) almost anywhere. Feed could be gathered for most of them even in or near most American suburbs. If you seriously anticipate acquiring animals, take some time to learn more about the species you plan to have. In addition, feed, disinfectants, medications, tools, and equipment to handle them could also be acquired.

Rabbits

First there were 2, then there were 4, then 8, then. . .

For meat production, rabbits are marvels of nature. The gestation period is about thirty days. A rabbit doe is generally ready to breed at the age of five or six months and can safely produce four to five litters of six to eight offspring per year. The young can easily grow to two to ten pounds in about ten weeks (depending on the breed). Rabbits produce a fine-grained, nutritious, very palatable meat that is high in protein and mineral salts. Rabbits are, however, low in fat—a necessary part of a survivable diet.

There are many breeds of rabbits. Some popular breeds of medium to large size are Chinchilla, Flemish Giants, American, Bevern, Rex, New Zealand, and Champagne d'Argent. New Zealand White is a very popular medium-sized meat-producing rabbit breed. They will produce about four-pound fryers in about eight weeks.

Probably the best source of stock and information is someone who is already a reliable grower.

Feed   Rabbits may be maintained on alfalfa or clover. If their diet is to consist mostly of what you can gather, some of the feed should be dried before it is given to the animals to prevent digestive upsets. Commercial feeds, when available, are formulated to provide excellent nutrition. Nutritive requirements may also be met by providing grains such as corn, wheat, barley, oats, or milo if available. These are better if rolled or cracked. They may also eat seeds or nut meal for increased protein intake. It is important to supplement a doe's ration with one or more of these grains, when possible, during gestation and lactation.

A compressed salt block is desirable for the rabbits to lick, or some other source of salt should be provided. Rabbits also need clean, fresh water.

Rabbits excrete hard and soft feces and reingest the soft type. This is normal and will only enhance their nutrition.

Housing   Rabbits need protection from the extreme heat of summer and extreme cold and wind of winter. Hutches should provide a dry area and a screen or wire mesh on at least the floor and walls to allow the dwelling to stay clean.

A medium-sized rabbit needs a pen approximately two feet high by two and one-half feet by three and one-half to four feet in length. Recommended sizes for smaller breeds are two feet by two and one-half feet by three feet and for larger breeds two feet by two and one-half feet by five or six feet. A pen where mating of large breeds takes place may need to be higher to facilitate the process.

In addition, a nesting box should be provided in the pen for a pregnant doe to give her seclusion and protection as she bears her young (kindling).

Breeding   Small breeds are mature enough to breed at four to five months, medium at six to seven months, and large at nine to ten months. Gestation periods are usually thirty-one or thirty-two days. Nursing periods are usually from four to eight weeks depending on conditions. In cold areas with little feed available, it may be desirable to skip breeding during extreme conditions. One buck is normally sufficient for about eight to ten does.

When a doe is in heat and susceptible to breeding she becomes restless, rubs her chin on her pen, and shows a red coloration of the vulva. When signs of female susceptibility appear or when the breeder's schedule dictates, the doe should be placed in the buck's pen. Mating takes place almost immediately. Ovulation occurs after copulation, and after the mating the doe should be removed. Some breeders allow two successive matings to occur before removing the female, but the pair may fight if left together very long. In two to seven days after the mating has occurred, the doe may be reintroduced to the male's pen to insure successful impregnation. If she rejects him, she is probably already pregnant.

Another pregnancy test is made by palpation. About ten to fourteen days after the doe has been mated she may be restrained by the ears and shoulders with the right hand while the area in front of the pelvis and between the legs is examined with the left hand. Gentle squeezing of the area should reveal marble-sized embryos. Does may be remated immediately if they are not pregnant.

At or before about the twentieth day of pregnancy the doe should be provided with a nest box and some nesting material. She will usually also add some hair pulled from her own body. A litter is usually kindled at night. After the litter has been born and the mother is quiet the nest should be inspected and the dead young removed. If one doe has a very large litter and another a smaller one, the litters may be equalized between the two mothers to provide better care. An average litter is seven or eight. It is good policy to observe new litters closely for a few days after birth to help maintain adequate care for all the young.

Records may be kept to monitor performances. A good doe should produce fair-sized litters for two or three years.

Other Care   Rabbits are susceptible to several viral, parasitic, bacterial, and fungal diseases. Respiratory difficulties, intestinal problems, and infections may appear in the animals. Precautions such as keeping the housing, feeding, and watering devices clean and free from fecal contamination will aid in prevention and suppress the spread of disease. The isolation of diseased animals may also help prevent the spread of especially highly contagious problems.

Goats

Goats are able to provide basic nutrition in meat and dairy products. In many countries, goats are relied on heavily for sustenance. A good doe (female) can produce two or more quarts of milk per day for about ten months out of the year. Kids (baby goats) are usually old enough to butcher at about four months and provide very good meat. Frequently those who raise goats for dairy purposes will sell the buck kids for butchering and retain the healthy doe kids for milking.

There are many breeds of goats and it is not necessary to have a thoroughbred to have a good milker. In a survival situation good stock may be hard to come by, but there are some things to look for. A healthy goat usually has a glossy, dandruff-free coat. If a doe's teeth are excessively worn, she may be too old to be a good milk producer. A less-then-fully healthy goat may have hard places in the udder and teats, lumps under the jaw, or sores on teats or ears. If possible, it should be ascertained that the animal does not carry tuberculosis or brucellosis.

Feed   Goats are browsers. They will eat weeds, brush, and small trees; and they will try almost anything else. If you have a small tree you are fond of or intend to keep, do not let a goat near it. Goats thrive on good mixed pasture and try many kinds of foliage while pasturing. For this reason, it is important to protect them from poisonous plants. Hemlock, azaleas, wild cherries, oleander, members of the laurel family, castor beans, buttercups, rhododendrons, philodendrons, some mushrooms, toadstools, mistletoe, milkweed, and some members of the lily family are poisonous. Some other garden plants containing toxins are the greens of potatoes, tomatoes, and rhubarb. Although a goat may eat some or many of the plants containing toxins, it may not show extreme signs of distress. On the other hand, it can die. Common signs of difficulty are diarrhea, constipation, skin problems, lameness, nervous disorders, and sores. Goats will eat and be nourished by a great variety of foods. They will generally eat citrus rinds with gusto. They also eat root crops—such as carrots, beets, turnips, and cabbage family members—and their greens. These and many other discarded culinary wastes are a good feed supplement. However, it quickly becomes evident in pasturing a dairy goat that what the goat eats has a very marked effect on how the milk tastes. Moldy or spoiled feeds should not be used.

Alfalfa or clover hay is a feed of choice for goats. They may leave the coarser stalks and will waste a great deal of the hay if it is not placed in a manger where they can get it a little at a time. If kept clean, the parts of the hay not eaten by the goats may be fed to other animals, such as cows or horses. Early grass hay is also good feed. Silage may also be substituted at the rate of about three pounds for one pound of dry hay.

As with other animals, a pregnant and/or lactating doe should receive a supplement to hay rations. Root crops can serve as part of this supplement, but some grain is very desirable. Oats, corn, milo, wheat, barley, and so on all make good feed. Corn should only be fed in combination with other grains. Cottonseed, soybeans, and other products high in protein and oil are also excellent feed sources. High-oil-content feed should be free of mycotoxins produced from molds in the feeds, since these contaminate the milk. It is desirable to crack or roll the hard grains and other feeds to increase palatability. A lactating doe will maintain good milk production on about three to four pounds of hay and one to two quarts of grain per day, or on a combination of pasture and other feeds. As with other yard animals, it is important to keep clean, fresh water before a dairy goat.

A good way to pasture a goat is to tether it with a stake, line, and leather collar. Goats are curious and love to climb, so be careful where you leave them. (A neighbor once left his goat untethered and later found it had climbed a ladder to the roof of his house.)

Housing   Goats need some protection from temperature extremes; shed or barn of modest but tight construction will suffice. A bedding area is a good idea. The bedding may be any dry material—straw, leaves, hay, and so forth. This becomes especially important in extremely cold conditions. When the bedding material is "soiled" sufficiently, it makes very good compost or fertilizer. Outdoors, solid four-to-five-foot fence will securely keep a goat enclosed. Gates should have good latches; goats are notorious gate-openers.

Breeding and Milking   A buck goat is a rather strong-smelling animal and should be kept away from milk goats; otherwise, the milk may be tainted by the smell. For this and other reasons, it is probably most desirable not to keep a buck unless you have a large herd. Taking the doe to visit a buck at breeding time is probably the easiest solution to reproduction for a small herd.

A doe is usually large enough to breed at ten to twelve months and comes in heat at approximately twenty-one day intervals in late summer through midwinter (usually between August and March). When in heat—normally for a period of one to three days—the doe is restless, acts oddly, and shakes her tail almost constantly. It is not necessary to join the doe to a thoroughbred, but using the best stock available is a good plan.

The normal gestation period is just about exactly five months (149 days). It is a good policy to place a doe in an undisturbed area with dry bedding and clean water when she is within two or three days of giving birth (kidding). It is normal for a doe to give birth to two kids, although she may have three or even four and occasionally only one. After birth, the young should be kept warm and dry. If the weather is cold, you may want to dry them. The afterbirth (placenta) will normally be expelled within an hour after birth and should be buried or burned.

Within three or four hours the kids should begin nursing. The colostrum they receive from the doe in early nursing is very important to their nutrition and resistance to disease. After a few days the kids should be trained to drink from a pan so that they may be fed only a portion of the mother's milk. Kids will begin eating hay, grain, and pasturage at a very early age.

Goats are milked in the same manner as cows, only it is desirable to make an above-the-ground stall to avoid bending over while milking. The milk should be strained to remove any foreign particles and cooled in a closed container immediately after the milking.

Other Care Buck kids may be castrated at about one week of age. They are frequently butchered at four to five months.

It is a good idea to remove horn buds (disbudding) from all goats at an early age.

Goats should have their hooves trimmed regularly (use a sharp knife). If not trimmed, hooves grow out of manageable size and shape and the animals become more susceptible to foot rot. Dry pens also help control foot rot.

Sheep

Raising sheep on a small scale can be farily simple and can be a source of both meat and textile fiber. Good healthy breeding stock from your area is usually well-adapted to living there. The health of the animal is probably more important than the breed.

Feed Sheep forage very well. Pasturing areas for sheep should be rotated to prevent complete loss of vegetation and to prevent disease. They do well on many types of grass and hay. It is also advisable to feed some kind of rolled or cracked grain during the last month of pregnancy, lactation, or fattening for slaughter. They need iodized salt in their diet as well; granular or block salt may be kept before them to provide this. Sometimes ground limestone is added to the salt, as is also an antiparasite agent such as phonothiazine.

In the winter a ewe may require five hundred to six hundred pounds of hay and fifty to seventy-five pounds of grain (if available) before spring pasture is again available (about six months). A daily winter ration for a ewe might be three to five pounds of hay and one half to one pound of grain. Fresh water is needed at all times.

Lambs are frequently sold at market for meat at four to six months. By then, they may weigh from eighty to one-hundred pounds. A one-hundred pound lamb is a culinary delicacy in my opinion.

Housing Normally, no indoor housing is necessary for sheep. However, good fencing is one of the difficulties with sheep if they are to be fenced in. A good design for a sheep fence is a strand of barbed wire on the ground, a three-foot section of four inch mesh field fence from the barbed wire up, and two strands of barbed wire above the field fence. Substitutes, of course, may need to be made. Dogs are among the worst enemies of sheep, and keeping the dogs away from them is one of the main concerns. The dogs will either maul them or run them to death or both.

In addition, when lambing is imminent, the ewes should be housed where it is dry and fairly warm and free from drafts. A bedding of straw or other material should also be provided.

Breeding A healthy male sheep (ram) can service as many as thirty to fifty ewes. In a small operation it may be easier to pay for the services of a ram to avoid having to feed him the whole year. It is customary to mix a nonpermanent but visible pigment with some type of oil and apply the mixture to the ram's belly at breeding time. After breeding, the ewe has pigment on her back and it can easily be determined which of them has been serviced by the male.

Female sheep (ewes) are usually bred at around eighteen months of age, although they can be bred as early as six to eight months. They will be in heat about every fifteen days in the late summer and fall. The gestation period is around 145 days, making them about two years (or one year) of age when they first give birth. They can be productive up to eight or ten years of age.

Some quick arithmetic applied to the above information reveals that lambs could, and frequently are, born in February and March. In most of North America this means it is cold when the lambs come. Cold weather discourages parasites in the young sheep and for this reason it is a good idea to breed sheep as early in the fall as possible; however, some precautions must be taken to protect the new lambs from the cold.

A week or two before lambing prepare housing and bedding for the ewes and forthcoming lambs, and it may be desirable to clip tags and some wool from the rear end and udder areas of the ewes. Healthy ewes usually lamb without much difficulty and give birth to one or two lambs. Newborn lambs should be dried and kept warm. If the lambing area is not very protected, frequent trips should be made to the area night and day during lambing or there will be undue loss of animals to the elements. The navel cord of a newborn lamb is usually dipped in or swabbed with an iodine solution (three to five percent) to prevent infections.

A lamb will usually get on its feet and nurse right away. It may need a little persuasion or help starting the process, but it should happen one way or another within about a half hour of birth. As with the young of most mammals, the colostrum received from the mother in early nursing is important to early development. The ewe and lamb should be kept together in some reasonable separation from the other sheep for about a day or so to give the lamb a chance to gain strength without getting knocked around by the larger animals and to assure a bonding between the ewe and lamb. This helps to eliminate the occasional tendency for a ewe to abandon her lamb.

A lamb creep is a good idea for their protection. A fence should be made in a corner of their feeding area with vertical slits in the fence sufficient in size for the lambs to enter and exit but too small for the larger animals (about ten inches wide). Inside the device the lambs should be fed a good ration of hay and grain or whatever is available. This eliminates competition with the larger sheep for feed. The lambs will begin to nibble at hay or grain at about ten or fifteen days. The young should be protected as long as convenient from extremes of weather and predators.

Other Care   Lambs should be docked (tails removed) and the non-breeding stock males castrated by the time they are one or two weeks old. Docking enables both males and females to stay cleaner and avoid maggot infestations, as well as facilitating breeding. Docking is done by facing the lamb, straddling its back, and grabbing the end of its tail with the left hand. Then, holding a knife in the right hand, and the tail taut with the left, cut the tail off near the ends of the caudal folds—about one to one and a half inches from the animal's body. Cut from the underside toward the top of the tail. If the animal bleeds profusely, a string may be tied near the end of the tail and left for a few minutes to stanch the flow; it should then be taken off. Most animals will stop bleeding shortly after the operation is performed.

It is a good idea to pull the skin of the tail toward the animal's body before cutting to allow an excess of skin at the end of the shortened tail. This seems to speed the closing of the wound. It is also a good idea to apply a solution of iodine or some other disinfectant to prevent infection. (Lysol is commonly used for this.)

In a survival situation there is probably no good reason to castrate a male lamb. If it is desirable to do, it can be done by cutting off the bottom third of the scrotum and gently pulling the testicles out until the cord pulls free and is severed. A knife may also be used to gently scrape the cords in two near the testicle(s) of more mature lambs. This process should also be done when the lambs are a week to ten days old.

Older sheep need their feet trimmed to keep them from getting lame and to help prevent and heal foot rot. Foot rot can be a serious problem. Overgrown hooves allow pockets in the foot where bacteria are able to grow and infect the animal. If hooves are kept trimmed so that pockets are eliminated, it will help protect against the disease. If the animals become infected, as evidenced in early stages by lameness, the hooves should be carefully trimmed to clean away all infected areas and the feet disinfected.

Sheep with foot rot may be made to stand in about four inches of a ten-percent solution of formalin or a twenty-percent solution of copper sulfate for at least four or five minutes several times over a period of a few weeks to destroy the infection. They are usually run through a clean-water bath before being treated to keep treatment solution usable for as long as possible. The feet of diseased sheep need to be inspected, trimmed, and treated often to do away with the disease. Foot rot may go dormant when very dry conditions prevail and reappear with wet weather. It is a good idea to keep infected and uninfected sheep separated. Sheep may also become lame by having overgrown hooves or other diseases or abnormalities.

Sheep should be sheared as soon as the cold weather is past. Shearing is normally done with power clippers, but hand clippers are available. Shearing with a pair of scissors would be a possible but very arduous task. The entire fleece is normally taken off in one piece and spread out. The tags and sweat locks are then removed and the fleece is generally cleaned up and the shorter areas are pulled away and separated from the rest of it. The fleece may then be rolled up from back to front and stored for future use or sale.

Cattle

Caring for cattle may present a logistical problem that is prohibitive to many. There are, however, places and circumstances where it may be possible in a survival situation to raise one or more of them for meat or milk. If a dairy farmer were temporarily unable to care for his herd due to lack of power or machinery, it might be easy to work out a "cheap" rental or lease agreement for one of them. In such a case, information from him on care of the animal could be very helpful.

A good dairy cow is healthy, easy to milk, and does not cause a lot of trouble. Its udder should not be lumpy and should shrink with milking. The milk may be examined for signs of an unhealthy cow. Clots, strings, or blood in the milk may mean trouble for you. Some of the diseases carried by cattle that can be transmitted to man are tuberculosis, brucellosis, and leptispirosis. Any good dairy herd would be free from these diseases at the time the milk was last sold for commercial purposes.

The most prominent breed of dairy cattle in North America is the Holstein. Guernsey, Jersey, Ayrshire, and Brown Swiss cows are also used to a lesser extent. The Jerseys and Guernseys are smaller in size than the others, eat less, and produce less milk that is higher in butterfat content. This makes them possibly more desirable for a small producer. Most beef cattle in the United States are of the Hereford, Angus, or Shorthorn breeds.

Feed   A cow may need two or more acres of pasture for about six months of summer. (A tame cow taught to be led with a halter and rope can be tied at ditch banks, roadsides, and other such places when pasture is not available in one place.) When pasture is not available the animal should be fed twenty to thirty pounds of hay per day plus grain or feed concentrate if lactating. Vegetable greens could be a supplemental feed including sweet corn stalks. Cattle need a source of iodine. This can be provided by placing a trace-mineral and salt block in an accessible area for on-demand use. Granular iodized salt could also be added to the animal's feed in small amounts to supply this need. In some areas, pasture available may be short on nutrients and the animals may need larger spaces to forage.

Housing   A four or five-strand barbed wire fence or a combination of field fence and barbed wire are frequently used to enclose pasture areas. It is wise to make a secure fence for cattle at an early age to teach the animal the habit of being fenced in. Otherwise, they learn that a fence may be broken to pursue "greener pastures." A cow that has this habit can be a lot of trouble.

Cattle can stand a great deal of cold if they are well fed, but it is advisable to provide at least a windbreak or a roof and windbreak in the cold and some shade in the summer. In addition, dry, protected areas for calving during weather extremes is necessary.

Breeding and Milking   The gestation period for cattle is about nine months. Cattle are usually bred artificially (which requires a trained and equipped technician) or by having a suitable bull for about every forty to fifty cows and heifers (a heifer is a cow that has not yet given birth to a calf). When a cow is being milked it is a good idea to make her go dry four to six weeks before calving by gradually discontinuing the milking and perhaps even by gradually reducing feed a bit.

Cows generally do not have difficulty in calving. During cold weather, however, shelter and dry bedding should be provided for the process. In nonextreme weather, a dry place with a windbreak may be adequate shelter for calving. Keep an eye on an animal in labor. Problems do arise that may require your help.

Calves will stand and suckle quite soon after birth. As with other animals, the colostrum received in early nursing is important to their early health. Keep calves warm when it is extremely cold. Calves may be weaned at about three months or partially weaned after around two months by separating the cow and calf, milking the cow, and then feeding some of the milk to the calf and reserving the rest for culinary or other use. Calves should be healthy before being weaned. Male calves not being used for breeding purposes may be castrated at about any time after two weeks but not during hot weather when there are flies present.

During lactation a dairy cow should be milked twice a day at evenly spaced intervals. (Some dairymen milk three times a day at eight hour intervals.) The udder and teats should be washed or wiped off before milking. A stanchion in a barn or shelter is a preferred site for milking. The cow may also be close-tied and fed during milking although many a gentle cow has been milked by just sitting down on a stool by its side in a corral or fenced yard and doing the job. Cows should be milked as rapidly as possible, using both hands to squeeze the teats and trying to keep discomfort to the animal at a minimum. Cut your fingernails! When milk no longer flows, a small quantity more may be "stripped" from the cow by gently pressing the udder above the teat between squeezes.

Pigs

Pigs produce disagreeable odors and require heavy feeding, but they can be a plausible addition to your sustenance in a time of difficult conditions and are generally easy to raise. A healthy sow or gilt (a female that has not yet produced her first litter) is able to farrow (give birth to) about eight to twelve pigs in a litter. It is probably more important to acquire a healthy animal than to look for a thoroughbred of some type. Crossbreeds frequently produce more and better than do thoroughbreds.

Feed   Hogs can forage in a pasture, but they also require other feed such as grain. Corn is a preferred feed. Grain should be cracked or rolled if possible. A pig will normally require around two to three pounds of grain for every pound gained. Vegetable scraps and surpluses and milk products are also good feed. Clean water is essential. A sow suckling a litter may drink as much as five gallons per day.

Housing   Pigs need protection from weather extremes. They must have a dry, draft-free pen in winter and shade in summer. A simple, well-constructed lean-to or A-frame shed are easily constructed shelters.

Breeding   The normal gestation period for pigs is about 114 days. A female will come in heat about five or six days after her litter is weaned. For a small operation it is best to have females bred for a fee with a larger producer to avoid having to feed a boar (male) for the whole year. If you only want one hog for meat in a year's time, you may choose to raise a barrow (castrated male) instead of a litter.

When farrowing is imminent, a sow or gilt may be washed with soapy water if weather permits to remove worm eggs and other disease sources from her teats. Young pigs are fragile and mortality is usually quite high among them. During farrowing, be alert to prevent the sow from inadvertantly lying on the young. Sometimes a baby pig will have a membrane over its face which may suffocate it. Remove such membranes, and rub or slap the sides of the baby pig to encourage respiration. Farrowing may take as long as five or six hours even under normal conditions.

The baby pigs should be kept warm. Their navels may be swabbed with a two-percent solution of tincture of iodine to prevent infection. Male pigs not being used for breeding should be castrated at two or three weeks of age; otherwise the meat will have a less desirable odor and flavor. Some bare, clean earth should be provided for the small pigs during the first two or three days of life or some other source of iron should be given them to prevent baby pig anemia. Once they are eating they will get sufficient iron. The young may be weaned at four to six weeks. Six month old pigs are usually a good size for slaughtering.

Other Care   It is desirable to deworm a sow and young pigs before or shortly after weaning if deworming preparations are available. External pests should also be controlled when materials such as disinfectants, sprays, and dusts are available to control with.

Many producers clip off the tips of the eight tusklike needle teeth of the small pigs.

Ducks

Ducks produce meat rapidly and may be an asset to gardening by controlling many garden pests. Meat ducks can grow almost a pound a week for the first couple of months (to seven or eight pounds). Although duck eggs are not popular in the United States, in some areas they are widely used.

Most commercially grown ducks in the United States are the White Pekin breed; the Rouen breed is frequently found in small flocks on farms. The Muscovy and Aylesbury breeds are also grown in the United States for meat. Khaki Campbells and Indian Runners are excellent egg producers.

Feed   Commercial feeds are obviously the best option; but if they are not available, ducks may be fed standard cracked or rolled feed grains (oats, corn, wheat, barley, soy, and so on) and/or allowed to forage. Late spring ducklings with adequate area to forage usually do fairly well foraging.

Ducks should have water available whenever they are fed. Although swimming facilities are not necessary, they are highly desirable. Watering is best done outside a laying house to prevent the interior from getting wet. If it is done inside, it should be over a grating or drain. Ducks "play" in the water and make messes.

Housing   It is a good idea to confine ducks at night, especially breeding stock. Most duck eggs are laid during the night or early morning. A general rule of at least five to six square feet of space per bird in a laying house is good to observe. Nesting material may be straw or any other similar dry bedding. Nests should be at floor level and in the neighborhood of twelve by eighteen inches. Ducks are nervous, especially in the dark, and they may run or stampede if disturbed. This becomes a serious problem with a large flock. A small night-light will inhibit or prevent such behavior.

Breeding   As a general rule, keep one male duck to six females for breeding purposes. From among a young flock, breeding stock may be selected when they are about two months old if you are planning to harvest the nonbreeding stock for meat at that time. In an ongoing operation, desirable breeding stock may be selected by using body weight, body conformations, hatchability of eggs produced, and fertility as measurements.

The males and females vary a great deal among some breeds, and not so much among others. However, females usually have a sharp quack or honk, while males make a more muffled "belch" sound. This will normally become noticeable at six to eight weeks of age. Egg production will begin when the ducks are several months of age. In commercial production this is done at about seven months by allowing the young ducks increased time in the light. (And, of course, they are retarded in attaining sexual maturity by being kept in "the dark.") At the proper age, they will mature in a few weeks if they are exposed to twelve or fourteen hours of light daily. If they are allowed to run free, they will obiously obtain at least that much light.

Some breeds of ducks will set on their eggs and raise a brood, but others frequently will not do so and their eggs are best artificially incubated. If eggs are collected, either for consumption or for incubation, it should be done in early to mid-morning and the ducks should be taken out of the area before gathering begins. Badly misshapen or extraordinarily small brood eggs will probably not hatch. Brood eggs may be stored at moderate temperature and relatively high humidity with the small end down for a week or two without losing hatchability, but they should be turned at least once per day after being stored for a week. Soiled eggs may also be washed in warm water (about 110 degrees fahrenheit) without damage, but do not use cold water.

When incubation is to begin, eggs should be gradually warmed to incubation temperature and turned three or four times per day. After about a week, the eggs may be candled and the "dead" eggs discarded. A live embryo appears as a dark spot in the large end of the egg, with threadlike blood vessels emanating from it. A dead embryo appears as a spot clinging to shell membranes; no blood vessels are visible. An infertile egg appears clear.

Sometimes bacteria will pass through the eggshells and spoil the eggs; these "bad eggs" can actually explode. To kill the bacteria and prevent this, the eggs may be fumigated by using forty grams (one and a third ounces) of potassium permanganate crystals in a Pyrex or earthenware dish with eighty cubic centimeters (two and two thirds ounces) of formalin for every one hundred cubic feet of space to be fumigated. The fumigated area should be kept closed for twenty minutes and then allowed to vent. Do not breathe the fumes. Fumigation should take place during the first eighteen hours of incubation only.

Some breeds, such as the Muscovy, will readily set and hatch their own eggs. The setting ducks should be provided with a dry bedding area and feed and water. When the birds make overly long departures from the eggs for foraging, an overcooling of the eggs can occur and they will not hatch. Chickens can sometimes be convinced to set duck eggs.

Ducks generally hatch in twenty-eight days. (The Muscovy hatches in thirty-five days.) Hatched ducklings should be given dry bedding (moldy bedding is a killer of ducklings), warm quarters, and feed and water immediately. Feed substitutes for duckling starters could be wheat bread or cornbread moistened with milk. They should be kept around eighty-five to ninety degrees Fahrenheit for a week and slightly cooler for about three more weeks. After that, they need protection only at night or in extreme conditions.

Other Care Duck down and feathers make good insulation in clothing and bedding. Birds killed for food should be plucked before they are cleaned and dressed.

Geese

Some common breeds of geese are Emden (white), Toulouse (grey), African, and White Chinese.

Feed Geese are excellent foragers. They can live entirely on pasture if it is adequate. A ration of grain is also desirable for the very young and laying females. Wheat, corn, barley, and oats are all good. Insoluble grit should also be available at all times.

Housing Geese do not like being confined to a house. The best arrangement is probably a yard with a house for shelter during weather extremes.

Breeding Geese usually mate best in pairs or trios of one male and two females, with years two through five bringing good breeding results. The mates do well staying together. Goose eggs hatch in twenty-nine to thirty-one days and until that time may be handled similarly to duck eggs. Like ducklings, the hatched goslings may be fed wheat bread or cornbread moistened with milk if no commercial starter is available.

Other Care Geese are frequently killed for home use at around six months. By then they should weigh ten to fifteen pounds. Their feathers and down are valuable for insulation also. Like ducks, they should be plucked before they are eviscerated and cleaned. Starve geese for about twelve hours before killing them but give water freely.

Geese are good weeders for some garden areas as long as they do not forage on the crops in the garden. They are frequently used to weed orchards and vineyards, corn patches, strawberry patches, sugar beet patches, and so on.

Chickens

One of our friends is so sold on raising chickens that she thinks every upstanding family should also do so. Of course, that is not always possible; but raising them is certainly a good survival activity.

The most common breeds of small farm chickens are New Hampshires, Plymouth Rocks, Rhode Island Reds (all of which usually lay brown eggs), and some varieties of the White Leghorn breed (which usually lay white eggs). A good laying hen such as a White Leghorn will usually start laying at five to six months of age.

Feed   Chickens can forage for some of their feed but also need to be given additional feed and water. Commercially prepared feeds are the best source of nutrition, but in their absence any available grains should be added to their diets. Corn, oats, barley, wheat, soy meal, alfalfa, dried fish, cottonseed, and peanut meal are all possible survival feed for chickens. However, the coarser grains in particular should be cracked or rolled.

Chickens also need magnesium and especially calcium in order to produce eggshells, which are largely calcium carbonate. Oyster shell is given to them commercially, as is ground limestone. Limestone, which may be fed as a grit, should be possible to acquire in most places. It can be fed in boxes or with other feeds. Finely ground eggshell or bone meal can also be used.

Caution should be observed with all animals to prevent wasting feed. Grains may be fed to chickens in feeders or spread on litter. If the feed is spread on the litter, the chickens scratch for it and stir up the litter, helping it to dry out—which is not a bad idea.

Housing   Chickens need housing—about three to four square feet of floor space is needed for each bird. The space should give protection from predators and extremes of weather, including moisture, and be well ventilated. Roosting bars are usually placed two or three feet off the ground and may be as close as a foot apart. A droppings board may be hung underneath the roosts to catch droppings and make cleanup easier. As with other fowl, a dry floor helps eliminate disease. Litter should be kept as dry and clean as possible.

A nest for every three to five hens is sufficient in a laying house. A mat of straw or similar material may be placed in the nest. Nests may be elaborate or very simple, as time and materials dictate. If nests are not provided, the chickens may begin laying on the floor or in out-of-the-way places. Eggs should be gathered at least once or twice per day.

Breeding and Eggs   One cockerel is needed to mate with ten to fourteen hens if chicks are desired. If they are intended to replace laying hens, plan for this brooding to occur five or six months before they are needed. Sometimes a hen will set and hatch her own eggs. If she does not, artificial incubation is done in much the same manner as described in the instructions on raising ducks. Eggs must be kept warm and turned several times per day; high humidity is desirable.

After the chicks hatch, they should be separated from the larger birds. It is usually advised that chicks be kept indoors at ninety-five degrees Fahrenheit for the first week and five degrees Fahrenheit less each week until they are at seventy-five degrees Fahrenheit. Baby chicks have been kept in a box by a wood-burning stove until they went outside in many houses in past years.

Male chicks may be castrated (caponized) at three to five weeks. The operation increases quality and quantity of meat production but is not necessary. If good feed is available, males may grow to two to three pounds in three or four months and be ready for the table.

Female chickens less than a year old are called pullets. They will usually start laying at five or six months, and culling them (separating the layers from the nonlayers) around this age may be a good idea. Good layers have large combs and wattles; a hen with a dry and shriveled comb and wattle will probably not be a good layer. A layer has a pink, white. or light blue vent (a chicken's multi-use exit orifice) and separated pubic bones. A nonlayer's pubic bones may be close together. A yellow vent is also an indication that the bird is not laying.

After about a year or so of producing eggs, a hen will molt (quit laying and drop and replace old feathers). The layers may be replaced at this time by new pullets if continuous egg production is desired. If not, the layers will begin producing eggs again after they have finished molting.

Other Care   Lighting has a powerful effect on birds. A rule given for chickens is: Do not increase light on growing pullets and do not decrease light on laying hens. This obviously requires some fairly sophisticated lighting programs and is not observed by most small, "backyard" producers

Chickens have a tendency toward cannabalism. Stress factors, such as insufficient feed, water, or space, and temperature extremes tend to promote this behavior. Birds with sores from being pecked should be removed from the flock and allowed to heal. Debeaking the birds reduces their ability to harm each other. They can be debeaked at any age; but it is quite common for this to be done when they are one or two days old, and again when they are four to five months of age. The very front part of both beaks or of just the upper may be removed. It is probably best to remove part of both beaks to insure proper ability to feed.

Assisting Animals in Giving Birth

Since some references have been made in the foregoing concerning assisting the birth of animals, a few general comments on this subject may be helpful. Cattle, goats, and sheep usually take about an hour to deliver and sometimes longer. The normal position of the young of these animals is front feet first with the head between the legs (figure 9.16). If all goes well, about all you need to do is paint the youngster's navel with an iodine solution and burn or bury the placenta, or afterbirth. Occasionally, one of the young may need to have some of the mucus cleaned from its face or get a stimulating rubdown with a cloth to get it going. Also, if the young animal does not begin nursing right away, it might need some assistance in finding the "spigot." The clostrum received in early nursing is important to its nutrition.

If a cow, sheep, or goat has not delivered in an hour or so of labor—starting after the water has broken—there could very likely be some difficulties. After three hours they definitely need help. To examine the animal when this abnormal situation occurs, have someone hold the mother while you scrub the area around the birth canal. Then scrub your hands and arms and roll up your sleeves; there's work to do. Lubricate your arm(s) and hand(s) and between contractions, slowly push your arm in through the animal's vulva into the birth canal. The common positions of the fetus that give problems are one or both legs bent back, head bent back, and breach (rump first) with legs bent back. Multiple births can also cause problems (figure 9.17). The first things that must be ascertained when feeling inside the mother is how many babies there are and what positions they are in. Keep your movements slow and deliberate. The onset of a contraction will probably make you hold still until it is over.

The first thing you encounter will probably be a leg. To find out whether the leg is a front or rear one, flex it at the joints. If both joints bend the same way, it is a front leg; if one joint bends one way and the other joint in the opposite direction, it is a rear leg.

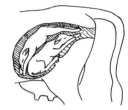

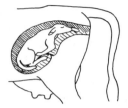

Figure 9.16   Normal Position   Figure 9.17   Twins

Figure 9.18   One Leg Front, One Leg Back

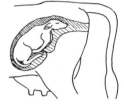

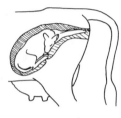

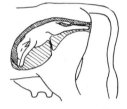

Figure 9.19   Both Front Legs Back   Figure 9.20   Head Turned

Figure 9.21   Breech, Back Feet First

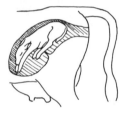

Figure 9.22   Breech, Rump First

When you have determined which end is coming first, follow the leg to the body and find the other leg. Watch for signs of more than one baby. If one or both front legs are bent back (figures 9.18 and 9.19), carefully manipulate the leg(s) to a straight-out position. If the head is turned (figure 9.20), it may be necessary to push the animal back into the womb a short distance to allow the head to straighten. Wait until the mother is not contracting to attempt to push the baby back in, of course; then tuck the animal's head between its front legs to allow a smooth exit.

A breech birth can go well if the legs come out first (figure 9.21). If the legs do not come first (figure 9.22), then apply the above-described technique to the back legs to straighten them out. In a breech birth the head (obviously) comes out last and may possibly leave the young animal trying to breathe and not able to for a longer-than-desired period of time. It is a good idea in these cases to help by removing the mucus and other material from the baby animal's nostrils as quickly as possible.

The helps for multiple births are largely the same as above. The young are usually smaller and the individual deliveries easier. If the mother needs help after the baby is in position, give the help by pressing down. The birth canal can only enlarge below the opening, not above. If the downward pressure alone does not do the job, then give a gentle pull while also pressing down.

The placenta should be expelled within a few hours after the birth. If it takes longer than twenty-four hours, see if you can find a veterinarian.

Most deliveries go pretty smoothly and no help is needed. When difficulties do arise, it is probably best to get a vet or someone knowledgeable in the matter.

USE OF ANIMALS

As with other resources, animal life must be protected and used with maximum efficiency, especially in a survival situation. Available game could be quickly used up without the restraints imposed by proper game management. Conservation is always important. Take only what is absolutely necessary and use it to the fullest. If large game were available in a survival situation, it would be easy to rely on the game totally—or nearly so—for food, without also turning to some of the other alternative food sources. However, if the situation were to continue for any length of time, this resource could soon be totally and irreversibly exhausted.

The proper use of animals for food can involve some somewhat laborious processes, including killing, bleeding, cleaning, dressing, skinning, cutting up, and preserving in a useful way. Hides may also be tanned and used for leather.

Slaughtering, Cleaning, and Dressing Larger Animals

Cleanliness is very important in slaughter and preparation for use. It is easy to allow yourself to be sloppy and thereby create a breeding ground for disease. Keep dirt out of the carcass, wash your hands and instruments, and keep flies and other vermin away from the dead animal.

If you have a choice, it is good to withhold food from especially a larger animal to be slaughtered for twenty-four to forty-eight hours before the kill. Give them plenty of water but no solid feed. After the kill, be sure to bleed the animal thoroughly; and after it is eviscerated, cool it as quickly as possible. Perform these kinds of duties on cool days if there is a choice, and at least during the cool time of day.

A larger animal can be effectively shot or bled to death. A well placed shot should enter the brain to bring immediate death to the animal (see figure 9.12). From the front, shoot between the eyes and slightly toward the top of the head. From the side, shoot about midpoint between the eye and the ear.

An animal that can be easily restrained such as a young sheep or a young goat can be bled to death quickly and with minimal trauma by severing the jugular vein. Restrain the animal and make an incision just back of the jaw with a sharp knife from about midpoint in the throat up one side. The blood should squirt out or at least flow very freely. If it does not, the incision must be enlarged to sever the vein. In just a few seconds the animal will become weak and lie down and die.

Instructions for slaughtering rabbits and fowl follow shortly.

To bleed the carcass, make an incision just under the chin, cutting completely through the jugular vein and the esophagus, and arrange the head lower than the body to allow the blood to drain out.

The animal should then be eviscerated. To do this, cut through only the skin from the tailbone to where the throat is cut (see figure 9.23). Cut around the anus and around the penis of a male being careful not to cut through the urethral tube. Also cut around the mammary glands of a female. (If the animal is going to be dragged over rough terrain before being skinned, cut the skin only from the tailbone to the bottom of the ribcage to help avoid getting dirt into the body cavity.) Be very careful not to perforate the tissue just inside of the skin bordering the intestinal cavity. Pull the skin away with your fingers and cut through it with the knife facing away from the animal.

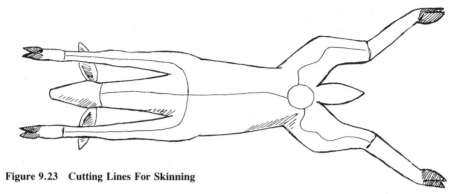

Figure 9.23   Cutting Lines For Skinning

The animal may be skinned before or after evisceration; but if there is no way of keeping the flies off the carcass and it is not going to be processed right away, the skin may be left on until just before the meat is cut up or introduced into cold storage. It is most convenient to hang the animal up with a gambrel attached at the hock joints of the rear legs before skinning or by the head before eviscerating, but both may be done with the animal on the ground.

The skinning operation is begun with the initial incision through the skin of the belly. Continue it by cutting the skin from the center incision up the inside of both front legs to near the bottom of the foreleg and up the hind legs from the anus area along the rear inside of the leg (usually where the long hair and the short hair meet) to just outside the hock joints. Cut the skin around the legs at the knee joints and around the neck in back of the ears. Pull and cut the skin away from the legs, neck, and belly, being careful to leave the meat on the animal and not on the skin. The skin can be removed mostly by pulling. The animal may be rolled from side to side to complete the skinning task if it is not hung up. Deer have musk glands on the inside of their rear legs near the hock joint. These may be removed to eliminate contamination, but wash your hands after removing them. More venison is probably tainted by the musk left on the hands of the person skinning the animal than would be if the glands were left alone.

Eviscerate the animal by carefully cutting through the membrane covering the intestinal cavity along the line of the original incision through the skin. Do not pructure the entrails. Cut completely around and through the anal aperture to free the large intestine so that it may be pulled back into the intestinal cavity and then removed. Be careful not to perforate the bowel. Remove contents of the intestinal cavity. Split the brisket, or breastbone; cut through the diaphragm; and remove the lungs, heart, esophagus, and all other organs, loose tissue, and fat. Again, if the animal is going to be transported in the outdoors, it may be better not to split the brisket to help avoid getting dirt in the body cavity.) If the gall bladder, urine bladder, or intestines are broken and their contents spilled on the meat, it should be washed with water. It is then important to cool the animal as quickly as possible and, again, keep it clean and away from flies and other vermin to avoid contamination and spoilage.

In a survival situation, edible fat from the animal should be saved. Fat is essential to life; in the absence of the oils used in modern diets, animal fat can supply this vital need. Fats can also be used to make soap and candles.

After skinning and eviscerating the animal may be quartered by cutting lengthwise through the center of the ribcage or brisket if this has not previously been done and then splitting the pelvic bone into the naval aperture. The animal is then sawed in two through the center of the spine. Next, make a cut between the first and second ribs to the spine and then through the spine. The front shoulders and hind legs may also be easily removed with only a knife, if necessary, by cutting through to the shoulder or hip joint. The meat should be kept as clean as possible. A loose covering of cloth will help keep away flies and other pests. The following diagram of various meat cuts is included to give some idea of what cut comes from where and to facilitate the cutting chores.

Since the foregoing descriptions are somewhat general, the following more specific information may also be helpful.

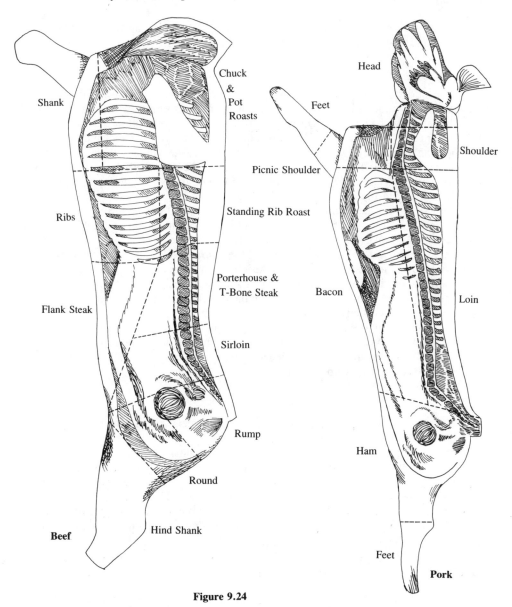

Figure 9.24

Rodents and Other Small Mammals

Small rodents such as mice, rats, squirrels, and such, should be skinned, eviscerated, and then pounded with a rock on a hard surface until all bones are pulverized. The animal should then be thoroughly cooked in the soup or stew pot. Such animals may also be skinned, eviscerated, and dried for future use. If they are to be dried, spread the body open to allow free air passage and dry them on a rock or other surface in open sunlight or by a fire. After the meat is dried, crush the bones by pounding them and continue drying the carcass another day or so to dry the bone marrow.

Animals such as badger, otter, and mink, have anal musk glands. Care should be used to avoid cutting into them when cleaning the animal.

Rabbits

A domestic rabbit may be made unconscious with a blow to the base of the skull or the dislocation of its neck. The neck may be dislocated as follows: with your left hand, hold the rabbit by the back legs across your right hip with the rabbit's back facing you. Place your right thumb on its neck just behind its ears with your fingers under its chin. Still holding the rear legs with your left hand and the neck and chin with the right hand, stretch the animal out by pulling (figure 9.25). Then give a quick upward movement of the head with the right fingers while pressing down on the neck with the right thumb. This will dislocate the neck of the rabbit if done properly.

Removal of the head must be done immediately after rendering the animal unconscious to allow it to bleed thoroughly. First, hang the rabbit by the hock of the right leg, then remove the head, tail, front feet, and other rear leg at the hock joint. Cut the skin loose around and below the hock of the suspended leg. Slit the skin down the inside of the right leg to the tail, then up the left leg to the hock joint, where the bottom of the left leg was removed. Pull the skin loose, keeping as much fat and meat *on* the carcass (and *off* the skin) as possible. Continue to pull the skin down over the rabbit so that it is inside out when it is pulled off the bottom of the carcass.

If the skin is to be used, it should be stretched with wire, a board, or whatever, to remove wrinkles. Do this with the hair side in, just as it came off the carcass, and let it cure out of the direct sunlight but with access to free air passage. Rabbit skins should not be salted if they are to be used.

Fowl

Birds are easily slaughtered by hanging them by the feet, severing the jugular vein of each behind the lower jaw, and allowing them to bleed completely, or chopping off the head on a chopping block and then hanging the bird upside down.

The small feathers and down of ducks and geese make valuable insulation. They should be plucked after the birds are dead and thoroughly bled and before further butchering. The down should then be separated from the small feathers. After chickens are slaughtered they can be immersed in hot water to remove their feathers. If they are scalded at about 130 to 140 degrees Fahrenheit (135 degrees is as hot as you can stand to put your hand in and then a bit hotter) for one to one and one-half minutes, plucking them becomes an easier job. A chicken can also be skinned by making an incision through the skin of the breast and pulling the skin off, feathers and all—a two-person job.

Once the fowls are plucked, the heads, feet, and non-meat-bearing wing extremeties

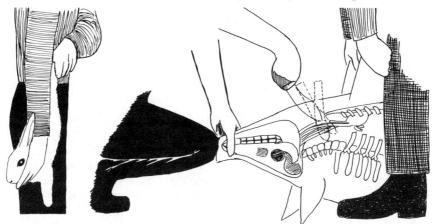

Figure 9.25
Correct Method of Minimizing Pain to
The Animal While Stunning A Rabbit   **Figure 9.26   Sticking A Pig**

can be removed, if desired. The feet should be removed first—at the hock joint—and then the head. To remove the oil glands from the tail, make a slit to the vertebra about one inch in front of the gland and then extend the slit to the end of the tail. Remove the glands in their entirety to avoid contaminating the meat. Next, cut the skin of the neck from the shoulders to the end of the neck where the head was (or is). Remove the esophagus, trachea, and crop and cut the neck off, leaving a flap of the skin. On the other end of the carcass, cut around the vent, pulling it out carefully until two or three inches of intestine show. Make a lateral slit about three inches across and about one and one-half inches below the point of the breast and pull the vent, intestines, and other viscera through the slit. Separate the gizzard, heart, and liver (giblets) from the viscera for consumption, but remove the gall bladder from the liver and the yellow lining from the gizzard. Completely clean the remainder of the carcass, removing the lungs and testes or ovaries from the back.

Hogs

Hogs are often killed by a blow on the head with a hammer or the blunt end of an axehead or by being shot (a .22 is sufficient) in the back of the head or between the eyes. After being killed, they should be immediately bled by "sticking" them (see figure 9.26). At this point, many people who butcher hogs dip them in a tub of hot water (about 145 degrees Fahrenheit) for a short time (three to six minutes) and then remove them from the tub and scrape the hair off. Scalding too long will set the hair and make it difficult to remove. The animal is then eviscerated.

To eviscerate the carcass, start at the place it was stuck and cut through the breastbone into the chest cavity back to (but not into) the stomach cavity. Then move to the back end and cut down in the middle between the hams to the pelvic bone. Divide the pelvic bone (a saw is necessary with older animals), being careful not to puncture the urinary bladder, which is just below. Split the skin on the belly down the middle with the knife facing out as you lift the skin, taking care not to puncture any internal organs. Remove all organs in a manner similar to the general instructions for larger animals previously given.

To prevent souring of the meat, the butchering process, in the absence of refrigeration, should be done in cold weather. A temperature between thirty-two and forty degrees Fahrenheit is ideal. The animal should be cut up and frozen or otherwise preserved immediately.

Preserving Meat

If meat is to be preserved in warm weather and no refrigeration is available, it can be dried or made into jerky and pemmican, stored in the cold, corned, or pickled.

Making Jerky   Modern "store-bought" *jerky* is usually made with brines, spices, and lots of sugar. It is so rich that it would probably cause sickness if used as a staple. "Real" jerky is simply dried meat. It is made by cutting lean meat in strips about one-eighth to one-quarter inch thick and drying them in the sun or over a low fire until they are hard. Use sticks, poles, racks, string, or whatever is available and seems appropriate to dry the meat on. Smoking with a nonresinous wood smoke (hardwoods—no fir or pine) will help preserve the meat. Jerky may also be rubbed lightly with salt or a brine solution before being dried to aid in preservation.

Take precautions to prevent flies from getting on the meat. Mold will sometimes form on the meat, especially if the weather is damp. Just remove the mold, rinse off the meat, cook and eat or redry it. Store the jerky in bug-proof containers.

Jerky is good boiled in soups and stews, fried, or lightly roasted over a fire.

Making Pemmican   Pemmican was used widely by North American Indians and is very nutritious. It is a mixture of dried meat, dried berries, and suet. The suet used is usually the hard fat from around the kidneys and loins of larger animals. The dried meat or jerky is pounded, the berries are smashed or pounded and dried, and the two are mixed together in melted suet. The mixture may be stored in lumps or balls in any good container. The Indians frequently stored it in the cleaned-out intestines of larger animals. If securely stored in a cool, dry place, pemmican will keep for a few months.

Cold Storage   In cold weather near or below freezing, meat can be stored for considerable lengths of time on a porch or other closed, unheated area where it can be protected from vermin and predators and used as needed.

Corning   Meat may be corned by curing it in a pickle brine. Corned beef is usually made from the round, brisket, plate, or chuck. Remove the meat from the bone, cut it to uniform size and thickness (about three inches thick), pack it in an earthenware or stainless steel container, and cover it with pickle. A pickle solution to do one hundred pounds of meat can be prepared by dissolving eight pounds of salt, three pounds of sugar, four ounces of baking soda, and four ounces of saltpeter (potassium nitrate or sodium nitrate) in four gallons of water. Meat pieces no thicker than three inches can be cured in about two weeks.

The pickle solution must remain below forty degrees Fahrenheit during the curing process (which may influence what time of year you begin such a process). If temperatures above this are encountered, a ropelike bacterial growth may occur in the brine. It may look a bit like ropey egg whites. If this occurs, remove the meat, wash it, and repack it with fresh brine in a clean container. After the meat is cured, remove it from the brine, wash it, and either dry it or smoke it with nonresinous wood smoke for three or four days. Store it in a cool dry area protected from pests.

Pickling   Pork may also be dry-cured or pickled in a manner similar to the corning process. Dry-cure it at temperatures below forty degrees Fahrenheit by rubbing a salt-sugar mixture on the meat: eight pounds of salt, two pounds of sugar, and two ounces of saltpeter in a mixture for about one hundred pounds of trimmed meat (use half of the salt mixture initially and save half for later). Place, roll, or pat a layer of the mixture on all meat surfaces—thicker (about one-eighth inch) on the lean meat surfaces. Place the coated meat in a crock or wooden barrel and allow it to stand at thirty-six to forty degrees Fahrenheit for at least twenty-five days (one to two days per pound for bacon and two days per pound for hams and shoulders). Resalt the larger cuts of meat after six to eight days with the other half of the curing salt mixture. Thin cuts probably need no more salt. Allow plenty of time for the salt to penetrate the meat. If the temperature is warmer than forty degrees Fahrenheit, bone and cut the meat in smaller pieces. Use lighter applications of salt, being careful to push it into all joints and holes, and do not pile the meat up.

Pickle Curing   Pickle curing can be done by stacking the trimmed meat cuts in a crock or barrel at thirty-six to forty degrees Fahrenheit and adding a brine made by dissolving eight pounds of salt, two pounds of sugar, and two ounces of saltpeter in four and one-half gallons of water. Do not allow the meat to float in the brine; weight it down with a clean rock. Every seven days remove the meat, remove the brine and stir it, and then repack the meat with the same brine as before.

Cure hams and shoulders about four days per pound or a minimum of twenty-eight days (a seven-pound or less shoulder for twenty-eight days; a fifteen-pound ham for sixty days). Bacon should be cured in fourteen to twenty-one days depending on weight. As previously described, if the brine becomes ropy, remove the meat and scrub it, discard the brine, thoroughly clean the container, and repack the meat with fresh brine. (The brine can, at this point, be made slightly more diluted by using five and one-half gallons of water instead of four and one-half gallons.)

After the pork is cured (the smaller pieces can be removed from the pickle and kept cold until the larger cuts are cured), remove the surface brine or salt from the meat and soak it in fresh water for fifteen to thirty minutes. Pass sturdy strings through the meat to hang it by in the smokehouse if it is to be smoked, then scrub the meat with a brush and hot water (110 to 125 degrees Fahrenheit). The meat may now be dried and stored (drying takes about one week) or smoked and stored.

Smoking dries, colors, and flavors the meat and takes two or more days. Some ideas for smokehouses are presented in figures 9.27 and 9.28. Smoking should be done at 90

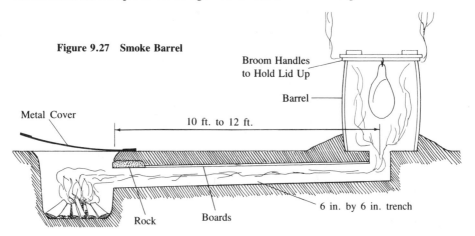

Figure 9.27   Smoke Barrel

Broom Handles to Hold Lid Up

Barrel

Metal Cover

10 ft. to 12 ft.

6 in. by 6 in. trench

Rock   Boards

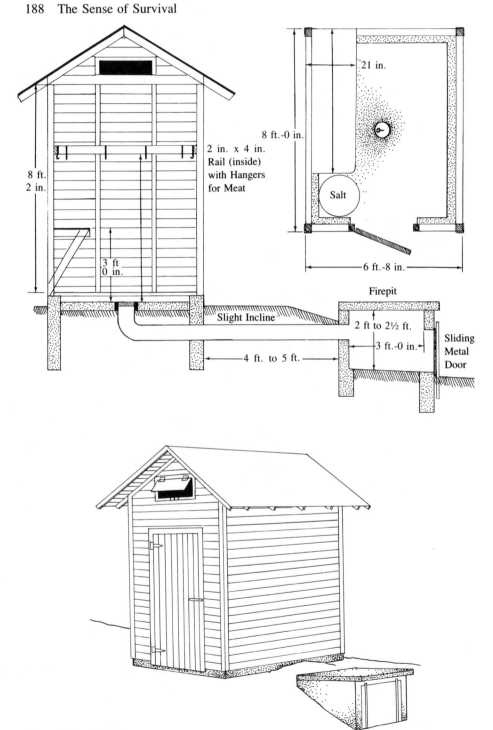

2 in. x 4 in.
Rail (inside)
with Hangers
for Meat

8 ft.
2 in.

3 ft
0 in.

21 in.

8 ft.-0 in.

Salt

6 ft.-8 in.

Slight Incline

Firepit

2 ft to 2½ ft.

3 ft.-0 in.

Sliding
Metal
Door

4 ft. to 5 ft.

Figure 9.28   Smokehouse

to 120 degrees Fahrenheit, with nonresinous wood smoke. No piece of meat should touch another. Be careful not to overheat the meat. After the meat has been smoked for two or three weeks, if it is not sour it will probably keep for a considerable length of time. To test a ham, stick an awl or similar instrument into the meat along the bones from both ends. A sweet smell should emanate from the hole in the meat. If the odor is not sweet, cut into the meat and examine it for spoilage. If it is spoiled, destroy it. Be cautious of insect infestations.

After the meat is smoked and cooled, it can be wrapped in heavy paper and stored in muslin bags. Make sure the paper is heavy enough not to become saturated with fat and coat the bag. Hang the bag in a cool, dark, dry, vermin-proof, well-ventilated area.

Ground pepper can be rubbed on the cured meat to add flavor. Mold may be retarded by periodically rubbing the surface of the meat with edible oil. If surface mold appears on the meat, trim or scrub it off. The meat is still edible.

Preserving Hides

Hides are very useful and can be preserved or prepared for use by tanning or by making them into rawhide. Other animal products can also be useful.

Tanning   There are a myriad of tanning methods and processes, many of which are specialized for certain types of skins or to produce definite kinds of products. For the purpose of this writing only three simple, general-purpose processes are described.

The hide is first soaked in water for about one full day. If it is soaked for more than one day it is a good idea to change the water each day. Next, stake the hide out flat and scrape all fat and flesh from the hide with a knife blade or other scraper. A long bone from the animal can be ground to an edge and used for this purpose, or a sharp-edged stone can be found—both these were time-honored scraping tools among the ancients, and they can save wear and tear on metal implements.

This "fleshing" operation can also be performed over a large smooth log or similar surface instead of staking the hide out or it can be stretched inside a rectangular frame by punching small holes around the edge of the hide and tying it to the frame. All flesh and fat must be removed, and the hide should be washed off as it is scraped. Scraping also helps to soften the skin.

Skins of animals with hollow hairs—such as deer, elk, and moose—may also have the hair removed before or during the tanning process, but it is not necessary. It is difficult to make hollow hairs set in the skin, and if they are not removed they will continually shed after tanning. For some uses this may not be particularly bothersome. If it is to be removed, the hair should be scraped off with a metal, stone, or bone scraper or with a knife blade, as in the fleshing operation. This process is facilitated by additional soaking or soaking the skin overnight in a solution of lime and water or wood

Figure 9.29   Staking Out A Hide

ashes and water. Be careful with such a solution; it is caustic. Rinse the soaking solution off before scraping. The lightweight skins of animals such as beaver, mink, or rabbit can be successfully tanned with the hair on them.

After soaking and fleshing, the skin can be tanned by using the brain of the animal from which it came, a soap solution, or a brine made with salt and alum (or any one of many other ways not mentioned here).

The brain of the animal is usually sufficient in size to tan its own hide. The brain may be preserved for a few days if necessary by lightly cooking it and storing in some cleaned intestine from the animal. Keep it as cool as possible.

When the hide is ready to tan, lightly cook the brain and mash it to paste. Rub the brain material into the staked-out hide with a stick, stone, or other instrument. Apply it to the smooth side if the hair is left on or both sides if the hair is removed. The hide should be nearly dry but soft and pliable before beginning this process. When the hide is saturated with the brain, roll the hide up for twelve to eighteen hours or soak it with the remaining brain material and just enough water to cover the hide. If the hair is left on do not soak the hide but roll it up by first folding the scraped sides together. Then stake the hide out again or otherwise spread it out and scrape the brain tissue off.

The hide should then be worked back and forth over the edge of a smooth narrow rounded surface such as a board, rod, or branch; some work it through a loop in a rope or wire. This process works the tanning material into the skin and dries it. The hide may also be pounded with a broadfaced rock over a soft surface, such as a grassy area, to soften it or scraped with a rounded stick or stone while it is staked out. The more it is worked the softer it becomes.

Smoking the hide over a smoky fire helps soften the leather and preserve its pliability when it becomes wet and is then redried. This also darkens it in color. Make a tripod over a smoky fire and wrap the hide over the tripod. Do not cook the hide; just smoke it.

An alternative to the brain tanning (although you will use the same soaking and scraping process before proceeding with this) is to dissolve a bar or bars of soap shavings in a bucket of warm water. Soak the hide in the soapy water for three to four days. Wring the skin out and allow it to dry, stretching it (by staking it out) as it does. Rub some animal oil, such as neatsfoot oil, into it. Soak it in soapy water (use real soap, not detergent) for another day or two. Rinse it with clean water and wring out as much water as possible. As the skin dries, work it as much as possible—stretch it, rub it, pull it, scrape it, wring it, twist it, and/or whatever else you can think of. The more you work it, the softer the tanned skin will be.

Hides can also be tanned using salt and alum (potassium aluminum sulfate, sodium aluminum sulfate, or other soluble aluminum sulfate). Prepare the hide by soaking and scraping as in the previously described methods. Then soak the hide for five to seven days or more in a brine made with two pounds of alum, five pounds of salt, and four gallons of water. I have seen many recipes for this method ranging from one pound of salt, one pound of alum, and three gallons of water to the one mentioned above. But the one above is the only one I can personally speak for. Make as much brine as necessary for the hide. After soaking, stake the hide out and allow it to dry until it is just damp and rub some light animal oil such as neatsfoot oil into it. Then work it as in the other methods (stomp, twist, rub, and so on) until the hide dries.

Alum can be purchased from chemical supply houses and when bought in bulk is *much* less expensive than in small packages—if you can get several people together who want some or you can use a lot of it yourself.

Many books on tanning are available at most libraries.

Making Rawhide   Much of the process for making rawhide is the same as for tanning. A fresh or ''green'' hide is scraped over a mound or log or staked out and scraped. A hide that has dried out should be soaked in water for twenty-four hours before scraping. Use a piece of bone, stone, or metal with an edge on it to scrape all fat and other nonhide tissue off the hide. As in the tanning process the hair may be removed or left on.

Urine is a material that will help dissolve fat and clean the hides (although its use may not be aesthetically pleasing). If you use it, apply it after the whole hide has been scraped. Let it soak in and then scrape any remaining fat off. Rinse the hide with water when scraping is finished.

Wash the scraped hide off and allow it to dry staked out for a day or two until it is stiff. If you have not removed the hair it can be removed at this point.

Rawhide is tough and hard when it is allowed to dry in place for lashings and fastening such as holding an axehead on or holding a shelter together.

If you are tying a structure together and the fastening is meant to remain in place, cut strips of hide to do the lashing before the hide dries. As the hide dries it will shrink and tighten up. If the hide is to be used for other purposes, such as making moccasins, it should be scraped and/or pounded over a soft surface until it is softer and more pliable. Use a blunt stone and moderate blows. A grassy area or a pithy or soft log make good pounding surfaces. Be careful not to make holes in the hide. If you make the hide into cordage start cutting the hide at the outside perimeter and cut the strip continuously around the edge toward the center. This leaves little waste and a continuous length of cordage.

Other Animal Products

Animals can provide much that is needed in a survival predicament. Oils (neat's-foot oil) can be removed from the hooves of some larger animals such as sheep or deer by crushing the hooves and boiling them in water. The oil will rise to the surface of the liquid during boiling and may be removed and used for waterproofing and preserving leather articles. The solution under the oil can be boiled down to thicken it and will make a pretty good glue that will do some general purpose gluing including sticking leather pieces together.

SELECTED REFERENCES AND NOTES

Belanger, Jerome D. *Homesteader's Handbook to Raising Small Livestock*. Emmaus, Pennsylvania: Rodale Press, Inc., 1974.

Benson, Ragnar, *Live Off the Land in the City and Country*. Boulder, Colorado: Paladin Press, 1981.

———. *Survival Poaching*. Boulder, Colorado: Paladin Press, 1980.

A booklet on trapping is available from Havahart, P.O. Box 551, Ossining, NY 10562. They manufacture traps.

Brown, Tom, Jr., with Brandt Morgan. *Tom Brown's Field Guide to Wilderness Survival*. New York: Berkley Books, 1983.

Churchill, James. *The Complete Book of Tanning Skins and Furs*. Harrisburg, Pennsylvania: Stackpole Books, 1983.

A Complete Guide to Home Meat Curing. Argo, Illinois: Morton Salt, n.d.

Hobson, Phyllis. *Tan Your Hide: Home-Tanning Leather and Furs*. Charlotte, Vermont: Garden Way Publishing, 1977.

Lockwood, Guy C. *Animal Husbandry and Veterinary Care for Self-Sufficient Living*. Phoenix, Arizona: White Mountain Publishing Company, 1977.

Macfarlan, Alan A. *Modern Hunting with Indian Secrets*. Harrisburg, Pennsylvania: Stackpole Books, 1971.
This book gets right to the nitty-gritty of the subject.

Park, C. Romney. *Raising Rabbits*. Provo, Utah: Value Write Publications, 1984.

Siegmund, Otto H., and others, eds. *The Merck Veterinary Manual: A Handbook of Diagnosis and Therapy for the Veterinarian*. 5th ed. Rayway, New Jersey: Merck, 1979.

Spaulding, C.E. *A Veterinary Guide for Animal Owners: Cattle, Goats, Sheep, Horses, Pigs, Poultry, Rabbits, Dogs, Cats*. Emmaus, Pennsylvania: Rodale Press, Inc., 1976.

U.S. Department of Agriculture from the Superintendent of Documents. *Beef Slaughtering, Cutting, Preserving, and Cooking on the Farm*. Farmer's Bulletin, No. 2263. Washington, D.C.: U.S. Government Printing Office, 1977.

U.S. Department of Agriculture. *Raising Geese*. Farmer's Bulletin, No. 2251. Washington, D.C.: U.S. Government Printing Office, April 1983.

Vuran, John. *The Manual of Practical Homesteading*. Emmaus, Pennsylvania: Rodale Press, Inc., 1977.

State Agricultural Extention Offices are a very good source of information about animals (see list of state offices at the end of chapter 8).

The outdoor magazines that are widely available, such as *Outdoor Life, Field and Stream, Sports Afield,* and *The American Rifleman* are also good sources of articles on hunting and guns. *The Mother Earth News* is a good source of general information.

SOURCES OF EQUIPMENT

M&M Fur Co., P.O. Box 15, Bridgewater, SD, 57319-0015, (800) 658-5554, (605) 729-2535. Good source of traps, calls, bait, equipment.

Wholesale Veterinary Supply, Inc., P.O. Box 2256, Rockford, IL 61131. (medicines and supplies)

Kansas City Vaccine Co., Stock Yards, 1611 Genesse Sts., Kansas City, MO 64102. (many kinds of supplies)

General Tractor Farm and Family Center, 3915 Delaware Ave., P.O. Box 3330, Des Moines, IA 50318, (515) 266-3101. (all kinds of animal medicines, supplies, and equipment)

CHAPTER 10

Water

One thing we cannot live without is water. It is essential in sustaining life and health, and during an emergency period it is probably more important than a food supply for the first period of distress. Yet, in many disasters the water supply is among the first conveniences (or should we say, necessities) to go. Even storage containers can be damaged or broken by flood, earthquake, and so forth. Storage water is a first source to use in an emergency. The toilet tank and water heater are also good sources of water. After these sources have been exhausted, it is necessary to find and make safe other resources.

SOURCES OF WATER

Lakes, ponds, ditches, and streams all have water, but it should be rendered safe before being used. It may become desirable to use some of these sources for washing and other uses while obtaining cleaner, better quality water for culinary use. This may create some special problems if radioactive fallout is present. To remove radioactive sediment, a filter such as that shown in figure 10.1 should be used. Springs, seeps, appropriately located cisterns, and wells can also be good sources of water. Even a small solar still could provide some drinking water until other sources could be developed.

Water in the ground drains by gravity. It moves to the level of least resistance. A zone of saturation occurs where the soil, rock, and spaces are filled to the maximum with water. Sometimes a stratum of impermeable natural materials will hold water above or below it, enabling the water to travel in its own stratum. A water-bearing stratum such as this is called an aquifer. An aquifer is sometimes sandwiched between two impervious strata. When the upper impervious stratum is perforated, water may leave the aquifer and travel to or near the surface. An artesian well results from this type of perforation when the source water of the aquifer is at a higher elevation, producing sufficient water pressure to drive it to the surface at the perforation (see figure 10.2).

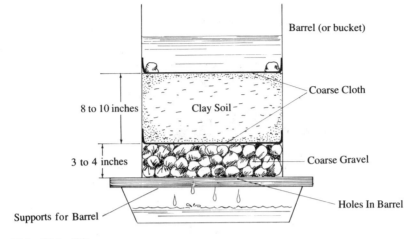

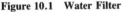

Figure 10.1   Water Filter

This filter will help remove radioactivity and other contaminants.

Springs And Seeps

Springs occur where ground water surfaces at some point due either to gravity or to artesian pressure; a seep occurs where ground water surfaces over a wider area. Seeps or springs may be developed by conducting the water to a catch basin or chamber. Such means as graveled ditches, perforated pipes, and hoses may be used to gather the water. A basin of grouted rock or even a wooden box placed in a shallow excavation may be used as a catch basin. The catch basin or cistern should be covered to prevent contamination.

Do not overdig in developing a spring or seep. The water is sometimes held near the surface by an impervious stratum. Penetrating the impervious stratum may allow the water to drop into a lower stratum. Also, be careful of surface contamination. Seeps or springs may drain from contaminated areas uphill from them. Stay away from swamps, sinkholes, corrals, outhouses, burial sites, and other unsanitary or polluted areas.

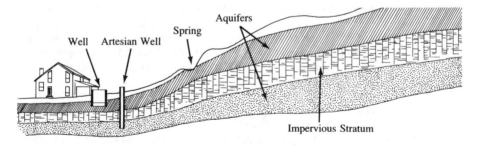

Figure 10.2   Underground Water Circulation

Springs and wells bring water to the surface from a water bearing stratum. An artesian well results when an impervious stratum holding water below it is perforated and the pressure on the water is sufficient to drive it to the surface.

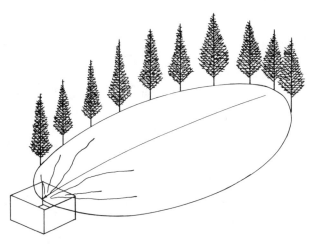

Figure 10.3   Developing A Seep

A seep can be developed by draining the water into a suitable container.

Wells

Wells are dug, bored, driven, jetted, or drilled by professional drillers; this last method is the only suitable way of drilling a deep well or one through hard ground formations. The other methods are pretty well limited to soft or relatively unconsolidated materials. A well should be constructed at least 150 feet away and uphill from cesspools, privies, corrals, and similar sites.

Making A Well   If a well is bored, jetted, or driven, some special equipment is usually needed to use the well such as pump, wellpoints, or casings. A well can also be constructed by just digging to a point below the surface of the saturation level (preferably this would be done during the driest time of the year, which in many areas is usually late summer or fall). The well may be "cased" by lining the hole with grouted rock or large pipe if available. A footing of larger rock at the bottom will add security to the structure. It is a good idea to build handholds into the wall of the well as a safety feature, in case someone or something is caught in it. The grouted rock well should then be backfilled.

If a fairly substantial bond is not made between the rocks used for the walls, the surface water will permeate the walls and seep into the well and possibly contaminate it. This is the reason for putting a layer of grout or plaster on the outside walls of the well before backfilling it. The well should be covered when not in use to prevent contamination and entry accidents. In the absence of any testing apparatus, water should always be purified for continual culinary use.

Improvised Well Pump   It may occur that no power would be available to pump water from existing cased wells. In this case you can use a can as a pump. (Use the largest volume can you can find that will fit into the well casing.) Remove one end of the can completely, and make a fairly large hole in the center of the other end (see Figure 10.5). Trim the lid that was removed from the one end so that it will easily fit down into the can or use a lid from a smaller can. In either case it must be substantially larger than the hole in the center of the bottom. Cover the lid with flexible plastic

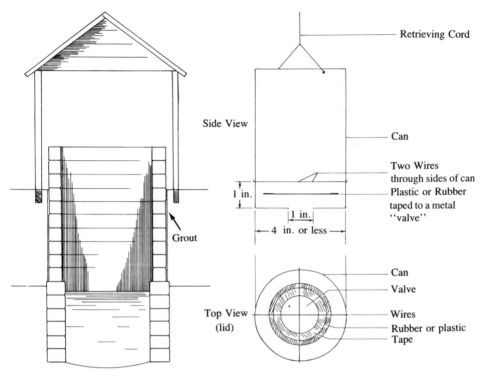

Figure 10.4 Well

A well should have a cover and keep surface water out.

Figure 10.5 An Expedient Device For Removing Water From Cased Wells

sheeting (plastic wrap, garbage bag, or whatever), and tape the plastic onto the lid so that the bottom of the lid (the part facing the bottom of the can) is as smooth as possible. (A piece of rubber tire tube can also be substituted for the plastic covered lid, but it generally will not work as well.) Drop the smooth side to the bottom of the can. About one inch above the bottom place two pieces of wire across the center of the can at right angles by making small holes in the side of the can. The wires should be slightly longer than the width of the can so that they can be bent up at the ends to hold them in place. Attach a cord to the top of the can through holes on opposite sides.

Drop the can into the well while holding the cord. The hole in the bottom will allow the can to fill and then the valve (plastic covered lid) will close over the hole and allow the can to be drawn to the top full of water with the cord.

Cisterns

Rain water may be caught from roofs or paved areas in a cistern constructed of wood, grouted stone, concrete, or similar material. The first portion of any runoff should be discarded to reduce pollution by means of a drainpipe in the bottom or by some other method. The rain water will be fairly pure but should still be treated before being used for culinary purposes. The cistern should be kept covered to prevent contamination or accidents.

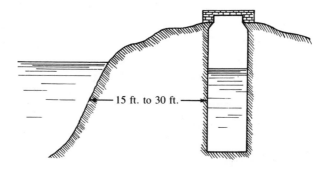

Figure 10.6 Relatively clean water can be acquired from a hole a few feet from a muddy stream or pond.

Solar Still

The type of solar still shown in figure 10.7 was developed by two U.S. Department of Agriculture scientists as a means of obtaining water. It performs to some degree even in dry areas; and all you need is a sheet of thin clear plastic (a specialty plastic is sometimes recommended but not necessary), a container (any small container will do but a plastic collapsible cup is lightweight and easy to use), a drinking tube, (surgical tubing) and a hole. Production capacity may vary, but a six-foot-square of plastic over a three foot wide by three foot deep hole can produce about a pint of water per day or more. The water will be essentially pure (providing the plastic is clean) and free from most radioactive materials. Water evaporates from the ground, condenses on the plastic, and runs into the container. A drinking tube is convenient, but one edge of the plastic can be lifted to retrieve the water if you do not have a tube. The edges of the plastic should be held down securely to prevent loss of water vapor around the edges. You can also add vegetation to the still to increase the yield. If you are in a desert, do not depend on just one of these to keep you alive for very long. Make several and look

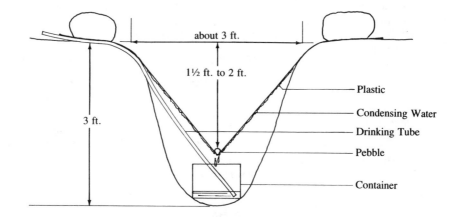

Figure 10.7   Solar Still To Get Water

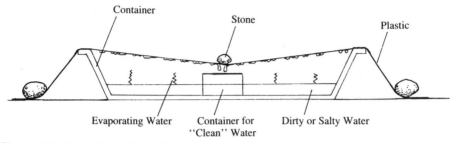

Figure 10.8   Solar Still to Clean Water

for other water sources. A solar still can also be used to distill salt (ocean) water or water that is otherwise contaminated (Figure 10.8).

Places To Look For Water In Dry Areas

Water in desert areas may be difficult to find, but some areas are more likely places to look than others. The base of a cliff or rock outcropping is sometimes found to be damp and may possibly be developed into a useable water source. Look for an abnormally green area on a hillside to signal the presence of water. The presence of birds may also indicate water. And the deep shady area of a dry streambed or bottom of a gully may be a good place to look.

A hole may be dug in a likely area to allow water to seep in and be collected. If the hole only becomes damp, the moisture may still be retrievable by using a piece of cloth or other dry, absorbent material to sop the moisture up. Then wring the cloth out into a container to drink.

This same process of "sopping up" moisture can be used on dew. Even in dry areas, substantial amounts of water can be gathered in the early morning by collecting dew from leaves, cacti, stones, metal, and other damp areas with a cloth. Moisture may similarly be gathered from a basin fashioned with a piece of canvas or plastic for lining and filled with stones or pebbles; moisture will collect on these overnight and may then be retrieved from the basin lining.

Cactus can be a source of water in extreme cases; it can be cut and the moisture sucked from its pulp. Larger cacti such as barrel cactus will yield larger quantities of moisture. Another way to get moisture from cacti is to cut the top off, smash and stir the pulp, and wring the moisture from it.

Water from a muddy stream may be cleared by digging a hole a few feet from the bank to a depth below the water level. The undistrubed water that seeps into the hole will be essentially clear.

Using Water In The Desert

In addition, remember these basic rules for desert water use: First, do not eat if you can not drink, too. Second, drink when you are thirsty; do not save water in your canteen. Third, travel at night when it is cooler; rest in the shade while it is hot.

PURIFICATION OF WATER

To be safe for culinary use, water must be rendered chemically and biologically harmless. A complete water purifier would remove harmful bacteria, protozoa, amoebae, fibers, sediment (microscopic particles of dirt, rust, scale, radioactive particles and so on), organics, and other agents such as chlorine and radioactive gasses. Many portable water purification devices are available today; most use systems of spun glass and/or charcoal for filtering and silver, iodine, or chlorine to kill bacteria. Many of these are only partly effective. Systems using iodine probably give the most complete bacteriological protection. Mechanical filters are also very effective.

In the event of nuclear war, when fallout is a problem, the dissolved and undissolved radiation particles should be removed from the water to the greatest extent possible *before* other purification measures are taken. Activated charcoal contained in many of the purification devices will remove most of these contaminants.

One of the principal radiation offenders is radioactive Iodine-131. It dissolves to some extent in water and if ingested can cause serious health problems, particulary in the growth of children. Removing it from the water and giving the body additional sources of "clean" iodine, will help prevent radiation-produced disease. For this reason, systems using iodine to render the water biologically safe are preferable for use *in a nuclear war situation.* In this case, the iodine should be added *after* the water has been thoroughly filtered, not before as most directions specify. (The filtering would also remove the "clean" iodine.) The taste may not be as good, but the end results are superior. If you were taking potassium iodide or receiving extra iodine from some other source, the filtering could be done last. Iodine-131 cannot be reliably boiled out of water.

In circumstances other than nuclear war, the directions may be followed to use iodine first and then filter to produce a good tasting drink. (Persons with thyroid trouble may need to avoid ingesting iodine.) Iodine destroys giardia organisms and viruses, whereas many other purification methods do not.

Municipalities treat water by allowing it to settle in clarifying ponds; flocculating with agents such as polymers, alum, and lime to produce sedimentation; filtering through sand filters; aeration; and treatment with chlorine. Appropriate combinations of these processes are used on given source waters. Similar procedures may also be used at home. It is best to allow any heavy sediment to settle and then to pour off the clear water as a first step. Alum may be added and thoroughly mixed to promote settling out of additional sediment. The clear water should then be poured off and further treated for purification by some appropriate method.

Radioactive sediment may be partially removed from water by mixing one part clay (removed from at least several inches below the surface so it is not contaminated) to four parts water, mixing them thoroughly, allowing the sediment to settle for several hours, and decanting the clear liquid for further treatment.

Purification treatments include bleach, iodine, boiling, and mechanical filters.

Bleach

Standard liquid household bleach containing about 5.25% (standard strength) hypochlorite will kill harmful organisms if added to water at the rate of two drops per quart for clear water and four drops per quart for cloudy water and allowed to sit for thirty minutes (contact time). Bleach older than a year or two may have become too weak to be effective, so use the newest bleach available or increase the quantity used. For safest and best tasting results use bleach that has sodium hypochlorite as its only active ingredient. A slight odor of chlorine should be detectable if the water is properly treated. If not, add a bit more.

Granular calcium hypochlorite can also be used to disinfect water. Dissolve about one half teaspoon in one gallon of water and then use one pint of this solution to treat ten gallons of relatively clear water.

Halazone tablets can be used according to manufacturers directions but they do not always do the job, have a short shelf life, and do not taste good.

Iodine

Iodine is an excellent purifying agent that is not affected by many of the conditions that chlorine is. A two-percent tincture of iodine can be added to clear water at the rate of four drops per quart of clear water and eight drops per quart of cloudy water with thorough agitations and a contact time of twenty or thirty minutes. Let the water stand with the iodine for about thirty minutes before using it.

Resublimed iodine tablets are relatively new on the market. They have long shelf life and are in my estimation an excellent storage item. For a few dollars you can have an antibacterial water treatment that will remain usable for years. Follow manufacturers directions, which should indicate using about 10 milligrams of a saturated solution (a solution which after an hour or more still has undissolved crystals at the bottom) to a quart of water with a contact time of around twenty minutes at normal room temperature (about seventy five degrees fahrenheit). Various combinations of more or less solution, longer or shorter contact times, and higher or lower temperatures can also be used. *Be careful with iodine crystals;* they are harmful if ingested, can cause skin burns, produce toxic fumes, and long-term use can cause other health problems. Purifiers using iodine resins, such as Tritek and PentaPure, are state-of-the-art.

Other purification tablets (tetraglycine hyperiodide) are also marketed as Coghlan's, Globaline, and Portable Agua. These generally do not have long shelf lives and are sensitive to heat and light. They turn color from grey to yellow as they become less potent. The usual dose is one tablet per quart of fairly clear water.

Boiling

Boiling water is usually a good way to sterilize it. Filter or allow any silt or other debris to settle, and decant the clear water before boiling it. Most water can be rendered safe to drink by simply bringing it to a full rolling boil; but to be on the safe side, boil water five minutes plus another minute for each thousand feet you are above sea level. After boiling, the water's taste may be improved by pouring it back and forth between two clean containers to aerate it.

Mechanical Filters

There is a growing number of mechanical filtering devices on the market. These units are, in my estimation, the purification devices of choice. They force water through microporous material to remove contaminants, and some have small amounts of silver or other material infused into the filtering material to kill microorganisms. Some also have charcoal filters. Examples follow; others are also available.

• The *Katadyn PF Pocket Filter* (Katadyn Products Ltd.) is imported from Switzerland. It weighs twenty-three ounces, is about the size of a flashlight, costs about $260, and delivers about one quart of filtered water per minute. A larger size is also available for about $660 and a smaller size for about $180. A fantastic siphon unit sells for about $100.

• The *Explorer and Scout* from PUR are filters with iodine resin and optional charcoal filters. The Explorer is larger and self-cleaning; cost is about $150, Scout is about $75. These units eliminate giardia, bacteria and viruses.

• The *First-Need* (General Ecology, Inc.) is a small (twelve ounce) unit that delivers about one pint of filtered water per minute. Cost is about $40. A larger unit from the same company and designed for home use is the *Seagull IV*, which costs around $300.

These units are available through mail-order catalogs, sporting goods stores, and so on. My own personal choice of purification devices for all occasions would be the Katadyn, but at two hundred dollars it may seem out of budget for many. One thing to consider, however, is that it could be used for many years on camping trips, trips to foreign countries, and so on and at the same time could be counted on in an emergency. The Katadyn has consistently taken top honors in independent testing.

STORAGE OF WATER

A short term store of water is essential to emergency evacuation as well as emergencies which allow the home "fires" to stay lit.

It has been proven many times that water may be safely stored for years when very simple precautions are taken.

How Much?

Authorities consistently recommend storing at least one half gallon of water per person per day for a two week period for emergency drinking purposes (seven gallons per person).

How To Store It

There is an endless variety of containers available for water storage. Glass or heavy polyethylene containers with good lids are very suitable. Glass containers are made less potentially messy and more durable when a layer of plastic sheeting is used to enclose the bottoms of the containers and they are placed in a cardboard box with dividers or in other solid, cushioning containers. Many emergencies involve movement of the ground (such as an earthquake), which can break glass and light plastic containers. (The plastic sheeting contains the breakage and the dividers help prevent it.)

One gallon opaque containers of heavy plastic with screw-on caps are ideal as are two-liter pop bottles. Most plastic milk bottles are too light and break easily. Sturdy metal containers can also be made acceptable for water storage with a polyethylene liner; but, unless a liner is used or it is a proper canning container, corrosion could occur and produce unpleasant taste and other contamination. Mylar liners in cardboard boxes is another popular option.

Heavy-duty plastic containers are available in most areas for a cost of about sixty cents to about a dollar and a half per gallon of storage space, depending on size. They come in five-gallon, fifteen-gallon, and larger sizes and are recommendable for water storage. Remember that water weighs over eight pounds per gallon; do not store more than fifteen gallons (about 125 lbs.) in any container meant to be portable.

Heavy containers should always be stored close to floor or ground level and be secured to prevent breakage or possible injury in the event of sudden movement of the earth. Also, be sure containers are clean. Containers that have residues of petroleum products or other unsafe chemicals or bacterial contamination must be avoided for all storage of culinary items.

What Water To Use

Normal family drinking water is probably the best source for storage; the family is used to its taste and mineral content, and the water is probably also bacterially safe. Bacterially safe water can normally be stored in clean containers without further preparations. Harmful organisms generally tend to die with storage. The longer the water is stored (if stored properly), the safer it will become from a bacteriological standpoint, generally speaking. Culinary water can be treated for storage, however, by adding one teaspoon 5.25% hypochlorite per five gallons of water.

No overall rule on the shelf life of water can be given, since water qualities vary so widely. It is noted from experience, however, that tap water placed directly in glass or polyethylene containers cannot be distinguished from freshly drawn water, even after considerable time (more than a year). Water stored in containers may, however, develop changes in taste, odor, or appearance. Though unpleasant, these changes will not normally be harmful. If the water is checked every few months, any undesirable changes may be noted and the water replaced.

Other emergency supplies of water may be obtained from water heaters, pipes, or toilet tanks. (Do not use water from a toilet tank that contains colored disinfectant; it is poisonous.) Also, storage of drinking liquids may be supplemented by canned juices, soft drinks, and water-packed foods.

An expedient method of storing water is in plastic-lined holes in the ground. A hole or narrow trench lined with plastic in the form of sheeting or bags can hold a considerable amount of water. Water can also be stored temporarily in a sturdy sack with plastic liners (e.g., a burlap sack or pillowcase with plastic sack inside). Just place the pastic sack inside the sturdier sack and fill with water. Watch for leaks and support the sides when setting the sack down. It is best to use at least two plastic sacks—one inside the other—for the liner. Do not use garbage sacks for drinking water.

If it is desirable to treat water for storage, use the methods described in "Purification." **Use only approved, food grade containers.**

SELECTED REFERENCES AND NOTES

U.S. Department of Agriculture. *Treating Farmstead and Rural Home Water Systems.* Farmer's Bulletin, No. 2248. [Washington, D.C.: U.S. Government Printing Office].

U.S. Department of Agriculture. *Water Supply Sources for the Farm and Rural Home.* Farmer's Bulletin, No. 2237. [Washington, D.C.: U.S. Government Printing Office].

Sources of Water Storage Tanks

Fiberglass Systems, Inc.
1080 W. Amity Rd,
Boise, Idaho

Steve Regan Co.
4215 So. 500 West,
Murray, Utah 84125

Robert Taylor & Sons, Inc.
381 Ironwood Dr.
SLC, Utah 84115

Domestic Water-Works/Growers Supply
P.O. Box 809,
Cave Junction, OR 97525

Equipment for Drilling Small Wells at Home

Deeprock
7067 Anderson Road
Opelica, Alabama 36802

Water Purification

Real Goods
966 Mazzoni Street
Ukiah, CA 95482-3471
(800) 762-7325
(Solar power well pumps,
seminars on solar power,
and other supplies.)

Nitro-Pak Survival Foods and Equipment
325 West 600 South
Heber City, UT 84032
(801) 654-0099, (800) 866-4876.

Major Surplus and Survival
435 West Alondra
Gardena, CA 90248
(800) 441-8855, (310) 324-8855.

Country Harvest, Inc.
325 West 600 South
Heber City, UT 84032
(801) 654-5400.

Emergency Essentials
165 South Mountain Way Dr.
Orem, UT 84058
(801) 222-9596, (800) 999-1863.

Recreational Equipment, Inc.
Sumner, WA, 98352-0001
(800) 426-4840

Intertech Trading Co., Ltd.
170 South Mountain Way Dr., #110
Orem, UT 84058

CHAPTER 11

Equipment

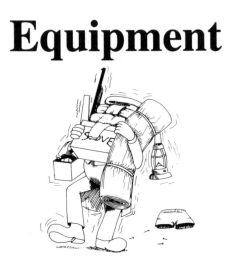

Life has become so specialized for most of us that many of the materials and methods we would need in possible survival situations have no part or function in our normal lives. Since it is always easier to spend money on things that will be used often than on things that may never be needed, some of these items can be easily incorporated into hobby or recreation activities or other practical purposes. Keeping this dual-purpose idea in mind may modify some purchases.

The needs addressed here are: Heat (including fuel and cooking); light; software (clothing, bedding, etc.); hardware (tools, materials); and miscellaneous. Sources for acquiring some items are also included at the end of the chapter.

HEAT

In almost any disaster those services providing heat and light are among the first to be interrupted. Life can become distressing without light and very uncomfortable and even dangerous without heat. Plans should therefore be made for alternative methods of heating and cooking. Stoves, fires, cooking in the rough, fuels, and lighting up are subjects of interest.

Stoves

Many kinds of stoves can be thought of as emergency gear. Wood and coal stoves, propane stoves, small camping stoves, kerosene heaters, and other miscellaneous units all have potential usefulness.

Wood and Coal Stoves   The efficient metal stoves that have become popular in the last few years are excellent auxiliary and emergency heating and cooking devices. Many of these are made to utilize both coal and wood. The really good stoves are expensive ($400-$1500); but there are small cast-iron or steel stoves of acceptable quality available that would take care of small-scale emergency needs at reasonable

Figure 11.1 Cooking Stoves

Small and large propane stoves, white gas, canned heat, and heat tab stoves (left to right) are all useful for the heating and/or cooking of food in emergencies.

prices. An oil-drum stove can be made for $75 or less, but EPA regulations now preclude using all but the most efficient stoves.

There are also some small, fold-up, light-gauge, portable metal stoves, made to burn wood, that are very effective for short-term heating and cooking. Among them are the Pyromid, the Raemco, and the Sims. These units weigh around twenty-five to fifty pounds and fold up to a few inches thick by less than two feet square. They have accessories such as ovens and various venting (chimney) options and most come with carrying cases (addresses at end of chapter).

A fireplace can also serve for heating and cooking. Masonry fireplaces are not as efficient as stoves and can eat a good supply of fuel in a hurry; metal fireplaces and inserts are much more fuel-efficient. Cooking with a fireplace requires a "hanger" or some type of grate to hold pans.

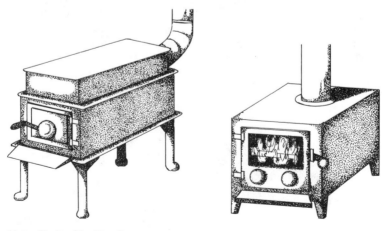

Figure 11.2 Heating/Cooking Stoves

Propane-Fueled Stoves and Heaters   Propane stoves and ranges are effective cooking tools. A range made to use natural gas can also burn propane, but it usually requires a conversion kit to accommodate the use of the propane. Even if you do not ordinarily cook with gas, enough propane could be easily stored in a small bulk tank to take care of modest cooking needs for months even on a propane camping stove. Altogether a propane cooking device is one of the most convenient and effective backup means of cooking. The small propane heaters are also very efficient.

Refillable bulk tanks are recommended to secure an adequate supply of the gas and are cheaper in the long run than disposable containers. Caution and good practice must be followed in the storage of any fuel, however. Store the tanks properly and watch for rust or other defects.

Small Camping Stoves   White gas camping stoves and good back packing stoves are very effective cooking units. Many people worldwide cook on small white gas stoves resembling the camping stoves. Fuel is relatively expensive, however, and although most types do contain stabilizers and may be stored safely for a few years, it is generally not recommended for extra long term storage.

One small stove that has great utility is the MSR model XGK backpacking stove. It weighs about a pound and will run on kerosene, diesel, white gas, gasoline, and other similar fuels. The MSR Internationale and the Coleman Peak 1 Multi-Fuel run on white gas or kerosene, are very lightweight, and work very well. Table 11.1 shows some characteristics of several good quality backpacking-type stoves and compares the relative merits of using various fuels. The table is graciously provided and allowed to be used here by Recreational Equipment Incorporated (REI).

Kerosene Heaters   Kerosene heaters are very efficient for heating and many are made so that they can easily be used for cooking. Kerosene is a storable fuel but should be rotated regularly. Store it in outbuildings in top-vented metal cabinets or other safe locations outside of the home.

Remember that any heater that uses a flame consumes oxygen and produces hazardous fumes. Use one only where there is adequate ventilation.

Other Miscellaneous Stoves   *Canned heat* stoves and *heat tab* stoves are also usable for heating food, inexpensive, and compact. Canned heat is good for minimal cooking chores and tab stoves are good for warming food; but it would be a bit of trouble to make a large stew from scratch on either one. These stoves can be purchased at sporting goods and military surplus stores; they are adequate to warm food from properly provisioned 14-day Emergency Kit.

Figure 11.3   Buddy Burner Stove

A tuna can filled with corrugated cardboard and covered with paraffin is a simple heating device. A cotton string wick should be included. Also, a stove can be made using a larger can.

Figure 11.4   **BUCKET STOVE**

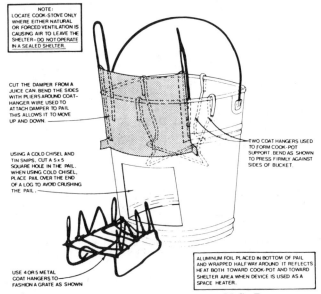

THIS COMBINATION COOK-STOVE/SPACE HEATER IS MADE USING A 10 TO 16 qt. METAL PAIL SOME COAT-HANGER WIRE, AND METAL CUT FROM A LARGE JUICE OR VEGETABLE CAN. WHEN ASSEMBLED AS SHOWN, THE STOVE WILL BRING 3 qts. OF WATER TO A BOIL USING AS FUEL ABOUT 1/2 lb. OF DRY, TWISTED PAPER OR DRY WOOD. PIECES OF WOOD ABOUT 1/2 x 3/4 x 6 INCHES ARE BEST.

NOTE:
LOCATE COOK-STOVE ONLY WHERE EITHER NATURAL OR FORCED VENTILATION IS CAUSING AIR TO LEAVE THE SHELTER– DO NOT OPERATE IN A SEALED SHELTER.

CUT THE DAMPER FROM A JUICE CAN. BEND THE SIDES WITH PLIERS AROUND COAT-HANGER WIRE USED TO ATTACH DAMPER TO PAIL THIS ALLOWS IT TO MOVE UP AND DOWN.

USING A COLD CHISEL AND TIN SNIPS, CUT A 5 x 5 SQUARE HOLE IN THE PAIL. WHEN USING COLD CHISEL, PLACE PAIL OVER THE END OF A LOG TO AVOID CRUSHING THE PAIL.

TWO COAT HANGERS USED TO FORM COOK-POT SUPPORT. BEND AS SHOWN TO PRESS FIRMLY AGAINST SIDES OF BUCKET.

ALUMINUM FOIL PLACED IN BOTTOM OF PAIL AND WRAPPED HALF WAY AROUND IT REFLECTS HEAT BOTH TOWARD COOK-POT AND TOWARD SHELTER AREA WHEN DEVICE IS USED AS A SPACE HEATER.

USE 4 OR 5 METAL COAT HANGERS TO FASHION A GRATE AS SHOWN

U.S. Department of Defense

A small stove can be made from a No. 10 can (figure 11.3) that will work well with canned heat or a tuna-can "buddy burner." (These tuna can heaters are made by coiling strips of corrugated cardboard tightly into an empty tuna fish can and then pouring melted paraffin onto it until the cardboard is completely filled and nearly covered. Lighting them is made easier by placing a wick of cotton string or paper in the center before the paraffin hardens.)

A stove may also be made from a metal bucket (figure 11.4). Cut a strip out of the side from bottom to top about eight inches to twelve inches from the bottom. (A grate may be rigged by making holes in the sides and running wire through them in an appropriate pattern.) These will burn newspapers and other small pieces of solid fuel.

Fires

Using a stove or fireplace to the best advantage requires some practice. If circumstances nullified the building inspectors' job and you had to rig something up in a hurry, keep in mind that good ventilation is a must for combustion and exhaust. A small fire is as good as a large one in most cases and takes much less fuel. Often a fire will only burn well in a fireplace if there are at least three sticks of wood burning together. When refueling it, bring the coals forward and add the new wood toward the back of the fire. This helps reflect heat and light toward the front. At the right point, a fire may be "banked" or fed and lightly covered with ashes to increase the length of time the glowing coals can be held sufficient to rekindle the fire (such as overnight).

TABLE 11.1
Backpacking Stove Comparisons

Stove	Fuel	Stove Wt. (oz.) (no fuel)	Fuel Cap (pts.)	Burning Time at Max. Flame (mins.)	Av. Boiling Time	Water Boiled per Pt. of Fuel (qts.)	Ease of Operation	Cold Weather Use	Compactness	Stove Stability	Pot Stability	Approximate Cost
Coleman Peak 1 Multi-Fuel	Kerosene	18.5	.52	83	4.1	36.7	F	G	G	E	E	$59.
Coleman Peak 1 Multi-Fuel	White Gas	18.6	.52	70	3.9	34.0	G	G	G	E	E	$59.
Coleman Peak 1	White Gas	29.2	.69	70	3.4	29.8	G	G	G	E	G	$39.
Gaz C-206 Bleuet	Butane	11.6	(5)	183	8.5	21.5(5)	E	F(4)	F	F	G	$23.
MSR Internationale	Kerosene	13.3(2) (2)		111(3)	4.0	29.3	F	E	E	E	E	$53.
MSR Internationale	White Gas	13.3(2) (2)		112(3)	4.1	28.8	G	E	E	E	E	$53.
MSR WhisperLite	White Gas	12.7(2) (2)		110(3)	3.8	28.7	G	E	E	E	E	$45.
MSR X-GK	Kerosene	18.3(2) (2)		116(3)	3.0	30.0	F	E	G	E	F	$80.
MSR X-GK	White Gas	18.4(2) (2)		128(3)	3.4	29.2	G	E	G	E	F	$80.

E = Excellent; G = Good; F = Fair; P = Poor. (1) Water at 70°F in covered pot, in still air at sea level. (2) MSR 22-oz. Fuel Bottle: wt.: 4 oz., cap.: 1.29 pt. (3) With MSR Fuel Bottle. (4) Fuel cartridge must be kept above 300F. (5) 6.7 fl. oz. cartridge (empty wt.: 3.3 oz.).

Fuel Comparison

Advantages	Disadvantages

White Gas
Caution: Do not use automotive fuel.

Spilled fuel evaporates readily; stove fuel used for priming; fuel available in U.S.; high heat output.

Priming required; spilled fuel very flammable; self-pressurizing stoves must be insulated from snow or cold.

Kerosene

Spilled fuel won't ignite readily; stove can sit directly on snow; fuel sold throughout the world; high heat output.

Priming required; spilled fuel does not evaporate readily.

Butane

No fuel to spill; no priming required; immediate maximum heat output.

Higher cost fuel; cartridge disposal is a problem; fuel must be kept above freezing for effective operation; cannot change. Gaz cartridges until empty; lower heat output.

These charts, taken from the 1989 Recreational Equipment, Inc. Camping Catalog, (REI), show the effectiveness of some of the popular backpacking stoves and the relative advantages of some fuels. Used by permission.

Cooking In The Rough

Cooking and baking may need to be done on a stove or over an open fire. A covered dutch oven or other heat-conductive, covered pot can be used as an oven by placing it on top of the stove. Flour sprinkled on the bottom of an oven will slowly brown at 350 degrees to 400 degrees fahrenheit. Bread can be baked in such an oven or in an integral oven on a stove if necessary by placing the loaf directly on the bottom without bread pans. A sprinkle of coarse flour or cornmeal underneath the loaf will prevent sticking. Many people in South America bake in square five gallon cans that have an opening in the side and a grate fitted to the inside. The device is set on a stove or near a fire.

A cast iron dutch oven can be placed on the glowing embers of a fire and some other hot coals placed on top of it to do the cooking, or the oven can be simply placed in the hot firebed if there is sufficient grease or water in the container. If you are baking some type of bread in a dutch oven you need to do some cooking from the bottom but it should be done mostly from the top. This is done by removing the oven from the coals in the firebed after some cooking time and then setting it on the ground and placing hot coals on the lid of the oven.

Cooking and heating with an open fire can be a "finger-warming" experience for the unprepared. Here again, a small fire is usually as effective as a large one. The hot coals from a fire are usually easiest to use for cooking. Generally, the flames do well for boiling and baking, and coals are good for frying and broiling. Cast iron is king when it comes to cooking on a fire, but other types of cookware are also usable and considerably lighter.

If no cookware is available, cooking can still be done. A green willow or branch (be sure to use a non-toxic wood—some woods such as oleander are very poisonous) can serve as a roasting stick similar to what would be used to roast a hot dog. Roots, meat, breads, and much else, can be cooked on a stick. An appropriately shaped rock can also be placed near the fire and used for a cooking surface. (Warning: Never place wet rocks, such as one from a stream bed or from wet ground in or near a fire. They can violently break apart or explode.) Also, water may be made to boil by adding very hot stones from a fire to the cooking container. Add food and cover the container until the food is done. Any pot of food brought to a boil may be surrounded with insulative material such as crumpled newspapers and kept sufficiently hot to do a lot of cooking after the heat is removed.

Food may be steamed under a fire by lining a shallow pit with leaves, grass or other green vegetation, placing food in the vegetation-lined pit, covering it with more leaves, covering that with one-fourth to one-half inch of dirt or sand, and then building the fire directly on top. Pull the fire away and retrieve the food when it is done. This process usually takes at least an hour—depending on what is being cooked, the size of the fire, and so on. Some foods can be cooked by simply wrapping them with leaves and placing them in the coals of the fire. Vegetables and some other food items may be coated with mud or clay and cooked in the coals or the flames.

Fuels

It is possible to store considerable quantities of wood and/or coal in many residential locations. If wood is not being used regularly it may be desirable to store most of it in log form with some cutting and/or splitting to be done when needed. This keeps the wood that is taken inside more dirt- and bug-free. It may be argued that in a survival situation the time and effort for cutting and splitting would be needed for other activities; this is an individual decision. An ax and saw, however, are "must" items. A bow saw is inexpensive and very effective for cutting firewood. Coal may be stored in a plastic-lined pit or in sheds, bags, boxes, or barrels. It should be kept away from air and light. Anthracite coal is easiest to use, and highly volatile bituminous is probably the trickiest to use.

Newspaper "logs" can also be made by soaking the newspapers in water overnight (or until saturated), rolling them up around a piece of pipe or dowel to a diameter of about four to six inches, tying them with string, removing the pipe or dowel, and standing them on end to dry. A small amount of detergent may be placed in the water to help "wet" the papers.

Hydrocarbon fuels (gasoline, diesel, kerosene and so on) deteriorate in the process of time to produce undesirable entities, such as gums and lacquers, that render them less useful or even dangerous in some cases. Micro-organisms can also grow in them. If you plan to store this type of fuel, you should add a stabilizer and biocide to help preserve its usefulness and keep it as cool and dark as is practical. One common fuel stabilizer is BHT (butylated hydroxytoluene). Also, Ethyl Corporation produces 733-PDA-75 for gasoline and EDA-2 for kerosene. Fuel Mate Plus and Diesel Plus are additives for gasoline and diesel respectively. One biocide made for addition to fuels is Bioban P-1487 (Angus Chemical Company). Retreatment of stored fuels is necessary after a few years and should be done according to manufacturer's directions. The addition of disodium EDTA can also prolong storage life especially in metal containers. Fuel Saver from Nitro-Pak is also effective. (see sources)

Gasoline normally contains some low-weight hydrocarbons that will easily vaporize and aid in starting a motor—especially in cold weather. These may be lost through vaporization by improper storage. The moderating temperatures of underground storage and vapor recovery systems can prevent or minimize this loss.

It is a good idea to filter any fuel that has been stored for a while through a fine cloth or other filter to prevent clogging or fouling of machinery. Mixing the old fuel with fresh fuel is also advisable.

Bulk liquid fuels are probably best stored in underground tanks, and the foregoing should help promote the idea that you can't just stick them there and leave them for a few years and expect them to perform flawlessly. Smaller amounts of these fuels should be stored in metal or heavy-duty plastic containers made for the purpose in cool, dark places outside the home. A locking metal cabinet which is vented at the top would add to the safety factor and keep children away from the fuels.

In the absence of something better, dried grain stocks, weeds, nut shells, fruit pits, wood and paper scraps, and whatever else will burn safely can be used for expedient fuels.

Lighting Up

Making fires can become rather difficult without matches. It is essential to store a good supply of matches in a solid, waterproof, preferably metal container in at least two locations around the house. Matches can be waterproofed at home by dipping them in melted paraffin. Lighters may, of course, be substituted for at least some of the matches. The disposable butane lighters are quite inexpensive but do not have an indefinite storage life. When building an outdoor fire the most important thing to do to preserve matches is to prepare the fire bed adequately before lighting the match. Gather a copious supply of fine, dry, combustible material as well as larger-size material. Fine branches, shredded bark (light bark such as sagebrush or cedar are very good for this), dry moss, dry shredded grass, deserted birds' nests, and so forth will help get things going.

Some methods of lighting up without matches are flint and steel, steel wool and batteries, and a bow drill. With these methods, as with matches, preparation of the fire bed is essential to success. It may seem a bit harsh to say that the flint and steel and bow-drill routines are not for the novice, but experience has shown it to be true. It is *not easy* to make fires by these methods unless you have made it a matter of substantial practice.

Flint and *steel* can produce sparks but making a fire from the sparks is difficult and when it is cold and damp, it borders on the ''not-likely'' category. To produce the spark a piece of ''flint'' (agate, quartz, etc.) is struck with the back of a solid hunting knife, the back of a closed blade of a pocket knife, or other piece of steel. It takes practice to make good sparks. Catch the spark in a *well-prepared* tinder bundle of the most flammable materials you can find, close the material around the spark, hold it up and gently blow it to a flame.

A *bow drill* consists of a bow made from a string (or leather thong, shoelace, or other cordage) and a somewhat flexible stick about two feet long and about a half-inch in diameter; a socket made from a piece of hardwood to hold the top of the drill; a drill made of a medium soft wood such as cottonwood or willow about a foot long and about a half-inch in diameter; and a wood fireboard a few inches by a few inches and about a half-inch to three-fourths inches thick. (see figure 11.5)

The bow should be unstrung when not being used to preserve its elasticity. Using a stick with a small fork in one end may facilitate the process of hooking and unhooking the string. The socket may be made by carving or drilling a depression in a small block of hardwood to receive the upper end of the drill. Lubrication is a help in letting the drill turn in the socket when using it. Grease or oil will lubricate the socket, or run the upper end of the drill through your hair to pick up some of the natural oil there.

The fireboard should be wood that is dead and dry; cottonwood is often recommended as are aspen, willow, sagebrush, or other relatively soft woods common to your area. A depression is carved and/or drilled near the middle of one edge of the board and then a notch is cut into the depression from the side of the board.

The string or other cordage is wrapped around the drill and strung onto the bow. Twisting the cordage tighter before stringing the bow can make the operation run more smoothly. The bow is then run back and forth rapidly to turn the drill in the fireboard. The socket holds the drill steady and allows pressure to be exerted on the fireboard. The

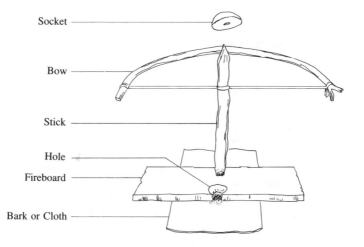

Figure 11.5   Bow Drill

fireboard should be placed on a piece of cloth, bark, or board to collect the black charred wood dust that accumulates at the notch when the drill is operated. When there is a small pile of smoking wood dust, cease operating the bow and fan or gently blow the pile to a glow. When it begins glowing, pick it up on the cloth, gently place it into a prepared tinder bundle and blow it into a flame.

A battery can be used to start fine *steel wool* (use size 00 or smaller) on fire. Simply stretch the wool out into a string long enough to reach from one pole of the battery to the other and connect two ends of the steel wool to the two poles of the battery. The steel wool will ignite and can then be placed in a tinder bundle to start a fire. A small nine-volt radio battery works excellently for this. Steel wool can also be used as tinder with matches or sparks.

Fires may also be started by using some *optical device* such as a magnifying glass or binocular lens to concentrate the sun's rays on some well-prepared tinder.

A *magnesium block* with a piece of "flint" attached is another device that works very well for starting fires. Simply carve some pieces of magnesium from the block into a pile with your pocketknife. Place the pile of shavings on some other tinder and then use the back of the closed blades to scrape the "flint" and produce a spark to ignite the magnesium. Keep the magnesium shavings together so they will ignite each other and use a good-size pile of tinder. These are great for cold or wet weather fire-starting and are recommended as an emergency device to carry with you on outdoor excursions.

Flammable materials in the form of pastes, strips, noodles, ribbons and so on are available to assist in making fires. These may be attractive addendums to the fire-starting equipment.

LIGHT

The absence of light can do strange things to people. An emergency source of light is essential. Flashlights, candles, lamps and lanterns, Cyalume light sticks, and electrical generators are all reasonably good auxiliary sources of light.

Flashlights

Most households own at least one flashlight. (The problem is that when it's not lost,

Figure 11.6   Some Emergency Light Sources

the batteries are usually dead or dying). It is a good idea to have a flashlight in every vehicle and also one in a bedstand or under your bed for nightime emergencies. Alkaline batteries are more expensive than carbon-zinc, but their shelf-life is longer and they last about five times as long in use. Keep the poles of stored batteries covered with non-conductive material to prevent discharge, and store them in the refrigerator or freezer for added shelf-life. Good batteries and a low-watt globe can allow a flashlight to give ten hours or more of continuous light. Krypton Star bulbs are among the brightest, longest living, and best on the market.

Lights and lanterns employing the larger six-volt batteries last longer than ordinary flashlights. The fluorescent lanterns are also long lasting and give good light.

Candles

Candles are easy to use and handy. For safety, a good holder should be part of the gear. The old-fashioned kind like Scrooge used are ideal. A three-fourth inch diameter candle will burn about one-half hour or a little longer per inch of length. An increase in diameter to seven-eighths or one inch should give you double the burn time or more. A two inch square candle should burn several hours (five to seven) per inch.

Candles can be made from animal tallow, paraffin, wax, or combinations of them. Beef tallow is especially good. They may be molded in any convenient container: milk cartons, cans, paper cups, toilet paper rolls, other cardboard or plastic containers, or anything else that seems suitable.

Wicks should be of cotton string. (If you were in a pinch, you could use some dried plant stems. Pithy ones such as the center of mature cattail stems work fairly well.) The quality of the wick material can be improved by soaking it (before making the candles) in turpentine or a solution of one or two tablespoons of saltpeter (potassium nitrate or sodium nitrate) and one-half pound of lime in a gallon of water for a few minutes and then allowing it to dry.

Tallow should be rendered and cleaned before making candles. Cut the tallow into small pieces and heat it over low heat until it is all melted. (Odors may make it desirable to do this outdoors if possible, but it is not necessary.) Strain the melted fat through a cloth (save the pieces of cooked meat to eat or feed to animals) and then put it back into the cleaned pot or kettle with some water. Boil it for a few minutes, then cool it and remove the clean tallow from the top. Allow it to dry and then make it into candles by melting it over low heat or in a double boiler and either molding it in containers or dipping.

Dipping candles is done by dipping the wicks into the melted fat and allowing them to cool. When the initial dips cool enough but are not yet hard, roll them between your hands or two boards to develop intimate contact between the wick and tallow. Dip them into the hot tallow again and allow them to cool. (This works very well during cold weather when they can be cooled quickly.) Dip them again and let them cool; and repeat the process until the candle is the desired size. Several can be dipped at one time by tying the wicks to a wire or stick or nailing them to the edge of a board about two or three inches apart for as many as can be dipped into your pot.

Lamps and Lanterns

Kerosene lamps of varying descriptions were very widely used for many years. They are fairly easy to use and quite efficient. According to my own calculations, ten gallons of kerosene would give a few hours of light per day for a whole year. In his book

Family Storage Plan (Salt Lake City, Utah: Bookcraft, Inc., 1966), Bob Zabriskie estimates that one quart of kerosene will burn forty-five hours in a lamp with a one inch wick. At this rate, a ten gallon supply would burn 1800 hours, or 5 hours per day for 360 days. A supply of spare wicks would also be necessary.

Lamps need not be fancy, but it is a good idea to make sure they are not too cheaply built. Some will not turn the wick up and down very well, and the mechanism for doing so is easily broken. Hurricane lamps, railroad lamps, plain lamps, fancy lamps, brass lamps, steel lamps, and glass lamps all give about the same light. The lamps with a mantle, such as those made by Aladdin produce a very bright light. However, the mantles break occasionally (when the kids knock the lamps onto the floor), and this makes the process of keeping and using them a bit more complex and expensive. But for a bright light they are the top of the line and very recommendable.

The black char that forms on the top of the wick should be gently pinched off before lighting a kerosene lamp, and the wick may occasionally need to be trimmed with scissors. The chimney should be kept clean inside and out. It can be wiped or washed clean. If it is washed, make sure it is completely dry before being used. One drop of water on it will probably cause it to crack when the lamp is lit. It is a good idea to keep a fire extinguisher or a container of sand or soda handy to extinguish a fire in case the kerosene lamp is broken and a fire starts.

White gas and propane lanterns give excellent light but consume more fuel than wick-type kerosene lamps. A single-mantle gas lantern will give approximately eight to nine hours of light per pint of fuel. A two-mantle lantern gives a little bit more light (about one-fourth to one-fifth more) but uses about one and one-half times as much fuel and is more expensive to buy and maintain. In my opinion the single mantle is the better choice. A propane lantern will use approximately two-thirds the number of ounces of fuel as a white gas lantern for similar results. Some small backpacking stoves can double as lanterns with the extra lamp attachments.

Carbide lamps are also used for emergency light and are inexpensive to buy and use. The drawbacks are smell, soot, and keeping them working right. Look for them at surplus outlets and sports shops.

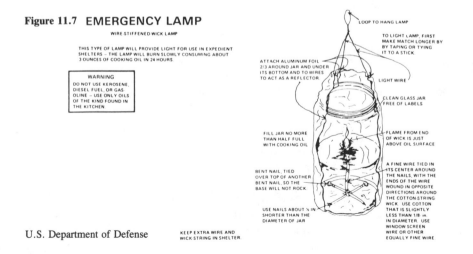

Figure 11.7 EMERGENCY LAMP

WIRE STIFFENED WICK LAMP

THIS TYPE OF LAMP WILL PROVIDE LIGHT FOR USE IN EXPEDIENT SHELTERS – THE LAMP WILL BURN SLOWLY CONSUMING ABOUT 3 OUNCES OF COOKING OIL IN 24 HOURS.

WARNING
DO NOT USE KEROSENE, DIESEL FUEL, OR GASOLINE – USE ONLY OILS OF THE KIND FOUND IN THE KITCHEN.

LOOP TO HANG LAMP

TO LIGHT LAMP, FIRST MAKE MATCH LONGER BY BY TAPING OR TYING IT TO A STICK.

ATTACH ALUMINUM FOIL 2/3 AROUND JAR AND UNDER ITS BOTTOM AND TO WIRES TO ACT AS A REFLECTOR

LIGHT WIRE

CLEAN GLASS JAR FREE OF LABELS

FILL JAR NO MORE THAN HALF FULL WITH COOKING OIL

FLAME FROM END OF WICK IS JUST ABOVE OIL SURFACE

BENT NAIL, TIED OVER TOP OF ANOTHER BENT NAIL SO THE BASE WILL NOT ROCK

A FINE WIRE TIED IN ITS CENTER AROUND THE NAILS, WITH THE ENDS OF THE WIRE WOUND IN OPPOSITE DIRECTIONS AROUND THE COTTON STRING WICK. USE COTTON THAT IS SLIGHTLY LESS THAN 1/8 in. IN DIAMETER. USE WINDOW SCREEN WIRE OR OTHER EQUALLY FINE WIRE.

USE NAILS ABOUT ½ IN. SHORTER THAN THE DIAMETER OF JAR

U.S. Department of Defense

KEEP EXTRA WIRE AND WICK STRING IN SHELTER.

Improvised Lighting

An oil lamp can be made by placing a cotton string or other wick material in a glass or metal container—a fruit jar makes a good container—of cooking oil or some other suitable oil. The wick material must be supported to keep it from flopping into the oil; a wire attached to the wick serves this purpose well. Also, a button lamp can be made by tying the opposite corners of a small, square piece of cotton flannel together over a button and setting it afloat in a cup or dish of oil and lighting the oil-soaked corners of the flannel. In a prolonged emergency, the cooking oil may be more valuable for dietary needs.

AUXILIARY POWER

Electrical generators can be very useful if the budget permits their purchase. In the absence of commercial power, such conveniences as furnace motors, refrigeration units, sewing machines, and other home appliances could be kept working enough to derive great benefit from them by operating the generator only a few hours a day. Make sure you have enough cords to operate all necessary appliances while the generator is running. If you plug into the house wiring, be sure to disconnect the main switch so that power will not be fed back into the lines where repairmen may be working. A small unit may not do everything that is normally done with power from the utility company, but it could be of great assistance in keeping warm (furnace fan), preserving food (refrigeration devices), or perhaps doing some work with a sewing machine.

Auxiliary power can also be derived either directly or indirectly from windmills, solar cells, alcohol stills, waterwheels, manure burners, and many other such products and devices. Solar cell-battery systems work very well.

SOFTWARE

Survival "software" includes clothing, keeping warm (including discussions on heat loss, insulation, layering, and an alternative cold-weather clothing system), bedding, and duffle. Temporary shelter (including tents) is discussed in Chapter 12, "Shelters."

Clothing

Many military leaders and parents of large families have recognized that provision for proper clothing is one of the foremost logistical problems. As a practical emergency preparation that requires no long-term extra expenditure, it is prudent to have at least one or two extra sets of sturdy clothing on hand. Where children are involved, it may not be as easy; but (speaking from experience) it is easily possible to buy many items that anticipate needs as long as a year in advance. This is especially true of basic clothing. (Styles may not be as easy to anticipate.)

Sewing is a practical skill to acquire and depend on for emergency clothing needs. Heavyweight denim, flannel, and other yardage can be bought and tucked away with needles and thread for future use. Patterns can be purchased or worn-out clothing can be used for patterns. Don't forget buttons, snaps, zippers, patches, scissors and other sewing aids. Some companies make treadle attachments for converting an electric sewing machine to manual use in case of loss of electrical power.

Special precautions are in order to protect against cold and wet. If ordinary conditions do not require you to cope with weather extremes, it may not be fresh on the mind; but with a little exposure to harsh weather, it does not take long to realize how uncomfortable life could be.

A poncho or raincoat and some waterproof foot covering will help reduce exposure in wet weather. This brings to mind an interesting experience with ponchos that our local scout troop had during a trip to a wilderness area in the high Uinta Mountains in Utah. Most of the boys had purchased the inexpensive plastic ponchos and had put them on for protection from a heavy rainstorm while hiking in to camp. The temperature dropped and the rain turned to hail. The colder temperature made the plastic stiff and brittle and the hail ripped the ponchos to shreds. Some of them were almost completely unusable after the storm. A good lightweight nylon poncho is more expensive but—as this experience proves—it may be a bargain in the long run. At very least, a good-sized piece of visquene is advisable for the purpose.

Rain chaps are lightweight and easy to use, and very effective for keeping legs dry. Rubber boots or galoshes are effective for keeping feet dry.

Good sturdy footware is practical, if not necessary, equipment for nearly every survival situation. Survival against the elements means work—largely outdoor work—and that kind of work is hard on shoes. The rubber-bottomed boots with felt inserts are great for keeping the feet warm, but those that are poorly made are difficult to walk in. If you are caught without shoes, make some with rags, several thicknesses of newspaper, or whatever is available and appropriate. Another covering of plastic can give protection from water.

Sandals can be made from old tires with pieces of rubber or leather used for straps. Shoes can be made from all kinds of leather scraps and have even been braided from pieces of cloth in the manner which braided rugs are made. To sew this kind of sandal and for many other heavy duty uses a sewing awl (such as a Speedy Stitcher) with heavy nylon thread and extra needles is the instrument of choice.

Many leather boots are water-repellent and will remain that way temporarily, but with use they lose their repellency. A good leather dressing appropriate to the type of leather in the boots will do as much to preserve and waterproof the leather as about anything, and a supply of it should certainly be part of a storage program. Shoe polish works very well if the leather is made to receive it. Most leather shoes and boots are made with chrome-tanned leather. For these, a dressing with wax or silicone for its base should be used. Sno-Seal, Leathe-R-Seal, and Ultra Seal are all good. REI (see end of chapter for address) sells a kit they call the Ultra Boot Care Kit which includes material to treat the leather, seams, welt, and inside of the boot. Oil-tanned leather needs an oil or grease dressing such as neatsfoot oil or shoegrease.

A couple of pairs of good heavy wool socks are a must for rough going. They give protection, comfort, good wear, and warmth. Some of the good quality outdoor synthetic socks are also durable and comfortable. For those not accustomed to heavy going, it may be good to know that changing socks every day is good practice. Rinse the dirty pair out and dry them while wearing another pair. This helps prevent foot disease and helps keep feet warm during cold weather. Even if it is not possible to wash the pair not being worn, just letting them "dry out" for a day helps.

In cold weather or when you must do a great deal of walking or work on foot, wearing a thin pair of socks under a thicker pair can help both to keep the feet warm and to prevent blisters. The inner socks made of poly-propylene or other synthetic material can help keep feet warm by conducting moisture away from the surface of the feet. The underwear of the same kinds of material can serve the same function.

Keeping Warm

In areas where it is possible to be caught in emergencies when it is very cold, warm clothing should be kept on hand. Without belaboring an obvious point, it seems fitting to recall again that there is a *great deal* of difference between going from a building to a car on the one hand, and having to work or sit for a few hours or sleep out in the cold on the other hand.

If you are stuck somewhere without heat, cover up all your body you can, including arms, head, neck and so on. Even though you may feel warm you still may be losing body heat from exposed areas. Don't wait until you feel cold. Stay dry; moisture is one of the greatest deterents to keeping warm. Another thing to remember is that in most cases bedding such as sheets and blankets, will keep you warmer if you wear it tucked into your clothing than if you try to cuddle up in it. Expedient materials such as newspaper and foam from chair cushions can be used very effectively in this manner. I have heard some amazing stories of transients and other indigent people keeping alive in extreme conditions using newspaper for covering and insulation.

When it is very cold, it is important not to get damp with sweat. Do not work hard enough to work up a sweat in these conditions—especially in an emergency situation where a warm shower or even a warm fire may not be close by.

Heat Loss   Several principles affect heat loss. Knowing something about them and some of the clothing available may be helpful. The body loses heat by evaporation of moisture at the skin, conduction, convection, and radiation.

Moisture is constantly being lost at the surface of the skin as an ongoing process sometimes called "insensible perspiration." Moisture also is lost through the sweat glands (which are most abundant at the feet). This is sometimes called "sensible perspiration." Heat is required to cause the moisture to turn to a vapor, and that heat is removed from the body. When the clothing becomes saturated, the clothing tends to become less insulative and more conductive, allowing the body to cool. And in spite of what is said about some types of clothing being better when wet, no clothing is really warm when it is wet.

Vapor barrier (VB) clothing can be worn in cold weather next to the skin to control the removal of moisture from the body. Specially treated polypropylene, polyvinyl chloride (PVC), and polyester underwear help serve this purpose because of the ability to wick the moisture away from the skin without allowing it to evaporate close to the skin. (The skin stays moist slowing further perspiration.) When you want to cool down and allow some of the moisture to be removed or when you are working hard, open the collar and cuffs and pull the shirttail out, allowing air to be drawn up through the clothing and permitting the evaporative heat loss to occur. When the work stops or you begin getting colder, simply tuck in the shirttail and close the cuffs and neck openings. As stated earlier, vapor barrier clothing is designed to be worn only in cold weather.

Conduction occurs when the heat is transmitted from more energetic molecules to less energetic molecules. This type of loss is minimal with any reasonable clothing system. Most fibers and even air conduct heat very poorly. The body has a built-in device to prevent conductive-heat-loss. The blood vessels in the outer areas of the body constrict in the cold to keep heat in the more vital areas. An exception to this is the head. A great deal of heat can be lost from the head. Keep it and the neck well covered in the cold.

Convection heat loss occurs when heated air rises and draws cold air in from below. This can occur around clothing layers of poorly fitting clothing or even *inside* poorly constructed clothing.

Electromagnetic radiation of body heat can be a major avenue of heat loss when the low-density synthetic fibers are used. This can be greatly reduced by including a wool or pile garment in the clothing system. Radiant barriers that reflect this radiant heat are also used in clothing and sleeping bags these days. The barrier must be shiny and must be faced with a non conductive material to work efficiently. Most heavier-weight outer garments, such as heavy winter coats, are substantial enough by themselves to prevent most of the radiant heat loss.

Insulations   Down is a premier insulation for outdoor cold-weather clothing. It is durable to washing, it has the best warmth for weight performance, and garments made with it for insulation drape well on the body—they look good and hug the body. Down is also usually the most expensive insulation material. If an insulated garment is not expensive, it is probably filled with feathers or something else; but it is probably not down-filled. Piles and fleeces made of polyester, nylon, and acrylic are all good insulators in medium-weight garments but are heavier than other synthetic alternatives for a given warmth. Wool is also very good.

Polyester insulations come in several packages and they are the closest thing (feature for feature) to down that has been made by man. They are even superior to down in some respects. The polyester insulations most widely used in sleeping bags and clothing are the lower density products. Some examples are Hollofil 808, Hollofil II, and Quallofil from DuPont, and Polarguard from Celanese. Hollofil 808 is a short (about two and one-half inch long) hollow fiber. (The hollow fibers provide good insulation and less weight.) Hollofil II is similar to Hollofil 808, but it has a silicon-based coating that makes it more compressible, more resilient, and—of course—more expensive. Quallofil is similar to the Hollofils but was more recently developed and has more holes running through the center. It is more thermally efficient, compressible, and resilient than most other synthetics. With recent improvements, Celanese's Polarguard nearly matches the thermal performance of Quallofil and is preferred by many people in certain applications. Polarguard is a continuous filament. Lamilite, also a continuous filament, is very effective. The new synthetic Lite Loft is bonded and is the loftiest and warmest for weight.

There are also two products which are more dense and give greater insulation for a given *thickness* than those mentioned above; these are Thinsulate from 3M and Thermolite from DuPont. They are made from very fine fibers and provide good insulation without high bulk. If there is a weakness associated with these two insulations, it is that they may be slightly heavier for a given warmth and may not conform to the body as well as the lighter insulations. They are very useful where bulk is a concern, however, and are widely used. Many cold-weather gloves are made using these for insulation.

All these synthetic insulations retain very little moisture. If a garment or sleeping bag becomes wet, just shake or wring most of the moisture out of it and most of the insulative quality will be regained.

Layering   Cold weather clothing should be worn in layers that can be added or removed to be able to stay at just the right temperature.

An outer garment to protect from wind and water is an essential element in keeping Old Man Winter out. This outer layer is sometimes a separate garment and sometimes just an appropriate facing on the insulated piece of clothing. Materials such as Gore-Tex are popular for this purpose. They are designed to allow water in vapor form to pass out through the fabric while at the same time not allowing liquid water to pass in through it. These materials are more expensive than most other fabric but are very effective. Better grades are more effective and cost more.

Poplin (cotton, cotton-nylon, and cotton-polyester) is also very durable and very popular for the outer shell, as is all-nylon. These fabrics are not waterproof, but can be made nearly so by treating with Scotchguard or Drifab (from Amway), or some other similar material.

For keeping the wet out, nonbreathable fabrics such as coated nylon are made into rainwear and are much less expensive than Gore-Tex and similar products, and in many uses they are just as useful and much more cost-effective. They do not allow vapor to go through the fabric, however, and provision must be made to prevent becoming wet from the inside. Proper design and construction of an outer garment, allowing for opening and closing and other features, can lend much to its usability.

Wear what is appropriate for keeping warm under the outer garment. In extreme conditions a well-insulated coat might be worn inside the outer shell with a pile jacket or wool sweater inside the coat and an appropriate shirt and underwear next to the skin.

The legs may not need as much insulation as the torso but underwear, a heavy pant and an outer shell are still necessary in extremes. The hands also need a protective outer shell with a fairly dense insulation underneath. Wool or polyester insulation with a nylon or leather outer glove or mitt are good combinations.

In many cases layers are combined using modern technology to produce some very effective products. An example would be a Gore-Tex glove with a leather palm, insulated with thinsulate and lined with a synthetic flannel.

An Alternative Cold Weather Clothing System   A very effective cold weather clothing involves the use of breathable, open-cell polyurethane foam. A friend named Jim Phillips and his father spent a couple of weeks on the ice north of the Arctic circle, and their clothing cost less than one-hundred dollars. In fact, Jim has cut a hole in the ice of a lake in freezing weather, jumped in long enough to saturate his clothing, got out and squeezed the excess water from the foam, and then spent the night outside. In Jim's system he uses a foam sleeping bag as well as head to foot foam covering.

The theory of the system is that the foam is placed next to the skin (except for maybe a loosely woven synthetic underwear) and allowed to breathe. No waterproof covering is added; the foam is just contained in oversize, breathable, non-absorptive synthetic clothing (big pants, big shirt, and so on). The body moisture is then allowed to be wicked to the outside while you still keep warm.

Jim tells a story of talking to a rather well-known arctic explorer on one of his northern outings. He learned that the explorer had turned his top quality down sleeping bag in for a new one when he was being resupplied during an expedition because the old one had lost its ability to keep him warm. He asked the equipment handlers to weigh the bag he had turned in. They did weigh it and found that it contained *several pounds* of moisture that had frozen into the down. (As moisture escapes from the body it is wicked to the outside by the down and then freezes the down into clumps, thereby causing the down to lose its ability to properly insulate. In the freezing weather the frozen clumps of down never have a chance to thaw out.)

The foam clothing system does not have this problem. In fact, it is my opinion, that this system has no peers for keeping warm in cold weather hour after hour; day and night. The muckluks made with the foam are an example; they actually keep your feet warm all day and/or all night.

After spending some time consulting with military cold weather outfitters Jim has developed some interesting training on the clothing and other aspects of cold weather living. They also have materials, patterns, and finished goods. He consults, provides training, and is *the* source on this subject. Jim Phillips is available at JP Associates, 3905 West 9850 North, Pleasant Grove, UT 84062, (801) 785-6027 (office), or (801) 785-5625 (home). Other companies are now also producing this type of gear.

Bedding

It is necessary to have adequate bedding to handle things in your climate area in case the furnace thermostat and the controls on your electric blanket are no longer able to do the job. If you are a camping enthusiast and have a good sleeping bag as part of your gear, that is great. But if you never camp and do not have an extra $100 to $200 each to buy good sleeping bags, then count on an extra two or three blankets per person for emergency purposes. Wool blankets are tough, warm, and very good for outdoor use. Synthetic blankets have many of the same qualities, are lighter, and retain very little moisture; and an inexpensive one can be purchased for as little as $5 at discount stores. Inexpensive plain wool blankets can be bought for around $25 to $30 new or from a surplus store for about $10 to $30, used or new.

Down sleeping bags are great for their weight; but they are expensive and the down clumps together when it becomes wet, making it much less insulative. Qualofil, Polarguard, Lite Lofts, and Lamilite (used in Wiggy's bags) are excellent, well-proven sleeping bag insulators and my all-around choices.

When you are bedding down outdoors, a waterproof ground cloth is essential—a function that is served well by a good poncho—and a tent is desirable. (If you do all of your sleeping indoors you may not know how much moisture accumulates as dew—not to mention snow and rain.) A piece of plastic or waterproof tarp will also make a good ground cover. An insulative pad is very desirable as well. Pads range from about $7 or $8 for a small, thin one to around $75 for deluxe models. Any of the pads are much better than nothing.

The plastic-aluminum foil laminate that has become known as "space blanket" or "emergency blanket" is very light and effective for holding body warmth. It reflects body heat back to the wearer. Put one of these around you or over your bedding in extreme conditions and they will help retain heat to an amazing extent. The drawback to using them over clothing is that they do not allow the body moisture to escape. As water vapor leaves the body (a normal, ongoing process), the vapor is trapped by the material, condenses to water, and dampens the clothing or bedding. Because of this, these devices should normally be used only when it is fairly cold or in the absence of something better. It may sound strange, but these emergency blankets would actually be more effective if they were worn next to or close to the skin to give a vapor barrier. This is also true of using one in a sleeping bag—put it close to your skin. If this escapes your logic, reread the above discussion on vapor barriers and also try it out for yourself.

Another similar product now on the market (Texolite) is a shiny, heat-reflective porous material which allows water vapor to pass through it. It is sewn into sleeping bags and reflects body heat back to the body but allows the water vapor to escape.

Expedient blankets can be woven from thatch and reeds. Even cuddling up in a pile of dry leaves or grass can keep you warm—but be sure to keep them dry.

Duffle

Keeping gear (especially a 14-Day Emergency Kit) contained and portable is a necessary part of organizing it. For that reason, a few ideas are included on the subject. Backpacks are excellent containers to use for some kinds of basic equipment if you have them. A wide variety of features, quality, and sizes are available. A backpack should be made with waterproof material.

If you are not a backpacker (or maybe even if you are), duffle bags are also very good containers. They are light, tough, inexpensive, and easily carried. The new military issue duffle bag and similar commercial models, have conveniently arranged shoulder straps which allow them to be carried like a pack. New bags can be purchased at sporting goods stores and from mail order suppliers, and new and used ones are available at most military surplus stores. Waterproof bags are a little more expensive but may be worth the cost.

In a pinch, if you had to leave home and pack things in a hurry, you can use the "Santa Claus" bundle mentioned in Chapter 5.

HARDWARE

The ultimate hardware kit would, of course, enable complete self-sufficiency. This may not be possible for everyone, but some hardware items should be part of every able-bodied person's personal equipment. In simplest form, an emergency equipment kit should contain saw, pliers, ax, hammer, nails, adjustable wrench, screwdrivers, mill file, hacksaw, sharpening stone, wire, knife or knives (at least a sturdy pocketknife), rope or cord, visquene, shovel, pick or mattock, and a pot and griddle suitable for outdoor cooking. Other gardening tools, hand tools, and materials are also desirable.

A *bow saw* is probably the most effective and the least expensive for cutting firewood. One can even be simply made if you have only the blade and some wire (see figure 11.9). Chainsaws are, of course, the best log-cutting machines, but they are expensive and require a good supply of spare parts and equipment to keep going. A carpenter's handsaw is also a useful tool and could be very helpful in rebuilding after a disaster. Suitable handsaws can be purchased for ten to twenty dollars each. A

Figure 11.8 Useful Emergency Equipment

Figure 11.9   Expedient Bow Saw

The blade is attached to two pieces of wood or metal for the handles. The cross-brace makes the frame solid and allows tension to be applied by twisting the wire at the top with a small stick. The tension makes the saw rigid and keeps the blade taut.

carpenter will usually have an eight point (eight points per inch) and a ten point (ten points per inch) crosscut saw for general coarse and fine (respectively) cutting. He will also have a five-and-a-half or seven-and-a-half point ripsaw for cutting along the grain of boards and for other uses. Power saws are, of course, much better for cutting when you happen to have electricity.

Pliers are very useful tools. A pair of plain straight pliers will bend metal, pick a pot out of a fire, cut wire, bend wire, loosen fasteners, and perform many other necessary functions. Poor quality pliers will usually break in a hard "pinch;" go for quality—not necessarily price.

An *ax* is useful for trimming and fitting timbers, splitting firewood, chopping firewood, driving stakes (single-bit ax), and so on. The rounded edge of the ax as it comes from the factory is okay for splitting wood, but it is poor for chopping. It may be reshaped with a slow-turning grinder or a file—a fast-turning grinder can overheat the metal and change the temper or hardening. Figure 11.10 shows the correct final shape of the blade. If a file is used, it should have a guard around the handle to prevent cuts if you file toward the blade or you can just file away from the blade. An old canning jar lid or a block of wood punched through the end of the file will work as a guard. Wear heavy gloves while filing. A vise is ideal to hold the ax, but any other device (or person) that can hold it securely will work. You can use a stone to keep an edge on the ax. Use a circular motion with the stone against the blade (figure 11.11).

Be cautious with an ax; many have been injured in a moment of haste. Clear obstructions above and around the target and on the ground before chopping. When the ax handle loosens from the head, as happens to all axes in normal use, small wedges are usually driven into the end of the handle to tighten it (figure 11.12), or the top few

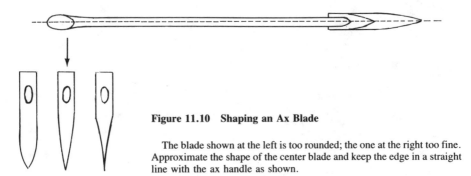

Figure 11.10   Shaping an Ax Blade

The blade shown at the left is too rounded; the one at the right too fine. Approximate the shape of the center blade and keep the edge in a straight line with the ax handle as shown.

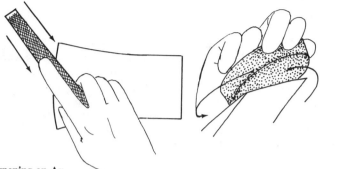

Figure 11.11   Sharpening an Ax

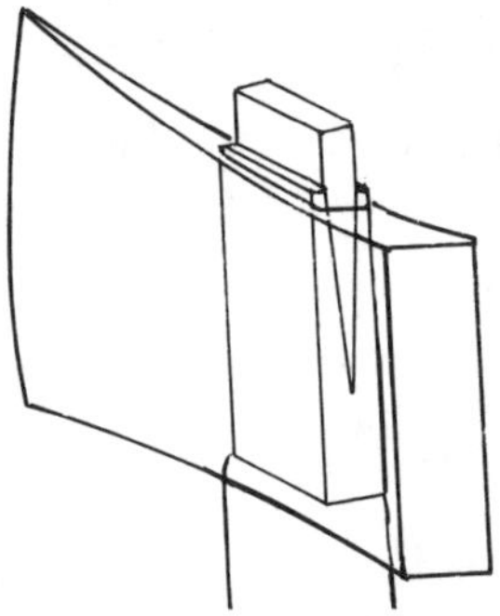

Figure 11.12   Tightening an Ax Head

Drive a wedge into the top of the handle to tighten the head of the ax when it becomes loose.

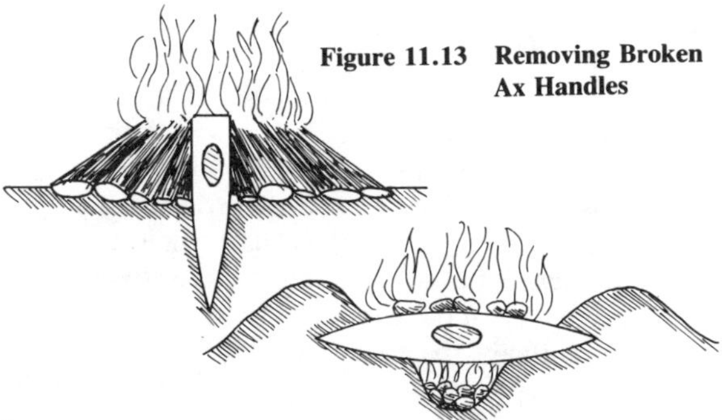

Figure 11.13   Removing Broken Ax Handles

When an ax handle breaks, it may be necessary to burn the old handle out of the head if it canot be easily removed by some other method.

Figure 11.14   Using an Ax

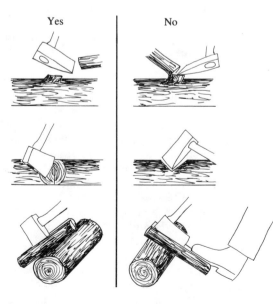

Yes   No

inches of the handle (including the head) may be soaked in oil to cause the wood to swell.

A good solid *hammer* and a supply of *nails* are tools of great use. By gaining permission to pick up dropped nails around a construction site, a variety can normally be gleaned in a little time; or a variety of small (such as drywall nails), medium (8d), and medium-large (16d) nails could be purchased.

An *adjustable wrench,* and several small and large straight-blade and Phillips' *screwdrivers* are minimal tools for taking care of many fastening and unfastening chores that are bound to occur and for turning off utilities. Socket sets, end wrenches, and locking pliers such as Vise Grips are also useful.

Mill files eight inches long or longer are good for shaping metal objects, including hoes, shovels, and axes. Store at least two or three. A sharpening stone with coarse and fine sides is necessary to shape and keep a good edge on knives and other tools. A hacksaw is used for cutting metal, and metal shears are used for cutting light-gauge sheet metal.

Knives are objects of fascination that some think of for self-defense. For most, the probability of using a knife as a self-defense weapon is probably about zero. Yet, a knife *is* an indispensible tool for survival. A couple of years ago I gave my wife a pocketknife for Christmas. She carries it in her purse and finds it one of the handiest things she carries there. (And that's saying something!) A pocketknife and a rigid or lock-blade knife are "must have" survival tools for skinning and cleaning game, gathering plant food and fiber, and ten thousand or so other miscellaneous uses. They do not have to be fancy in shape or anything else—just good steel and construction, and proper edge. The blade of the rigid or lock-blade knife should be at least four inches long. Features such as hollow-handles storage compartments and serrated back edges for sawing may or may not be desirable. A plain, good quality custom-made knife is about $80.

A dull knife can do some tasks, but a sharp knife is needed for many others. A carborundum, aluminum oxide, or natural stone can assist with the sharpening chores and need not be expensive. A sharpening steel also works well. In the absence of a commercially produced stone, a knife or other cutting tool can usually be sharpened on a piece of flat sandstone or other appropriate rock. A flat sandstone-clay stone works very well; quartz rocks such as granite will also do the job. Look for flat-surfaced rocks. If they have rough surfaces, rub two rocks together for a few minutes to grind a flat, even surface.

An all-around-use sheath knife should be sharpened to an angle of about forty degrees. A finer edge with a twenty degree angle may be desired for skinning animals and some other purposes. A twenty-to-thirty degree angle is good for a pocket knife (see figure 11.15). The wider-angle blade does not dull as easily as the finer blade. A circular motion on the stone alternating the sides of the blade will form the edge. For putting the final edge on the blade, draw the blade across the stone as though you were cutting a thin horizontal slice of the stone away. Maintain the blade at the desired angle with the stone during all sharpening motions. Keep the stone wet or oiled during sharpening procedures. Do not use a file on a knife blade unless a broken blade or mishap causes a need for major reshaping. Even then a good stone may work better.

Rope and *string* are useful for lashing, clothesline, tethering animals, and many other purposes. Strong, inexpensive nylon string and a little rope should handle most needs. If you live in a rural area, it should be easy to pick up a few pieces of used baling twine for nothing but the effort. Most farmers are glad to get rid of it.

Wire is one of those forgotten entities that can do wonders in a makeshift situation. Remember the old saying, "We keep things going with spit and balin' wire." Baling

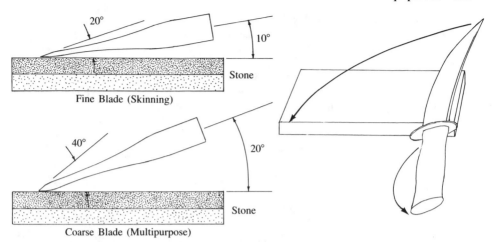

20°

10°

Stone

Fine Blade (Skinning)

40°

20°

Stone

Coarse Blade (Multipurpose)

Shown with semi-hollow-ground knife blade

Figure 11.15   Sharpening A Knife

Final sharpening should be done as though you were shaving a piece of the stone away. The edge of the blade should form approximately a forty degree angle for all-around use or about a twenty degree angle for skinning and other finer cutting chores.

wire or a small roll of form wire (used with concrete forms) is very inexpensive. Pliers are a must for working with wire.

Plastic sheeting can keep you and other things dry, make a solar still (see Chapter 10), help keep a child's bed dry, or serve any one of many other needs. Rolls of plastic are available at hardware stores and other outlets. An insulation contractor or supplier may be able to save you some money on a whole roll. A four-mil thickness is common and a good all-around weight.

A *shovel* is an absolutely essential survival tool for every able-bodied person, and more than one would be good for barter or prolonged use. A shovel can help in constructing ditches, wells, reservoirs, shelters, and latrines; cleaning house (when it gets to that point); harvesting crops; and much else. A pick and digging bar are also useful for digging in hard ground. (And most of it is!) Absolutely do not buy a cheap shovel if you plan to do much work with it.

A *pick* and/or a *mattock* are almost tools of necessity for digging. If your nuclear war survival plans include the possibility of building an expedient shelter, or if you can foresee any other substantial digging project, then acquire at least a pick. Hard or rocky ground is almost impenetrable without one. A mattock has a broad face which can also speed up the digging chores.

A *cooking pot* and *griddle* suitable for cooking under adverse conditions are near necessities. Cast iron is unequalled for outdoor cooking. Household wares may be adaptable for this under necessity, but they may never be the same again after being used over an open fire. Other important kitchen hardware might include a can opener, kitchen spoons, pancake turner, dishpan, bowls, and a bucket or two.

Other gardening tools that could be considered are a hoe, a rake, and a mattock or grubbing hoe. A grubbing hoe is, to me, one of the most useful hand gardening tools available. A large-wheel hand cultivator is also a useful gardening item.

Other hand tools useful for construction are a plane (a jack plane or a block plane are good for all-around use in planing wood articles), a chisel (for making holes, cutting, and shaping), a miter box (for cutting true angles), a level, a carpenter's square and a tri-square (for cutting true angles, rafters, and stairways), a brace and bits and/or a hand drill (for boring in wood), a chalk line, a measuring tape, and a carpenter's pencil and chalk or crayon. Some materials that might be added to the list where they are practical to store and might be used could include a sheet or two of plywood, chicken wire (for animal cages), caulking, screws and bolts, and lumber.

MISCELLANEOUS

A number of miscellaneous items should be considered as good survival equipment. One of the most important is writing tools. Pens, pencils, and notebooks are good not only for recording important details but also for recording important thoughts and feelings, writing letters, and for recreation. (It may be hard to think of writing as recreation, but under some circumstances it could bring a welcome relief from other duties and provide an outlet of expression by putting awkward moments to useful purpose.) In addition, important papers and other valuables should be kept available and "found." (See Chapter 5 for suggestions.)

A supply of cleaning agents and toiletries are important. This should include laundry detergent, all-purpose or dish detergent, germicide, hand soap, bleach, rubbing alcohol, razor and blades, and deodorant.

A sewing kit such as mentioned earlier in "Software" (this chapter) is something that could be easily overlooked. There is a story of a noted church leader who was touring Europe shortly after World War II to give comfort and help initiate assistance to some very deprived people. He was passing out oranges to one group when a woman—speaking German, which he did not understand—began trying to give hers back. Finally an interpreter told him that she wanted to trade the orange for the needle he had in his lapel so that she could make clothing for her child. The woman, of course, got the needle and the orange. Needles and thread are very inexpensive and easy to store. As mentioned earlier, some manufacturers make treadle attachments for sewing machines so that they can be made to function in the absence of electricity. A "Speedy Stitcher" for sewing leather and other tough jobs is also desirable.

As previously mentioned (14-Day Emergency Kit), dust masks could be popular items in more than just a nuclear war scenario. Earthquakes, volcanic eruptions, storms, and other natural disasters can also cause excessive dust in the air. Fairly good rubber masks with dust filters and valves can be purchased at safety appliance stores for as little as ten dollars. Better ones cost more. Paper or disposable masks are much less.

It has already been mentioned in the sections on gardening and preparing grains that a supply of seed and a hand-operated grinder could be valuable emergency storage items. Even more important is experience in using these and other survival measures. Why not start growing a garden now? Why not practice cooking with whole grains now? It can increase health, save money, and promote a feeling of independence.

EQUIPMENT CHECKLIST

Heat:

☐ Backup stove(s)

☐ Fuel

☐ Lighters, matches

Light:

☐ Flashlight

☐ Lamp and/or lantern

☐ Candles

Software:
- [ ] Sturdy, warm clothing, including shoes
- [ ] Bedding
- [ ] Duffle

Hardware:
- [ ] Saw(s)
- [ ] Pliers
- [ ] Ax
- [ ] Hammer, nails
- [ ] Wrench(es), screwdrivers
- [ ] Mill file, sharpening stone
- [ ] Wire, rope, cord
- [ ] Knife
- [ ] Plastic Sheeting

- [ ] Shovel
- [ ] Pick and/or mattock
- [ ] Pot and griddle
- [ ] Other hand and gardening tools and materials

Miscellaneous:
- [ ] Writing supplies
- [ ] Valuables ("found" and available)
- [ ] Cleaning supplies
- [ ] Sewing supplies
- [ ] Garden seeds
- [ ] Hand-operated grain grinder
- [ ] 14-Day Emergency Kit
- [ ] Experience

SELECTED REFERENCES

Brown, Michael H. *Brown's Alcohol Motor Fuel Cookbook*. Cornville, Arizona: Desert Publications, 1979.
How to Make and Use Alcohol.

Clark, Christine Lewis. *The Make-It-Yourself Shoe Book*. New York: Knopf. 1977.

Ovens, William G. *A Design Manual for Water Wheels*. Mt. Ranier, Maryland: Volunteers in Technical Assistance (VITA), 1975. VITA's address is VITA, 3706 Rhode Island Avenue, Mt. Ranier, Maryland 20822.

American Survival Guide. McMullen Publishing, Inc., 2145 West La Palma Ave., Anaheim CA 92801-1785.

Staff of Mother Earth News. *Mother Earth News: Handbook of Homemade Power*. Hendersonville, North Carolina: Mother's Bookshelf, 1974.

Fry, L. John, and Richard Merrill. *Mother's Methane Package*. Hendersonville, North Carolina: Mother's Bookshelf, 1973.

Stoner, Carol, ed. *Producing Your Own Power: How to Make Nature's Energy Sources Work for You*. Emmaus, Pennsylvania: Rodale Press, Inc., 1974.

Thomas, Dian. *Roughing It Easy*. New York; Warner Books, Inc., 1975. A classic on ways to ease into the inconveniences of the out-of-doors.

SOURCES

A list of some outdoor equipment suppliers can be found at the end of Chapter 12, page 237.

CHAPTER 12

Shelters

Knowledge of how to select and construct basic shelters can be a survival necessity. Hence, the design and construction of a miscellaneous variety of shelters, including tents and adobes, are discussed and/or illustrated in this chapter, excluding shelters pertaining to nuclear war which are discussed in Chapter 4. Logs and other wood products are not discussed specifically, but they are convenient and useful building materials.

It is a common saying among realtors that there are three factors to consider when building: location, location, and location! Even temporary shelters should be built observing these three important rules. A shelter is not just protection but should give a sense of security and comfort as well. Even though building a suitable shelter may take some long hours of work, it is probably worth the investment. Pick a dry area away from gullies or high ridges or peaks to reduce the dangers of flash floods, high winds, and lightning. Be observant of possibilities of flash floods or rock or snow slides. Locate near a usable water supply, firewood, and building materials for the shelter. Insulate yourself from the ground with bark, boughs, brush, wood, or other dry insulative materials. Makeshift shelters may make good use of gear such as plastic or ponchos and natural features such as banks, rocks, and vegetation: a simple lean-to built against a tree or some natural formation can get you by in a period of short-term need. Remember that breezes usually blow up canyons in the morning and down canyons in the evening. By moving up the side of a canyon even a few yards and/or taking advantage of natural barriers, the wind factor can be reduced considerably. Also, facing the entrance of a shelter toward the winter sun (south to southwest), or away from the summer sun (north to north-east), or away from the wind can make it much easier to keep the interior at the desired temperature.

SNOWCAVE

A snowcave is a shelter that could save a life by its considerable protection from temperature extremes. To construct one, pick a good snowbank in a stable location away from avalanche danger. Avoid wide sloping areas. Dig into the bank by forming a small opening and enlarging it inside. Make it big enough for comfort but small enough to be sturdy and practical to build. Use a piece of plastic and/or some pine boughs or other insulative material as a floor covering. Even the body heat of the occupants will warm the air within the cave somewhat and the snow will act as insulation to retain this heat. There must be enough air circulation to provide air for all occupants and a burning candle (if you use one); this air can come through the doorway and/or some other opening. If the snow is deep enough, you can dig the entrance below the level of the cave floor and then just leave the entrance open. The low area acts as a cold trap, not allowing cold air to move into the cave. (Figures 12.1 and 12.2) After a snowcave or igloo has been heated for a period of a few days to a few weeks the snow becomes icy and looses its ability to insulate and thus becomes a ''cold'' shelter. Other winter shelter ideas are shown in figures 12.3 through 12.6.

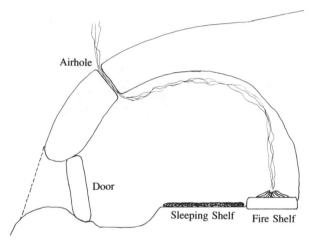

Figure 12.1 Snowcave

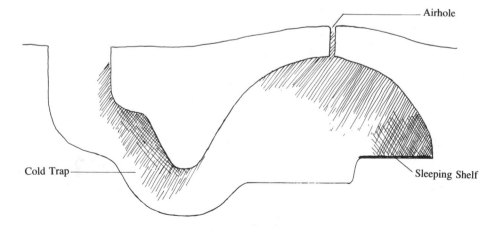

Figure 12.2 Snowcave with Coldtrap Entrance

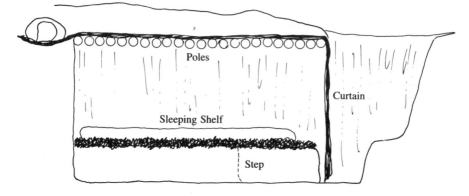

Figure 12.3   Snowcave Made From A Hole

Figure 12.4   Snowpit

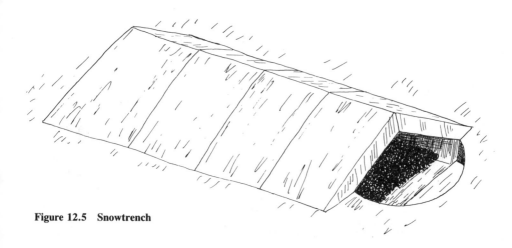

Figure 12.5   Snowtrench

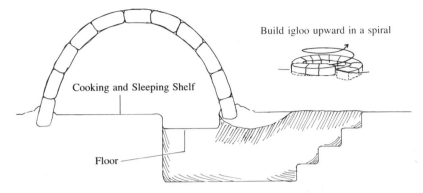

Build igloo upward in a spiral

Cooking and Sleeping Shelf

Floor

Figure 12.6   Igloo with Coldtrap Entrance

EARTH-SHELTERED LODGE

An earth-sheltered teepee is a small, well-insulated structure which is usually heated with rocks. Build a teepee frame starting with poles six to eight feet long. Cover them with smaller branches or boughs, bark, leaves, or other material. Then cover the entire structures with earth or sod, except for the entrance. The entrance should be small, and a small fireplace can be made just outside the door. Heat two or three good-sized rocks and bring one inside for heat while the others are warming in or near the fire.

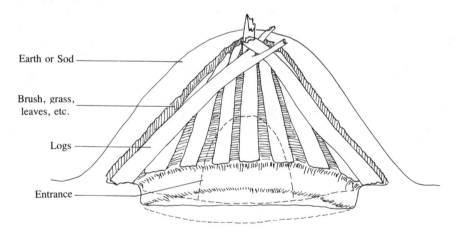

Earth or Sod

Brush, grass, leaves, etc.

Logs

Entrance

Figure 12.7   Earth Sheltered Teepee

TEEPEE OR WICKIUP

A teepee-type shelter can be made from scratch or a tree may be used by leaning a pole against the tree (and lashing it to the tree if possible) and then leaning smaller poles against larger pole and covering these with boughs, brush, thatch, or other materials.

In the open, a teepee may be made by using branches or willows for a framework and covering it with thatch (grasses, reeds, etc.), boughs, barks, or other vegetation. Leave

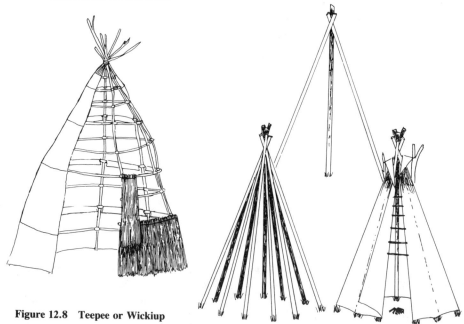

Figure 12.8   Teepee or Wickiup

an opening twelve inches to eighteen inches in diameter at the top for a smoke hole. Tie the framework together with green willow bark, rushes, string, or other cordage.

Start the thatching at the bottom and work up, tying it to the framework with the stems up, and overlapping each row with the next to allow water or snow to run off. Attach the siding material as firmly as possible. Make the opening as small as practical in cold weather or larger if it is warm. Make a suitable firepit in the center and take precautions, such as covering the inside bottom two or three feet of a thatch siding with bark (or mud) and being cautious with the fire, to avoid catching the sides on fire. Canvas, skins, or other suitable materials are, of course, also possible coverings for the framework.

TENTS

Tents may not seem a good investment as emergency gear for someone who would never otherwise use one, but if it would be otherwise used it could also serve as a temporary shelter in an emergency. Many sheepherders and others have spent time fairly comfortably during even extreme conditions in a good tent with a stove to huddle up to. Indians also made teepees with hides and built other similar structures that served well as shelter.

An inexpensive tube tent or a piece of visquene fashioned into a tent or lean-to is probably the simplest and least expensive approach. The next step is possibly a small two- or three-person tent. There is a great variety in prices in this category. The design, weight of material, quality of construction, and name all dictate price. It is often possible, for example, to see two three-man tents of nearly identical design in a sporting goods store where one is half the price of the other.

My own opinion is that if you are willing to make a few simple repairs down the road

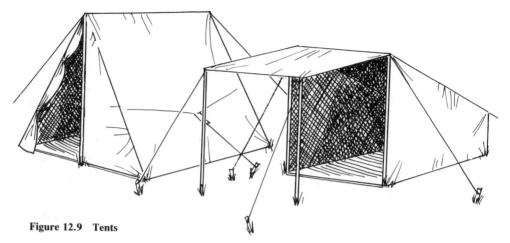

Figure 12.9   Tents

and are not going to be using the tent more than a few days a year, an inexpensive one will serve as well as an expensive one except in very specialized applications.

A rain fly is a must in a small nylon tent unless some special design feature otherwise provides the needed protection, and my experience causes me to suspect the usefulness of many of the "special designs." The nylon alone may hold out the wet for a time if left untouched; but if you touch the inside it will leak, and in a small tent it is next to impossible not to touch. Seam sealer seals the tent seams against moisture leakage.

Larger tents should be made largely of canvas. Otherwise, most are not stable in even light wind. Few things are more nerve-racking than trying to sleep in a tent that is continually blowing down. A treatment called "sunforger" is the best all-around for canvas to insure its performance and longevity.

ADOBE CONSTRUCTIONS

Adobe is a practical building material and is being used increasingly for homes in some areas. The main enemy of adobe is water. Hence, it is important to start an adobe structure with a sturdy, well-drained foundation and finish with a widely overhanging roof. The foundation and footing may be of stone, block, concrete, or whatever is available and appropriate; but remember that adobe is not light and needs a substantial base to support it.

Adobe is made by mixing sand and other coarser materials (such as fine gravel and straw) with clay and water. Clay can usually be found in and around streambeds or by digging below topsoil (sometimes a few feet), where it exists in many areas in layers. Some areas are more blessed (or cursed, as the case may be) with clay soil than others. Determining the amount of clay to sand and other material needed takes some trial and error. If the adobe has too much sand, the dried bricks will crumble. If there is too much clay, the bricks will shrink and crack. A good brick can usually be made with twenty to thirty percent clay.

If you want to make a more scientific study of ratios of materials, fill a glass jar about two-thirds full of what seems like a good adobe mixture and then add water to within an inch or two of the top. Mix all this gently but thoroughly and let it stand for a few hours. The coarser sand or gravel will collect on the bottom; the lighter, finer sand on top of

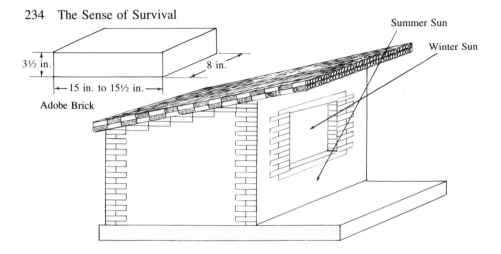

Figure 12.10   Adobe Structure

that; the clay next; and silt and the water in a layer at the top. Measure the amounts with a ruler and make calculations and adjustments accordingly by making a few bricks of differing ratios of materials and then letting them dry and testing their structural stability.

The adobe is traditionally mixed in a "pit" with a hoe (or, if you have nothing else, use your feet). A mixture with about the right amount of water will settle into the corners of a mold with a little persuasion but will not splat all over the ground when a handful of it is dropped.

The traditional size for adobe bricks is eight by fifteen or fifteen and one-half inches long and four to eight inches deep. The bricks are laid to make a wall that is eight inches from interior to exterior. A brick mold made of two inch by four inch lumber (three and one-half inches wide) produces bricks about three and one-half inch thick that dry in reasonably good time and are easy to handle. The wet adobe should be worked into the corners of the form and scraped off with a trowel or board. Try to keep the block material as homogenous as possible. Multiblock forms can be made for forming two or more blocks at one time.

Lining the forms with paper or plastic or oiling them can make removing the blocks easier. Wetting them will also help. The forms should be only sides—no top or bottom. Lay them on flat ground, fill the forms, poke the mud to get it to settle in the form and release air pockets, scrape off the excess, and then lift the form carefully off the adobe block. If the water-to-solids ratio is about right, the block will stand. If it is too wet the block will slump over. Let the blocks sit in the sun for a day or two lying flat, and then turn them on edge for a couple of weeks or longer until they are dry and hard.

The mortar for cementing the blocks together may be the mix the blocks are made of, or some cement may be added to it for greater strength and hardness. As in bricklaying, the seams of one layer of blocks should be overlapped by the next course. A plaster coat of the same material inside and out adds stability and better appearance to the walls.

Lintels, door frames, and other opening supports should be made of planks or other sturdy material. One easy way to roof an adobe structure is to raise one side of the structure progressively higher than the other (figure 12.10). Then lay peeled logs or planks sturdy enough for the span across the walls. Cover them with boards and some good roofing material such as shingles or galvanized sheet metal (if any is available).

Sod or dirt can also be used if the roof structure is made sturdy enough to hold it. If the structure faces south or southwest, a proper roof overhang in front can provide shade from the high summer sun and allow the lower winter sun an entrance.

FINISHING TOUCHES

Some shelters may need a few final touches to make them suitably livable. They may need more insulative capacity, better water tightness, better ventilation or a better heating system. Again, it is probably worth a little extra effort to make it comfortable.

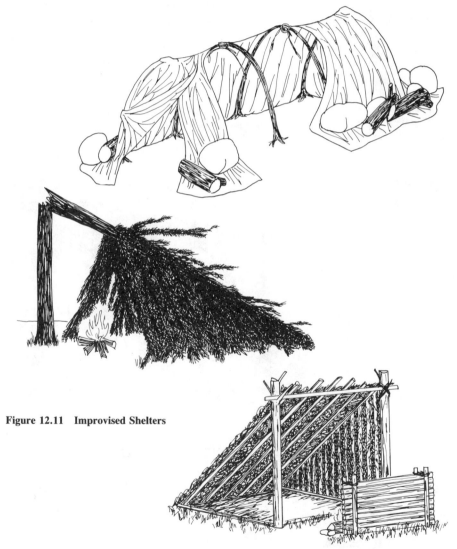

Figure 12.11   Improvised Shelters

Leanto With Fire Reflector

A quick-fix patching material that will work on some kinds of shelters is a mixture of mud and grass. When it dries, the grass holds the mud together to make a pretty tough material. This mixture can be used for patching holes, chinking log structures, mortar to stick rocks together in masonry projects, and so on.

Heating in small shelters can be accomplished by bringing in rocks that have been heated in or by a fire, as mentioned in "Earth-Sheltered Lodge." Firepits and fireplaces can also be very useful for cooking and heating. Remember that often a small fire can serve as well as a large one. Figure 12.13 shows a few designs for fireplace construction.

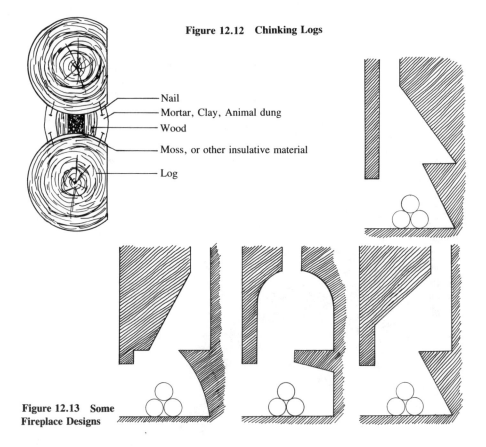

Figure 12.12   Chinking Logs

— Nail
— Mortar, Clay, Animal dung
— Wood
— Moss, or other insulative material
— Log

Figure 12.13   Some Fireplace Designs

SELECTED REFERENCES AND NOTES

Brown, Tom, Jr. with Brandt Morgan. *Tom Brown's Field Guide to Wilderness Survival*. New York: Berkley Publishing Corporation, 1983.

Langsner, Drew. *A Logbuilder's Handbook*. Emmaus, Pennsylvania: Rodale Press, Inc., 1982.

Olsen, Larry P. *Outdoor Survival Skills*. Provo, Utah: Brigham Young University Press, 1972.

Leckie, Masters, Whitehouse, Young, eds. *Other Homes and Garbage: Designs for Self-Sufficient Living*. San Francisco: Sierra Club Books, 1975.

U.S. Department of Agriculture. *Building with Adobe and Stabilized Earth Blocks*. U.S. Department of Agriculture Leaflet No. 535, A1.35:535/3. Washington, D.C.: U.S. Government Printing Office, 1972.

Equipment

Real Goods, 966 Mazzoni St., Ukiah, CA 95482-3471. Alternative energy products including all kinds of solar products. (800) 762-7325.

Lehman Hardware, Box 41, Kidron, OH 44636. Products from the past all kinds of non-electric appliances, etc.

Jade Mountain Import-Export Co., 254 Edgewood Drive, Boulder, CO 80302. Alternative energy products and information.

Edge Tek, 324 South 950 West, Orem, UT 84058. A good source of custom knives. Owner: Marvin Kay Christensen.

The Sportsman's Guide, 411 Farwell Ave., P.O. Box 239, So. St. Paul, MN 55070-0239. Clothing and equipment, good prices. (800) 888-6933.

Cumberland General Store, Route 3, Crossville, TN 38555. Wide range of old-time hardware and goods.

China Diesel Imports, 15749 Lyons Valley Rd., Jamul, CA 92035. Small diesel powered electric generators.

Campmor, 810 Route 17 North, P.O. Box 997-P, Paramus, NJ 07653 0997. Great variety, good prices.

Indiana Camp Supply, Inc., P.O. Box 211, Hobart, IN 46342. Great variety, good prices.

Gander Mountain, P.O. Box 248, Wilmot, WI 53192-0248. Wide variety of solf and hardware.

Recreational Equipment Inc. (REI), P.O. Box 88125, Seattle, WA 98138-0125. Quality gear and clothing. A co-op, they return a percentage of purchases as cash or credit yearly.

Eddie Bauer, P.O. Box 3700, Seattle, WA 98130-0006.

Cabela's, 812-13th 13th Ave., Sidney, NE 69160.

L.L. Bean, Inc., Freeport, ME 04033.

Vendors listed on pages 79, 80, and 88 are also good equipment sources.

CHAPTER 13

Sanitation

Even in the event of disaster or displacement, certain basic needs are always present. Recall, for example, the last family vacation: the car is packed, the family drives down the street and onto the freeway and gets almost out of town when someone needs a bathroom. Multiply this by the fears, anxieties, and dietary changes that occur when an abrupt change in routine takes place such as might be part of an emergency situation, and it is easy to see the everpresent need of toilet facilities. Other facets of sanitation and waste disposal also quickly become aggravating problems when our normal methods are interrupted. The disposal of wastes and the manufacture of cleaning agents from basic ingredients represent some major elements in the preservation of basic hygiene.

WHAT TO DO WITH WASTES

Daily human activity produces wastes that must be dealt with. Modern technology has placed in the hands of the highly industrialized nations some very wonderful and convenient devices. One's garbage is whisked away out of our sight, smell, or worry; we have seasonal pickup of bulky items such as tree limbs and leaves; we have trash compactors, garbage disposals, and automatic washers that pump the waste water quietly away; we shower or pull the plug on our tub and see the water no more. And last, but far from least, we are served by our wonderful toilets and sewage systems.

The usual waste produced by Americans, including waste water, exceeds one hundred pounds per day per person. (By some measurements it exceeds 200 pounds per person per day.) In emergency situations, where the normal removal of these wastes breaks down, the consequences can be very troublesome. Flies, rodents, and other pests are readily attracted to wastes and become roaming reservoirs of some of the deadliest diseases known to man. Military records show that during World War II there were nearly one million hospital admissions in U.S. Army camps for diseases associated with unsanitary conditions (Dept. of the Army, *Field Manual*, FM 21-10, p. 73). Further, it has been shown many times that during major disasters such as earthquake, flood, and violent storms, more death and misery are caused by starvation and disease than by the actual disasters themselves. Such ugly diseases as plague, dysentery, typhoid, and cholera can quickly reintroduce themselves in places where they have nearly been forgotten.

Wastes can be divided into three categories: human wastes; liquid wastes (wash, kitchen, and bath water); and solid and semisolid wastes (food wastes, waste paper, waste metals, waste glass, and so on).

Human Wastes

The excrement (primarily feces and urine) of even a small group contain disease-producing germs and must be properly cared for immediately. Germs from human wastes are readily transmitted by the classically stated "five F's": feces, fingers, flies, food, and fluids. Food or drink contaminated by improper handling, flies, or other vermin can spell big trouble. Close physical contact with an "unclean" person can also transmit germs.

Proper personal hygiene must be practiced continually to prevent disease, even though it may be inconvenient. Hands should be properly washed and finger habits watched closely. Clothing should be kept clean and eating utensils disinfected chemically (using soap or some other safe disinfectant) or by boiling. (Utensils may also be disinfected by soaking them in a solution made by adding one teaspoon of liquid bleach—five and one-quarter percent hypochlorite—to one quart of clean water.)

In most major emergency situations, there is a good chance that sewer and/or water service will be interrupted. (It is, in fact, likely that the main water supply will need to be shut off. Every responsible member of a household should know how and where to turn off all utilities, including natural or bottled gas, water, and power.) "Cat hole" latrines can be used in the backyard for a few days, but this system breeds disease very quickly and for other reasons is undesirable for more than a day or two.

Remember that a shovel is an absolutely essential storage item!

Portable Emergency Toilets   In case of a nuclear incident where all individuals may be required to remain continually sheltered for at least several days, wastes must be stored until they can be properly disposed of. In these cases, the obvious choice for storage would be a leakproof container with a tightly fitting lid. The preferred container fitting this description—and the most convenient short-term method of emergency human waste storage—is the portable chemical toilet. (If one were an avid camper he just might have a portable toilet or two around the house with a liberal supply of treatment chemicals.) Portable chemical toilets can be purchased at sporting goods stores, discount stores, and the large department store outlets.

If no portable toilet, then almost any leakproof container with a tightly fitting lid will do. A small garbage container or plastic bucket with a supply of sturdy plastic sacks for liners will serve the purpose. Liberal use of the same chemicals used in portable chemical toilets will control odors. If you can not afford the portable toilet, buy some of the chemicals and use a plastic bucket, or a small amount of household disinfectant added after each use will also make the make-shift toilet much less unpleasant. In a closed area a small hole can be made in the side of the bucket near the top to receive a piece of garden hose for venting to the outside. Another type of toilet that is practical, inexpensive, and simply designed consists of a toilet seat with a plastic bag attached to the bottom. If a place is provided to put the plastic bags after use, as described below, these can be quite effective.

If possible, every household should have on hand a large garbage can or other waterproof container (twenty or thirty gallon size). This should have a tightly fitting lid and a plastic bag or waterproof paper liner. This can hold the contents of smaller receptacles when they need to be emptied. A substantial amount of waste can thus be stored until it can be permanently disposed of by burying it under at least two feet of dirt. Of course, plastic sacks do not readily degrade in the soil and for this reason some recommend using a bucket toilet lined with a large paper sack and several layers of newspaper instead of plastic. If there is no room for the larger container, two or more smaller containers can meet the need.

Stationary Toilets   In urban and suburban America, the outdoor privy would undoubtedly become a convenient human waste disposal method in an extended emergency situation. (This excludes, or course, the time immediately after a nuclear incident in which nuclear fallout hazards exist outdoors.) For this reason some basic ideas on their construction and use are included.

It is desirable, of course, to build a privy as far away from the kitchen and the water supply as possible. It is undesirable for the latrine to be dug below the level of the ground water or where it would drain into (uphill from) a water supply area. It is also desirable to keep insects and/or rodents from spreading contamination by preventing their entrance into the latrine as much as possible. Precautions should be taken to divert runoff water caused by rain or snow so that it does not flood a latrine and scatter its contents.

TRENCH LATRINE   A latrine should provide shelter and privacy, but in an emergency situation it need not be as formal as the ones used widely in less urgent situations. A straddle trench is an easy temporary latrine to build. The U.S. Army (Field Manual, 21-11, p. 77) says that a trench latrine should be sufficient to accommodate eight percent of the individuals it serves at the same time and that a one-foot-wide, two-and one-half-foot-deep, and four-foot long trench will accommodate two people at once. So, if a group had twenty-five males and twenty-five

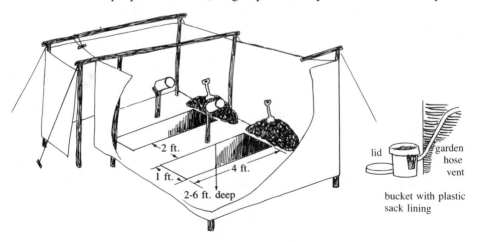

Figure 13.1   Army-Style Trench Latrine, Bucket Toilet

females, one trench this size on the men's side and one on the women's side would suffice—that is, if they could survive "army style" for a while.

A crude framework of sticks, boards, or poles with a covering of cloth, canvas, or even paper would provide privacy and some protection. A more substantial protection can be built for longer term use.

A sprinkle of dirt and/or lime over an open latrine after each use will provide some degree of protection from odor and flies. A bit of disinfectant will also help. When a trench latrine is filled to within twelve inches of the surface, it should be closed. It should be compacted somewhat as it is filled and sprayed with insecticide if possible. It should also then be mounded over with about twelve inches of dirt.

DEEP PIT LATRINES If a more substantial latrine is to be built (like a two-holer or deep-pit latrine) there are a few basics to keep in mind. A hole four to six feet deep that measures about two feet square per seat gives sufficient space for a seating structure and the capacity to last a while. This means that a two-seater hole should be four feet wide, two feet long, and four to six feet deep. A solid base should be built around the edge of the hole with a solid seating structure that will not collapse. Such a collapse would be not only embarrassing but dangerous. In some soils a six-foot deep hole will easily support itself. In other types of soil, such as loose sand, an interior structure of available materials such as wood or stone must be made to keep the sides from collapsing, even in a hole only two feet deep. Standard toilet height is approximately sixteen inches and the hole in the seat is usually an elliptical shape that is about nine inches wide by twelve inches long. The seating structure should ideally be tight with a lid to prevent entrance of flies into the pit. Seats should be scrubbed regularly with soap and water and the interior should be sprayed with insecticide if available. Disinfectant introduced into the pit can help control odors but may not be a "best use" if disinfectant is in short supply. A possible plan for a seating structure is provided in figure 13.2 for the readers and/or builders' convenience.

BARREL LATRINE An alternative seating device is useful if an auger is available to dig the pit. A drum, barrel, tub, or large bucket which is open at both ends may be placed partially beneath the ground so that about sixteen inches of it is above ground. It

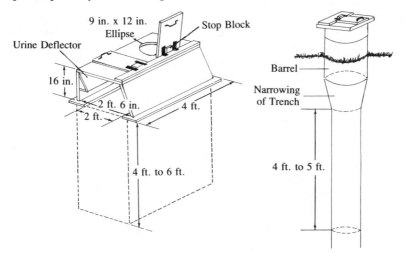

Figure 13.2   Deep Pit Latrine   **Figure 13.3   Barrel Latrine**

should be placed only over a hole that is smaller in circumference than the device. It may then be fitted with a proper seat.

OTHER ITEMS   In areas where the water table is too high to allow a pit of sufficient depth, a pit can be built above ground using wood, stones, or other available materials. Dirt should then be mounded around the pit structure and a seating device and housing mounted on top. The dimensions described for a deep-pit latrine can be followed in construction of this type of unit. The closing of this or any other type of latrine should be done in approximately the same manner as the ''straddle trench'' mentioned earlier.

A suitable holder for toilet paper—if only a large nail, a piece of sturdy twine, or a stick—is quite necessary. When this precious commodity is at a premium, it would be discouraging to drop it in mud or dirt or get it soaked by rain. A large can or plastic sack placed over it could prove effective in keeping it dry when it is not otherwise protected. Another idea is to place a roll of toilet paper in a plastic sack, remove the cardboard tube from the center of the roll, and then pull the paper out from the *center* of the roll. This works great for camping and backpacking trips. A quantity of toilet paper and sanitary napkins is a must for storage. (There is a story of a family who literally bought their way out of a German-occupied area of Europe at the beginning of World War II with a case of good old T.P.)

Liquid Wastes

Almost all of the liquid wastes produced by cooking and washing contain soap, grease, food particles, and/or other more mysterious contaminants (depending on who is cooking). In an emergency situation of short duration most wash water that is not too heavily laden with particulate and/or soap can probably be used for irrigation or just ''thrown out.''

In an extended situation, most fat and grease should be saved for making soap. However, there are methods of letting these waste liquids be absorbed by the ground without leaving an open ''compost pit'' at the dumping site.

In most soils, a relatively small soakage pit will absorb the water. A soakage pit may be constructed by digging a hole and replacing the dirt with rocks or gravel or other solid material. The size of the hole will depend on the size of the need, but a hole that is four feet square by two feet deep will accommodate considerable use. If such a pit is not constructed, the inconvenience faced with such waste water—especially from the kitchen—is the clogging of the soil with grease and so forth. The soil becomes like the ''ducks' back'' and will not longer absorb moisture. Waste water containing large amounts of solids or grease should be run through a grease trap before it is poured into the soakage pit (or onto the ground if you choose not to make the pit).

A grease filter or trap can be constructed by using an old barrel, bucket, or other device with no top and holes bored in the bottom of the container. The bottom six or eight inches is filled with gravel and the next twelve inches or so with sand or wood ashes or a combination of the two. The top may be covered with burlap or some other loosely woven fabric. If this is not available, dried weeds or grass or straw may be placed in the top portion. The device is placed over a soakage pit or elevated to allow the throughput to run in a desired direction.

The fabric or dried vegetation strains the large lumps of grease or solid material out of the liquid. If cloth is used, it should be cleaned regularly; the dried vegetation can simply be burned and replaced. The ashes and sand will absorb much of the particulate

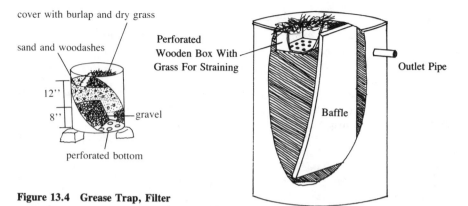

cover with burlap and dry grass

sand and woodashes

Perforated
Wooden Box With
Grass For Straining

Outlet Pipe

12"

8"

gravel

Baffle

perforated bottom

Figure 13.4   Grease Trap, Filter

and organic material; when the trap becomes slow in running the waste water through, the ashes and/or sand should be removed and buried.

A baffle trap may be made as an alternate to the above by using nearly any container that will accommodate entrance, vertical baffling into two chambers, and an exit. The waste water is poured in through a filter such as cloth or vegetation to remove large particulates; a baffle separating two chambers keeps the grease and floating material but allows the passage of the liquid into the other side where it may exit. Other particulates may settle to the bottom. Since this device needs to be cleaned periodically, access to the inner chambers must be provided so that the grease and sediment can be removed (see figure 13.4).

Solid and Semi-solid Wastes

The wastes produced while cooking or preparing meals, including containers, ashes, etc., must all be dealt with in order to avoid attracting undesirable houseguests. Much food waste is useable as animal food and/or compost. Vegetable and fruit scraps may be composted or fed to animals that will eat them. Hogs and goats will eat almost any of these fruit or vegetable scraps; chickens and rabbits are more selective but will eat many vegetables. That which is not composted or fed to animals should be removed as far as possible from living quarters and stored for proper disposal or buried.

Meat scraps should be fed to animals. These can become breeding grounds for flies if not cared for.

Trash that can be burned may be most easily disposed of by burning. Other solid material should be composted, buried, or stored, depending on circumstances. Containers such as cans should be reduced to their smallest volume by crushing. If garbage is to be buried, a hole about four feet deep and of suitable width and length should be dug to receive it. When it is filled to within about one foot of the surface, it should be closed or filled-in and mounded over about one additional foot.

CLEANING AGENTS

Agents for cleaning and disinfecting are a must at any time for preservation of health and morale. The divisions here are: "Making Soap" and "Some Ideas on Keeping Clean."

Making Soap

Making your own cleaning products can be a satisfying endeavor. It is also an activity that could easily become useful in a survival situation. There are a myriad of methods and materials for making soap; however, space here allows only a few basic instructions on soapmaking.

Three ingredients are needed to make plain ol' soap: fat, lye, and water. Other ingredients may be added to give desirable smell, texture, and other qualities. The process involves saponification of the fat with the lye water to produce the soap.

Animal fat or scraps with fat are the usual source of fat. Beef, mutton, or pork fats make satisfactory soap. Mixed fats work well too. Even rancid fat can be used—although not as desirable. Cook the scraps and fat over low heat until the fat is melted, then strain it through a coarse cloth or a piece of screen to remove pieces of meat and other debris. Squeeze the cloth to remove all the fat you can.

Many people who make soap boil the strained fat in salted water (about one heaping teaspoon per gallon) for ten or fifteen minutes, using about twice the volume of water that there is of fat. This process "cleans" the fat. When it cools after boiling, the fat forms into a crust on top and may then be removed for making soap. Scrape the grey layer off the bottom of the fat or tallow and discard it.

The easiest way to obtain lye is to buy it at your neighborhood grocery store. If the store is out of commission, lye can also be obtained from wood ashes or ground limestone can be used. Wood ashes are common and a fairly good source of lye in the absence of the store-bought product. Fill a plastic, wood, or porcelain bucket with the ashes (generally the lighter in color, the better they perform), add boiling water to near the top of the pail (the ashes will be greatly reduced in volume), stir them, add more ashes to fill the pail, stir them again, and let them stand overnight or until the sediment has settled and the liquid is clear. Decant the clear liquid and use it in making soap. This liquid may be boiled to remove some of the water and make a stronger solution. Old timers say that when a raw egg will just float it is about the right strength.

To make potash (solid lye) the clear liquid may be boiled down until all the water is gone. Heating is then continued until the black residue turns to a grayish-white color. This material is potash and may be cooled and collected for use in soap. (A heavy iron kettle is the traditional container to make potash in.)

A note of caution is in order here: All lye is hazardous. It can cause severe burns to skin. Eyes can be quickly damaged by it. Keep some clean water handy to flood any body area that may get some lye or lye water on it. Also, if you use concentrated lye to make lye water, add it to cold water. The chemical reaction causes considerable heat of its own in the mixture. It is also important to do the mixing where the fumes can be removed by adequate ventilation. And keep it away from children.

Soft Soap   The simplest soap to make is soft soap. It may be used for dishes, laundry, or about any other cleaning purpose. Keep it covered and away from small children.

SOFT SOAP RECIPE NO. 1   Stir two quarts melted grease into two quarts lukewarm water in a plastic or porcelin bucket. Add one can (thirteen ounces) of lye and, if you happen to have it, one-half cup of borax. (The borax moderates odors and makes a better laundry detergent.) Mix thoroughly (by stirring) with a wooden implement and repeat the stirring every ten or fifteen minutes for about an hour. Add one and one-half quarts more water and continue the mixing by stirring every half-hour for a couple more hours. Let stand overnight. If the mixture is too thick add more water and stir. When it turns white it is ready to use. Leave a lid on the container to keep the soap soft—you may also need to add water intermittently—or leave the lid off and let it dry, then grate the hard product and use as a powder.

This soap is especially effective as a laundry detergent if you have fairly soft water.

SOFT SOAP RECIPE NO. 2   Place an appropriate amount of fat in a soap kettle and add enough hot lye water made from wood ashes to "melt" the grease. Heat the mixture and add more lye water. A froth rising to the top indicates too much water; this can be removed by continued boiling and intermittent stirring. When the frothing action ceases, the soap level will lower in the kettle and become darker in color with white bubbles rising to the top.

One test that can help evaluate if the soap is done or not is to place a drop of the soap mixture on a plate. If too much lye has been added, a grey scum or skin will form over the drop and it will not be sticky. If this is the case, add more grease. If the drop is grey and has very small solid pieces in it, it is probably deficient in lye and more lye water should be added. A clear homogenous drop indicates that the soap is about right.

Hard Soap   Hard soap can be made by boiling soft soap down after adding salt (the salt helps the soap to harden) and water to a finished batch of it. Remove the soap from the top of the kettle as it cools and pour it into molds., Molds are usually wood (never aluminum), and a piece of thin cloth or plastic can be used to line them to prevent sticking. Cover the filled molds with several layers of newspaper or an old rug to hold in the heat and let the bars harden for a few days. Then remove the soap from the molds in a sink or on an outdoor surface to prevent any separated lye from damaging person or property and cut it into the desired size of pieces, using a string or thread to "saw" them with. Be careful of lye that may have separated from the soap—it is very caustic. Let the soap "cure" for a few weeks before using it; unless it is cured, the soap can be caustic and unpleasant to use. The aging allows more of the free caustic material to incorporate into the soap.

Here are a couple of scratch recipes for hard soap:

HARD SOAP RECIPE NO. 1   Add a hot mixture of one pound of potash (or one can of lye) and one-half gallon of water to five pounds of melted grease, stirring constantly. Let the mixture sit overnight and then add an additional gallon of boiling water and heat it until the grease is saponified and it is homogenous. Pour it into molds as described above.

HARD SOAP RECIPE NO. 2   Heat a mixture of one pound of potash (or one can of lye) in two gallons of water (or the same amount of approximately the same strength wood ash lye water) and slowly stir in four pounds of hot melted grease. Stir until it becomes an homogenous emulsion, and then simmer it for three or four hours. Dissolve a cup of salt in a gallon of hot water and stir this in. Simmer the mixture for a few more minutes, then cool it. Remove soap layer and put it into molds.

Additives   Various other entities can improve the quality of the soap. Borax is frequently added to improve the odor and color of especially soft soaps used for laundry—not hand soap. Fragrances may be added by such simple means as adding flower petals or pine needles to the soap mixtures and removing them again before the soap hardens, or add a drop or two of your favorite perfume. The addition of some glycerine can help produce a more pleasant-to-use hard soap.

Some Other Ideas on Keeping Clean

When things get right down to basics there are a few plants that can provide some germ-killing or cleaning power. Another thing that should be considered is storing a few basic cleaning and hygiene items.

The bark from most hardwood trees will provide a solution of tannin by boiling it gently for an hour or two. The solution can be used like soapy water.

Myrrh gum is antiseptic and is sometimes used as a gargle.

The pitch from lodgepole pine is a mild disinfectant.

A few basic cleaning products that it would be provident to keep a bit extra of around home might be laundry detergent, handsoap, dish detergent, and disinfectant.

Personal hygiene is also an important consideration to a scenerio of the survival mode. To be able to personally clean up a bit, brush the teeth, and maybe even smell good could bring a little breath of courage in a difficult situation. If you shave with an electric razor, keep a few razor blades and a razor or some disposable units around too in case the power goes out. Keep at least a jar of petroleum jelly, if not something more sophisticated, around to deal with chapped skin. Keep a little extra of the personal items it takes to keep clean.

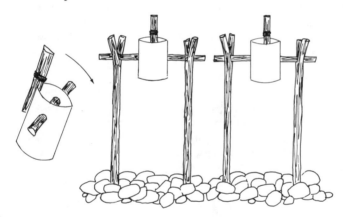

Figure 13.5   Handwashers

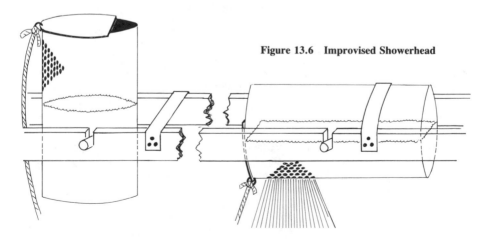

Figure 13.6   Improvised Showerhead

SELECTED REFERENCES AND NOTES

Krochmal, Connie. *A Guide to Natural Cosmetics.* Springville, Utah: Thornwood Books, 1973, 1980.
Everything you never even knew enough to ask about making cosmetics and many other household items from natural ingredients, Write to the publishers at Thornwood Books, 1680 South Main, Springville, Utah 84663 for information.

U.S. Department of the Army. *Military Sanitation,* Field Manual 21-10. Washington, D.C.: U.S. Government Printing Office, 1957.

Ellis, Marietta. *Introduction to Soap Making.* VITA Technical Bulletin, No. 510003BK. Mt. Ranier, Maryland: A VITA Publication, 1981.

Bramson, Ann. *Soap: Making It/Enjoying It.* 2nd ed. New York: Workman Publishing Company, 1975.
Very good.

SOURCES

Sun-Mar Corporation, 900 Hertel Avenue, Buffalo, NY 14216. Composting toilets. This is an excellent alternative to the big sanitation problem.

CHAPTER 14

Weapons

In a survival setting weapons are normally thought of as tools for procuring food and for protection. Proper weapons can be effective for both. First, it is understood that the use of modern weapons gives a great advantage for killing animals to provide food and other necessities. Also, in a catastrophe a natural concern of the survivors is how they can defend themselves from molestation while the normal law-enforcement and defense systems of their society are in disarray. This is especially true of some areas of the country where chaos is easily promoted. In event of war, it is even conceivable that the collective individual weaponry of our society could be a small, but real deterrent to enemy invasion.

In modern perspective personal weapons are largely firearms, and a discussion of firearms should at least include how to choose them, what types are available and the philosophy of when to use them.

PHILOSOPHY OF WEAPON USE

Using weapons as survival tools is a subject that is breached here with great caution. The unwise or impulsive use of weapons has prematurely taken many people to places they never intended to go. When the going is tough and weapons are readied, there is no room for lost tempers or impulsive actions, the consequences of which could be regret and perhaps tragedy. But weapons can help provide for those in need when used for hunting; and, as a last resort, they can help protect self and loved ones from the ravages of those who would destroy life, limb, or virtue.

When picturing life after a large scale disaster such as a nuclear attack, some imagine every household being prepared to repel invaders with high-powered weapons. Indeed, in most states, with proper licensing, one could legally buy a .30-caliber machine gun, a .50-caliber machine gun, and more. And some groups do have hideaways or retreats equipped for self-sufficient living which are provided with the firepower to see that no outsiders are allowed in. But, could you imagine an average person living on your street holed up in a foxhole outside his front door, keeping all others at bay day and night with his rifle as he hoards his last sack of wheat and can of tuna fish? In some areas this may represent an all-too-real possibility, but for most it should be labeled "nonsense!" Our neighbors are an asset to us.

Realistically speaking, the best means of survival is for reasonably self-sufficient people to help each other. The division of labor is always an integral part of independent survival. In a catastrophic collapse or degradation of present convenience and commodity, the best avenue of approach in renewing provisions is to organize and cooperate. Selfishness breeds contempt, but a few charitable deeds can usually turn others' hearts toward reason and cooperation. Even self defense (one of the most basic reasons for government) is best accomplished through sensible organized effort. And this includes the firm attitudes and actions sometimes demanded by tough times.

If things collapse, some people will be unruly, some uncooperative, some unprepared, some violent. But look at your neighbors—realistically. It would be hard to watch your neighbor's children starve when you have plenty. The important thing is to do all we can for ourselves. If enough people are prepared when disaster strikes, provision can then be made to meet the needs of those who could not or did not prepare adequately, and weapons have no part in this.

A church leader once told a story of a man who sat by him during a plane ride, really haranguing the "establishment" (of which he considered the church leader a representative member) for its unfair provision for its citizens. The man said, "If I asked you for your watch, would you give it to me?" The church leader answered, "No, the watch is a luxury; you don't need it. But," he continued, "I'll tell you what: if you came to my house in need of a meal I would surely feed you. And what's more, if I had to feed you again I'd make you work for your meal."

There in a nutshell is a welfare plan for all seasons: those who have not can work for those who have, with service that is mutually beneficial. Consequently, no one's feelings are hurt, no one is freeloading, and the situation of both should quickly improve. There is no reason to start giving out your provisions, however, until you establish a real need in those to whom you are giving. And, above what is necessary to meet immediate basic needs, nothing should be given without an agreement for mutual benefit.

Even in the most severe circumstances, some law enforcement agencies would still be intact. They could be inadequate, but still be able to solve some problems. In instances where a group of uncontrollable, marauding individuals (or an individual) without reserve begins vandalizing (or otherwise seriously threatening) a neighborhood or occupied area, the residents should stand together to try to avoid conflict, to try to acquire legal police protection, and—as a last resort—to take minimal necessary steps deemed appropriate to repel the menace (which, in the worst case, could all happen in about thirty seconds). Almost any imaginable situation from mob action to protracted war would be resisted best by organized, well-educated effort. For this and other reasons it is my opinion that the breadth of participation in our National Guard and military reserves could well be expanded.

Even so, there may be instances in times of breakdown when, in either isolated or general incidents, there is clearly no alternative but to resist or die. At times such as these the human bosom can be filled with some strange and terrible feelings. In these circumstances most people would choose to resist and would desire appropriate tools of resistance. Each person should consider what he would be willing to do in such circumstances and plan accordingly. The use of weapons takes care and practice.

The next step is to consider some practical aspects of weapon choice for both hunting and protection.

CHOOSING WEAPONS

When choosing any weapon, three basic questions can help discern the best approach. First, what do I want the weapon to be able to do for me? Second, what kind of weapon is used by those who are already doing what I want to do with it? Third, what is the least expensive and most effective alternative?

There is also a basic rule that has been borne out countless times: you are better off using a weapon that you can handle and are not afraid of than you are using a bigger, more powerful weapon that you can not handle as well. This is true for almost any purpose. For this reason it is highly recommended that, if you are going to obtain weapons for hunting, protection, or whatever, you should use them enough to know what you want and do not want and what you can and can not do with them. It is unwise to tie up money on something that is of no use to you, and it is unsafe to try to use something you can't handle. It is also a very important safety feature to know the limits of your weapon, whatever it may be. Many rifle bullets will carry for miles beyond the muzzle of the weapon and could cause harm to the innocent. Hunter safety courses can provide excellent instruction for learning firearm safety.

Another important consideration in choosing weapons is to limit the variety of calibers of firearms. Four different sizes of .30-caliber rifles (e.g.: .308, .30-30, .30-06, and .300 Mag.) would necessitate stocking four different types of ammunition and can somewhat limit portability. Sticking with common calibers, especially rifle calibers used by the military (namely .308 and .223), would more likely assure your being able to find ammunition when you run out because there is alot of it around. Finding ammunition for a .239 Blunderbuss Special may not be very easy when the going gets tough.

Similarly, it may be wise to limit the variety of weapons. If you have two weapons of the same make and model, one set of spare parts could probably serve for both. If they both break, you may be able to fix one with parts from the other.

TYPES OF GUNS AND ACCESSORIES

It is not necessary to consider every kind of weapon available to cover this subject. An adequate discussion should cover at least low-powered rifles, high-powered rifles, shotguns, air guns, barrel and chamber inserts, handguns, ammunition, spare parts, and care of firearms.

Low-Powered Rifles

Of all survival firearms I would consider a .22 rimfire rifle the one with the most utility (although it may not be the most indispensable). With minimal practice almost anyone can be a reasonably good shot with one. The ammunition is light and inexpensive. (You could purchase five-hundred rounds for $15 to $20 and carry it out in your coat pocket.) Small game and game birds (only in emergency; it is otherwise illegal to use a .22 on birds) can easily be taken with a .22 rifle. Domesticated meat

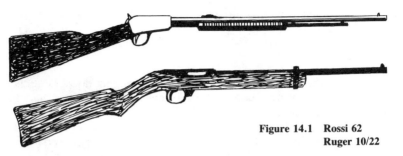

**Figure 14.1   Rossi 62
Ruger 10/22**

animals can be slaughtered with one. The recoil is minimal and the noise is small compared with larger weapons.

There are many reliable and inexpensive .22 rifles on the market. Ruger, Remington, Winchester, Mossberg, Marlin, Stevens/Savage, Harrington and Richardson, Browning, Charter Arms, and many other companies all make acceptable products. The most common type is semiautomatic. There are also pumps, bolt-actions, single-shots, combination guns, and take down models, such as the lightweight Charter Arms AR7 and the Springfield Armory M6.

The Ruger 10/22 is a much-talked-about .22 for which a large array of accessories is available. These accessories include folding stocks, fancy stocks, large magazines, and flash suppressors. There are those who speak for these expanded conveniences and capabilities, but it remains a matter of individual choice how important they are to your forseeable purposes. There is something to be said for large, easily removable magazines, but there is also something to be said for simplicity. For example, if you have a tubular magazine the only extra you need to carry is ammunition. As for the rest of the accessories, I have my doubts about their usefulness for a .22. Let's face it; if I am fighting a war with someone and I have a choice, I am not going to be using a .22 rimfire. Most of the fancy features cater to an improbable situation. If money and practicality are of no consequence, then a machine-gun rimfire .22 with laser sight and large drum-type magazine is available.

There are some expanded dimensions to the .22 rimfire these days in the increased-velocity ''Stingers'' from CCI and similar offerings from other manufacturers. CCI also makes a ''CB cap,'' a small, low-power .22 rimfire cartridge that makes very little noise and is ideal for pests and very small game. These cartridges cannot be reliably used in a semiautomatic weapon; however, a pump, lever-action, or bolt-action rifle or a revolver (pistol) can handle them. One of my favorite alternatives to a semiautomatic is the Rossi Model 62 pump. This is a copy of the Winchester that was so popular for many years before it was discontinued. It is dependable, has an exposed hammer, a tubular magazine, and comes with sixteen- or twenty-three inch barrel. The Remington Fieldmaster pump is also good, and there are some very nice lever actions on the market.

Durability, ease of repair and cleaning, and reliability are features to look for in making your selection. Aluminum receivers are not as durable as steel receivers. A solid .22 rimfire rifle, including the Ruger, can be purchased for $100 to $200 by prudent shopping.

A .22 rimfire is not suitable for taking larger game such as deer. It is suitable for taking rabbits and other game of a similar size.

Several manufacturers make guns chambered for the .22 Winchester Magnum cartridge. The ammunition for this is more expensive, but it does have greater firepower than the standard .22 long rifle. The magnum is especially useful for game that is larger than rabbits.

High-Powered Rifles

A hunting rifle capable of long-range shooting and bringing down large game is considered by some to be the most universal survival weapon. A standard hunting rifle chambered for any high-powered cartridge (.30-06, .308, .270, etc.) would provide hunting capability for just about any large North American game. Reasonably priced rifles of current manufacture by reputable manufacturers are usually as serviceable for hunting as more expensive models. As previously mentioned, rifles chambered in .308 are desirable in that there is a lot of military ammunition around that will fit them. At a time when ammunition may be generally hard to come by this could be very important. (This is also true of .223 ammunition.) Other non-military calibers that are very common, such as .30-06, .30-30, and .270, are also likely choices because the ammunition for them is so widely available.

Figure 14.2　Bolt-Action Rifle

Recommendations have been made that the designs of the actions of the Ruger Model 77 and the Interarms Mark X make them inherently more reliable than some of the other popular, competitively priced bolt-action rifles. However, with reasonable care and a reasonable quality of ammunition any good bolt-action rifle should function well through years and maybe even generations of use. A stock of basic spare parts and a little knowhow is advisable, though.

In recent years single-shot rifles have filled a growing niche in the hunting rifle market. Some hunters find an added feeling for making the first shot count and find that they shoot more carefully and accurately with a single shot. The Ruger No. 1 and subsequent models are well made and popular. Harrington & Richardson makes a line of singe-shots that sell for under $150.

It is possible to pick up survival literature that recommends having a military-type semiautomatic rifle and four or five thousand rounds of ammuntion. There are many types of these "assault" type rifles on the market. (A true assault rifle is fully automatic or semi-auto by selection.) Most are chambered for the standard NATO .308 caliber (7.62 x 51 mm), .233 caliber (5.56 mm), or 7.62 x 39 (AK47). They cost anywhere from around $450 to over $1500, and the ammunition for a .308 costs at least $250 per thousand rounds for remanufactured and around $450 for new and weighs about ninety pounds. Some quick arithmetic shows that a "heavy" expense could be involved here.

The question is: what do you want to do with it? If you are concerned with protection in case of extreme circumstances yet want the added capability of hunting, then an assault type rifle might be just what you want. These are the pieces used by those fighting the wars, and some of them are every bit as accurate as the popular bolt-action rifles used for hunting. (In Switzerland, where every able-bodied man is part of their national guard fabric, there is a military rifle in nearly every home.) If your planning includes the possibility of a firefight, then by all means buy an assault rifle. (Imagine, for example, being attacked by sveral individuals at about 100 yards with only a bolt-action rifle with a four round magazine to hold them off.)

If you are looking for protection of the assault rifle type that can be used by the unaccustomed and the small of stature, consider a rifle in .223 caliber. The big advantage of this over the larger .308 is that the recoil is much less; hence the weapon is easier to handle. Experience has shown that even a ten-year-old youngster or his grandmother, with limited experience in handling guns, can become confident and proficient with one of these rifles with minimal exposure and training. The .223 ammunition is less expensive than .308 and weighs considerably less, making it easier to buy and carry—the reasoning of the military. The 7.62 x 39 mm round is used in Russian and Chinese military weapons (primarily the AK-47 rifle). It has less punch and also less recoil than the U.S. .308 (7.62 x 51 mm), more punch than the .223, and is slower than either one. Recent legislation has made AK-47 and other semi-auto rifles much less available than previously. The sad part of this is that it will most likely have little effect on criminal use of weapons.

The main disadvantage of the .223 is that it does not hit with as much force as the .308 and the effective range is somewhat less. Larger game would require well-placed shots at close range with a .223. The standard U.S. infantry rifles fire the .223 round. Military sniper rifles are mostly .308. Some of the most knowledgeable writers on survival weapons recommend that the .308 will do anything the .223 will do and do it better. Except for the above considerations, I agree. But the above considerations can be very important in some cases.

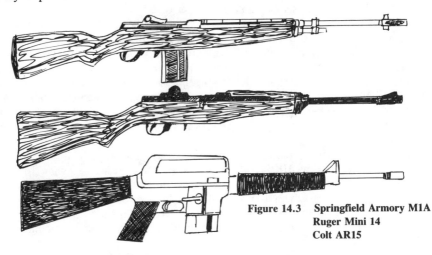

Figure 14.3   Springfield Armory M1A
Ruger Mini 14
Colt AR15

All of these military-type rifles come in semiautomatic (some can also be purchased, with proper licensing, in fully automatic) and have large-capacity magazines. There are twenty-, thirty-, forty-, fifty- and even seventy-five-round magazines available for some. Other accessories such as flash suppressors, bipods, folding stocks, are also available. The really recommendable accessories, in my opinion, are slings, extra magazines, and spare parts; flash suppressors and bipods may also be useful. An inexpensive nylon sling can be purchased for $5 to $8, or you can spend much more for a deluxe model. Many military and hunting rifles come equipped with quick detachable swivels to attach a sling. These are a must. For many reasons it is a good idea to have a few extra magazines—if nothing else, it is a convenient way to carry ammunition to practice sessions and saves time on the practice range. How many do you need? If you are planning to fight a war, a supply of extra magazines would be nearly indispensable. If you had six or eight thirty-round magazines and were having a local battle, the chances are very great that you would either have won or lost by the time the ammunition was used up. Ammunition for .308 and .223 can also be purchased packed in "stripper clips," which can be used to load the magazines very quickly.

Heckler & Koch (HK) of West Germany produce the HK93 (.223) and the HK91 (.308-7.62 x 51mm) which are used widely as military weapons. They use a delayed recoil roller lock bolt system that eliminates the inherent problems of gas-operated semiautomatic weapons and are very accurate. Some are still around but their importation is now prohibited.

The Colt AR-15 A2 (.223) is a commercial semi-automatic model of the U.S. military issue infantry rifle. This tried and true weapon is now copied by other manufacturers. The Ruger Mini-14 (.223) and Mini-30 (7.62 x 39 mm) are less expensive and very recommendable. Springfield Armory manufacturers the M1A (.308).

If you want a semiautomatic rifle but do not want a military type, they can be found. Remington's Model 7400 is popular and a fine hunting rifle and comes in a variety of calibers, as does the Browning automatic rifle. The Browning is more expensive. Many sporting rifles are not designed to endure sustained fire, and after several rounds fired over a short period of time some may become less accurate or otherwise malfunction.

There are other rifles available such as the .30-caliber M1 carbine. This retired military semi-automatic, lower-powered weapon is still manufactured and has a substantial following. A semiautomatic sporter rifle is available from Ruger (Model 44) in a .44 magnum. Many calibers are also available in lever and pump actions. Some of these are smooth and functional. The Marlin Model 1894 is a lever-action rifle that comes in .22 magnum, .357 magnum, and .44 magnum. The new Winchester lever-actions from U.S. Repeating Arms come in a variety of calibers. Browning makes a good lever-action rifle. If you opt for one of the lower powered guns and also have a handgun, you may want to stick with the same caliber to avoid having to stock another type of ammunition. For example, if you have a .44 magnum pistol you may want to expand the arsenal with a .44 magnum lever-action or semiautomatic rifle.

Semiautomatic rifles patterned after submachine guns are available, such as the UZI, from Israel and the HK94. These are mostly chambered in 9 mm. In spite of their popularity, they are costly and in my mind much less useful than other alternatives. For those who buy them they are usually either toys or close-range antipersonnel weapons which purposes they seem to serve quite well. Some of these pieces produced abroad are now on the endangered species list due to the fact that their importation has been halted.

There is also something to be said for buying new equipment. A case in point is the M1 Garand rifle. There are many on the market that may be (or have some parts in them that are) in the neighborhood of thirty-five to forty years old. A survival situation is no time to have to cope with a worn-out, difficult-to-repair, no-parts-available, semi-functional shooting iron. Although the M1s may still be superbly functional for years to come, other models may not be. It may be better to buy something new that fits into the budget.

Table 14.1 lists some rifles and their approximate retail costs. This is far from an exhaustive list but it does give an overview. Prices can be drastically altered by options. Most firearms can be purchased for less than the retail price with a little shopping and/or negotiating.

Shotguns

For bird hunting and even hunting some small land animals a shotgun is an obvious

TABLE 14.1

Some Currently Available Firearms

Firearm	Caliber	Approx. Retail Cost	Comments
Ruger 10/22	.22	$220	Semi-auto, many accessories available, dependable
Rossi 62	.22	$180	Pump action, good value
Ruger 77	many	$450	Bolt action
Winchester 70	many	$500	Bolt action
Remington 700	many	$470	Bolt action
Browning A-Bolt	many	$550	Bolt action
Browning Automatic Rifle	many	$570	Semi-auto
Colt AR15 A2	.223	$700	Semi-auto civilian version of U.S. military weapon
Ruger Mini 14	.223	$440	Semi-auto, many accessories available
Springfield Armory M1A	.308	$1,045	Military-type semi-auto
Browning Pump Shotgun	12 ga.	$400	Bottom-ejection - 20 ga. good for left and right handovers
Remington 870 Pump Shotgun	.12 ga .20 ga	$400	Very dependable
Smith & Wesson 19	.357	$340	Revolver
Browning Hi Power	9mm	$480	Semi-auto pistol
Springfield Armory 1911A1	.45	$487	Semi-auto pistol
Beretta 92F	9mm	$550	Semi-auto pistol currently used by U.S. armed forces

best choice. The expanded pattern of the shot makes a kill much more likely. If you are not already experienced with shotguns and have no preference, I would recommend a pump-action shotgun in either 12- or 20-gauge. The 12- and 20-gauge ammunition is common and competitively priced (especially around hunting seasons), and the pump action is reliable and easy to master and use. A 20-gauge has less recoil than a 12-gauge but is a bit less effective. The 20-gauge ammunition and the weapon are substantially lighter, however. The 12-and 20-gauge ammunition comes in

Figure 14.4   Remington Shotgun

two-and-three-quarter inch and three inch magnum and some guns will only function correctly (or at all) with one or the other. Some, such as the Browning pump, will handle either type. My personal feeling and experience is that the two-and-three-quarter inch is the right choice for most needs, but there is a movement in the market for that ''little bit extra.'' The magnums produce more recoil and slightly better performance.

For close range protection a shotgun is very effective and has the added safety feature that the projectiles do not carry far enough to penetrate multiple walls as do pistol and rifle projectiles. This means that in a building it is possible to disable an assailant without creating undue danger to those who may be on the other side of the wall or even several rooms away. At close range a 12-gauge shotgun with 00 buckshot can put out an incredible number of pellets in a hurry; but although they do scatter, the fact is that they still have to be pointed in the right direction to make a solid hit. And a buckshot pellet can not be counted on consistently to penetrate even a leather jacket at more than forty yards.

Several manufacturers make ''riot'' type guns with a variety of handle configurations and short (usually about twenty-inch barrels.) Most are made specifically to use slugs or buckshot and are basically for close-range hunting of large animals or antipersonnel purposes. For all-around practical use the twenty-eight inch barrel in modified choke is a good combination, or choose one of the variable choke systems that allow a variation of shot patterns by adjusting or changing choke tubes. Many manufacturers produce some type of choke system—among them are Browning, Winchester, Mossberg, Remington, and Harrington & Richardson.

There are many good pumps to choose from. Two popular models with survival writers are the Remington 870 and the Smith & Wesson 3000. Others—the Browning, for instance—have some attractive features such as bottom ejection, making them easy to use left- or right-handed. Other models from Ithaca, Winchester, Mossberg, Stevens, or some other manufacturer may also meet your foreseeable needs.

Other action types may also be of interest. Single-shot shotguns are available from under $100. (Harrington & Richardson or Stevens/Savage). Double-barrels are around for a variety of prices. The semiautomatics are also popular with many hunters.

A combination gun incorporating rifle and shotgun in mutiple barrels is considered by some to be an ideal survival weapon. There is a wide variety of such pieces available at a wide range of prices. The kings of the group are called drillings, and as far as I know are made only in Europe. They frequently come with two shotgun barrels and one rifle barrel and are made with a variety of calibers and gauges. They are also expensive, starting at around $2,000 and going up to over $5,000.

Many over/under combination rifle/shotgun pieces are available at more reasonable prices. Savage makes one that retails for about $160; the Harrington & Richardson model sells for a little less. Both come in a wide variety of rifle and shotgun chamberings. Springfield Armory makes a lightweight combination gun (the M6) which comes in .410 and .22.

The advantage of the combination type of weapon is that it gives the instant choice of either a rifle or shotgun. The disadvantage is the slow loading. Each barrel has to be reloaded manually after it is fired.

A good all around pellet size for shotgun ammunition is No. 5 or No. 6 shot.

Air Guns

Another survival weapon that is worthy of special mention is the air gun. Several manufacturers produce these useful pieces. They come in .177 and .22 caliber (there is some discussion as to which is better, but both are very acceptable in performance) and better ones operate by a spring piston. With a single pump (in which the piston is cocked), a good one can spit a pellet out at six hundred to twelve hundred feet per second. They are quiet, durable, accurate, and powerful enough to down small game and rodents. The ammunition is very inexpensive compared to firearm cartridges and with a few spare parts they will last for years.

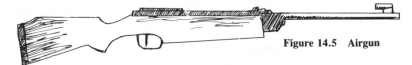

Figure 14.5   Airgun

Be aware that the better quality air guns approach and even exceed the cost of a good high-powered rifle. Beeman makes some superb air guns. RWS air guns are a little more competitively priced. Other models are available from Mendoza, Gammo, Daisy, Crosman, Benjamin, and others. Some of these manufacturers make pieces that can be pumped up in air pressure or use CO_2 cartridges to give a BB or pellet a pretty good ride. They can also be useful but are not in the same class as the spring piston models.

If you have plenty of money and are a full-sized adult, buy a Beeman 124 or a similar offering from a reputable manufacturer. If you have less money, buy an RWS Model 45 or something similar. If you are really poor, either stick with .22 shorts or C.B. Caps for your .22 or buy one of the air guns that can give fairly good velocity and go on sale at the discount houses for about $40 to $60.

A safety note: The more powerful air guns can be dangerous when used for backyard pest control. The pellets will overtravel and ricochet and can be potentially harmful to nearby innocents.

Handguns

Handguns have the advantage of being lightweight, maneuverable at close quarters, and easily concealed. These features give handguns substantial utility. They are, however, less accurate (generally speaking), more difficult to aim, and less powerful than rifles. If you are not experienced with handguns, it will probably be a substantial task to become proficient at hunting with one. Even some of the real handgun enthusiasts I know readily admit that for almost any all around survival situation they would rather have a rifle than a pistol. A handgun is just a "band-aid" weapon for many foreseeable survival purposes.

If you do not have a pre-fixed preference and experience and want a handgun, there may be a few considerations to make. First, stick to common calibers, namely .38 special, .357 magnum, 9 mm, .45 ACP, .40 S&W, .44 magnum and/or .22. Some may argue for others, but these are common and available in a wide variety of guns. Second, in my experience in teaching men, women, and children to use handguns, I have found that most have trouble with "cannons" such as a .44 magnum and big automatics. However, even small women and youngsters seem to be able to handle a .38 special, 9 mm, or even a .357 magnum with a lighter

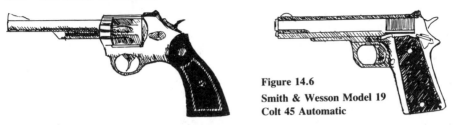

Figure 14.6
Smith & Wesson Model 19
Colt 45 Automatic

weight bullet. (A favorite load of many handgunners and police is a 125-grain bullet in .357. It shows good ballistics and has less recoil than the larger bullet loads.

The 9 mm is the most widely used handgun and submachine gun cartridge in the world. There are many nice automatic pistols chambered for this cartridge from reputable manufacturers. The Beretta model 92F 9 mm is the current U.S. military sidearm.

The .357 magnum revolver is a recommendable all-around handgun. There are some very good pistols made in this caliber, and the .38 special ammunition can also be used interchangeably in a pistol chambered for .357. Some good, competitively priced revolvers chambered for .357 include the Smith & Wesson Models 13, 28, and 19, and the Ruger Security Six. Both of these companies and several others make a wide variety of other models. One drawback to revolvers is that they can be difficult to repair. Automatics are usually simpler in construction and easier to fix but may need repairs more often.

The .45 automatic was for decades the U.S. military issue sidearm. It is effective and revered by those who become proficient at its use, but for beginners it is probably not a good place to start. Many of the knowledgeable writers on survival weapons count the .45 ACP Colt automatic (and similar offerings from other manufacturers) as the most valuable handgun for antipersonnel, sidearm protection. It is fast, powerful, and easy to reload and repair, and ammunition is widely available.

If you ever plan on shooting a bear with a handgun, buy a .44 magnum (at least) and learn to shoot it. But, as already mentioned, the .44 magnum recoil is hard for some to handle. The Ruger Redhawk or a Smith & Wesson Model 29 (or 629 in stainless steel), and Dan Wesson are good .44 magnum offerings. Others are also available.

A .22 pistol does not cover the same ground as the larger weapons. It can be lethal, but it has low shocking power. In my estimation a .22 rifle is, for most legitimate purposes, a much better device and a much better value; a good .22 pistol costs more than a rifle. Still, a .22 pistol may be a good, easy-to-carry anti-varmint weapon, and if you become proficient with it you may be able to bag some small game. The Ruger Mark II semi-automatic is a fine .22 pistol that sells for around $250 retail and is competitive with the more expensive ones. I am not aware of a bad or even questionable Smith & Wesson .22 in either automatic or revolver. Many other good .22 pistols are manufactured at a wide variety of prices.

Thompson Center makes very fine single-shot pistols in many calibers, some of which are actually rifle calibers.

One final note of personal prejudice on handguns: The reasons for using a handgun are such that a repeater of some sort is desirable. Also, because of the ease and speed of reloading, stick with either a semiautomatic or a revolver with a swing-out cylinder. If you choose a revolver, there are also speed loaders available for rapid reloading. There are also still some top-break revolvers around that were made by Harrington & Richardson, although it is my understanding that they are not currently manufacturing them. Some of the single-action revolvers are of very good quality, but the empty casings must be removed one at a time before reloading each chamber—also one at a time. The new 9 mm semiauto pistols carry as many as 15 rounds in removable magazines.

Ammunition

If you were hunting and could bring home one hundred pounds of meat with every bullet you would not need much ammunition to survive for a long time. On the other hand, if you could not hit the broad side of a barn or if you were fighting a war, you might run out rather quickly. The first consideration should be learning to be proficient.

The best way to decide how much ammunition to store is to become familiar with your guns and then project how many rounds you would use if you had to rely on it for the purpose you intend for it. It is probably not a good idea to store a lot of handgun ammunition. If you have a rifle, the reasons to use a handgun are few indeed. Another consideration is the amount of money you have to put into it. In his book *How to Prosper in the Coming Bad Years,* Howard Ruff recommends commodity storage not only to provide for one's own needs but for use in barter. He suggested that ammunition would fall into this category. I agree that this could be a possibility. If you store for this reason it is a good idea to store common calibers. And, again, it is best to keep your firearms to a minimum of varieties.

Much of what is reasonable can be dependent on where a person lives. For many it seems reasonable to have two hundred to three hundred rounds of hand gun ammunition and the same amount or more of shotgun ammunition if it can be afforded and stored. It will safely keep for many years. Several hundred to a thousand rounds of rifle ammunition could also be a reasonable possibility.

The .22 ammunition is easy to "burn up" and very useful and inexpensive. For these reasons a couple of thousand rounds or much more might be in order. For ease of cleaning and keeping weapons functional, buy copper-jacketed .22 ammunition. The plain lead bullets tend to leave deposits in the chamber and barrel. The same is true of other calibers.

Remanufactured ammunition is somewhat less expensive than regular factory loads. Handloading and storage of handloading components in a safe place appeals to many as a practical alternative to storing all their ammunition preloaded. Lee and Lyman make simple, inexpensive handloading tools that might be good items to consider buying if you do not plan to do a lot of reloading or if you don't want to tool up with the more expensive and more functional equipment. Dillon equipment also is good.

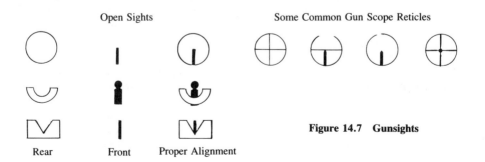

Figure 14.7   Gunsights

Ammunition should be stored where it is cool and dry and away from fumes of things like gasoline, paint thinner, and so on. A water tight container in a cool place is ideal. Also it may be of interest to store a certain portion of the rifle ammunition in spare magazines and in a carryable container for quick portability, as some writers suggest.

Barrel and Chamber Inserts

To give added utility to many weapons, inserts are made that can be placed inside a larger weapon to allow smaller caliber ammunition to be used. For example, a barrel insert can be fitted to a .30-06 rifle that will allow a .22 long-rifle round to be fired in it.

Some adaption can also be made to allow a smaller cartridge of the same bullet diameter to be fired in the weapon. These devices, some of which are called auxilliary cartridges, fit into the rifle's chamber. An example of this is using a .22 long-rifle round in a .223 rifle. Others are available.

Spare Parts

The parts most likely to break are not always the same on different firearms. In some cases, the owner's manuals give suggestions. An elaborate spare parts kit would include firing pin, extractor, springs, small screws, pins, iron slights, an extra barrel, an extra bolt, and a bluing kit. The simplest spare parts kit, in my opinion, would be an extra firing pin and an extractor. A manual of instructions on how to use and service the weapon is essential.

Care of Firearms

Firearms must be cared for properly to preserve their function; those that are will last many years. A good cleaning kit should never be far from a gun. It should include a rod (of appropriate size), lightweight oil, cleaning solvent, wire brush, patches, and patch holder. A protective gun case is also a must.

Use the gun-solvent to clean the barrel with the brush and/or a patch. (Light flannel makes very good cleaning patches.) Then run clean patches through the barrel until they can go through it and come out clean. Run a lightly oiled patch through the barrel after it is clean. (In the absence of a ramrod, a patch tied to a string works fine.) Clean and lightly oil the other metal parts. Some wood finishes also benefit from a light coating of oil (and some will not). Store the gun in a horizontal position. Before using the weapon again, it is a good idea to remove excess oil from the barrel by passing a clean patch through it.

If no cleaning equipment is available, boiling hot water may be poured down the barrel to clean it. Use a cloth to remove water from the action and other parts. The water will quickly dry from the barrel if it is left in a vertical position. Do not pour water down the barrel of a semiautomatic or automatic rifle with a gas-operated system unless the barrel is removed from the rifle first.

When brought into a warm area from extreme cold, a rifle will "sweat". This sweat can freeze upon re-entering the cold and stop the function of a weapon; it can also cause rust. To avoid these problems, leave the rifle on a cold porch or wipe the sweat off (and out of) the rifle as soon as it appears.

Always follow manufacturer's directions for using and cleaning a firearm.

There is still some ammunition around that was manufactured many years ago with corrosive powder and/or primers. This type of ammunition is especially detrimental to firearms unless they are properly cleaned soon after use.

OTHER WEAPONS

There are many other devices that can be used as weapons: sticks, knives, bows, crossbows, blowguns, slings, slingshots, BB-guns, and so on. Many of them are very powerful and useful as weapons. If you are interested in any of these it may be recommendable to take them up as a hobby so that you can become acquainted with their capabilities. Most of the more primitive weapons require a considerable amount of skill to use them properly. But they can be very effective; a stick may be every bit as useful as a shotgun to kill mice or rats.

Whatever weapons are chosen it is probably best to have a complete battery that will serve all of your foreseeable needs.

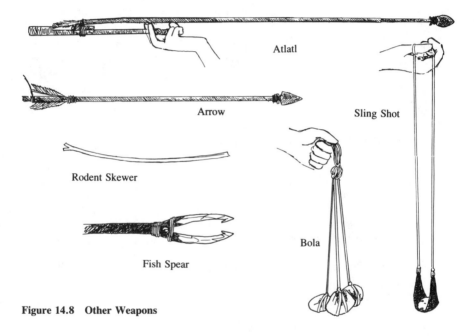

Atlatl

Arrow

Sling Shot

Rodent Skewer

Bola

Fish Spear

Figure 14.8   Other Weapons

SELECTED REFERENCES AND NOTES

Clayton, Bruce D. *Thinking About Survival*. Boulder, Colorado: Paladin Press, 1984.
   A discussion on defense of the old homestead is included.

The Shotgun News.
   This is a wholesale advertising catalog that will give you plenty of ideas about what firearms and airguns are available and how much you should pay for them. Published every ten days. Subscription price is about $15.00 per year. Address is P.O. Box 669, Hastings, Nebraska 68901.

Tappan, Mel. *Tappan on Survival*. Rogue River, Oregon: The Janus Press, 1981.

U.S. Department of the Army. *Combat in Fortified and Built-up Areas*, Field Manual 31-50. Washington, D.C.: U.S. Government Printing Office, 1964.

Cuthbert, S.J. *We Shall Fight In The Streets*. Boulder, Colorado: Paladin Press, 1965.

Ayoob, Massad F. *In The Gravest Extreme*. Published by Massad F. and Dorothy A. Ayoob, 1980. Available from Police Bookshelf, P.O. Box 122, Concord, New Hampshire 03301.
This book tells the things that need to be considered when using firearms for protection.

The Second Amendment Foundation is an organization that is battling for the rights of gun owners in America. They distribute information and support pertinent legal causes. They are at 12500 N.E. Tenth Place, Bellvue, Washington 98005.

SOURCES

Brownell's, Inc.
Route 2, Box 1
Montezuma, IA 50171
(Gunsmithing, books, tools, supplies)

Sherwood Distributors, Inc.
18714 Parthenia Street
Northridge, CA 91324
(Accessories)

Choate Machine Tool Co.
Bald Knob, Arkansas 72010
(Supplier of custom parts and work)

Sarco, Inc.
323 Union Street
Stirling, NJ 07980
(Accessories)

Sporting Arms and Ammunition
Manufacturers' Institute, Inc. (SAAMI)
420 Lexington Ave., New York, NY 10017
(They have a pamphlet about ammunition and its components that they will send you if you send a self-addressed stamped envelope.)

Rapp, Burt. Armed Defense, Gunfight Survival for the Householder and Businessman, Port Townsend, WA: Loompanics Unlimited, 1988.

National Rifle Association
1600 Rhode Island Ave., N.W., Washington D.C. 20036

Gun Owners of America
8001 Forbes Place, Suite 102, Springfield, VA 22151, (703) 321-8585.

Center For Action
1106 North Gilbert Road #2114, Mesa, AX 85203
Self defense seminars and more. Headed by Bo Gritz.

DRILLINGS

Paul Jaeger, Inc.,
211 Leedom St.
Jenkintown, PA 19046

H. Krieghoff GmbH
Boschstrasse, ULM
West Germany 7900

CHAPTER 15

Communications and Signaling

Survival communications and signaling may be easily overlooked but should certainly not be minimized. They can be very important. Both radios and other signals merit consideration.

RADIOS

Receiving (and in some cases giving) information via radio could bring not only information but also a sense of security that would be very important amid the uncertainties of disaster.

As inexpensive and potentially useful as a small portable AM-FM radio can be, it is reasonable to have one tucked away for every one to two responsible individuals in the family.

Besides the regular AM-FM equipment, other radio equipment could also be valuable. Some multiband receivers handle AM, FM, UHF, CB, public service, and weather station signals. There are also weather-station radios that have automatic alarms that signal you to listen for incoming emergency weather information.

A public service band (Police band) radio receiver could be a valuable aid in an emergency situation. A good set should cover VHF high-band (148-174 MHz) and UHF high-band (450-470 MHz) and should have "phased-lock loop" so that you do not need a different crystal for every frequency you want to use.

Citizens' band (CB), military, commercial and ham transceivers are all alternative avenues of communication. The least expensive, but that with the shortest usable radius, is the CB. Many varieties are available, and they are good for communication at close range. There are so many car batteries around that even in a major disaster there should be power for CB operation for some time. The other alternatives mentioned are more specialized, more expensive, need more elaborate power supplies, and are heavier, but they are more effective. A two-meter (146-148 MHz) hand-held or mobile ham radio could be very valuable and may well be a best bet for emergency communications. Numerous repeater stations are set up to relay signals from these units, and there are ham radio operators around the world. (A novice ham license is relatively easy to acquire.) Another alternative for interpersonal communicado that could be useful in a survival scenario is the "walkie-talkie." Some of these little units are very reasonably priced, use small and inexpensive batteries, are light in weight, and have good sound at short distances.

In a shelter or enclosed area, radio signals may be weak. To improve reception, antennas can usually be extended by connecting any insulated wire to the existing antenna. AM radios are an exception; they usually have an internal ferrite antenna, and connecting an external one to it is not always simple for the layman. (If you were stuck in a shelter and you couldn't get a good signal, you might, however, set or hang the radio outside a shelter to listen to it.)

There is a worthy note to be made concerning radios here. As mentioned earlier (Chapter 3) one of the known Soviet strategies for damaging America in time of war is to detonate a large nuclear weapon at a high altitude. The electromagnetic pulse (EMP) created by one or two such large bursts is picked up by wires and cables and could damage electrical equipment all over the nation. This could serve as a warning prelude to nuclear attack. Radio equipment may survive EMP by being disconnected from plugs and antennas and placed in a refrigerator or oven or other metal box. All electrical equipment including appliances should be unplugged from power and/or antennas, to prevent or reduce damage from EMP. According to the experts, a small battery-powered transitor radio with a short (somewhere between ten and thirty inches depending upon who the "expert" is) antenna should survive any probable wave of EMP even without such protection.

One final note on radios is power. A good supply (at least two or three sets) of the best batteries available should be stored for each radio and regularly rotated. Most batteries can be stored in a freezer for increased shelf-life.

OTHER SIGNALS

Circumstances arise in emergency situations when communication with those who are searching for you becomes quite important. There are myriad accounts of stranded people being able to see the searchers; but the searchers, being unable to see the stranded, pass them by. This is especially true of searches from the air. Air searches are difficult at best. If you want to be found, make yourself known in as obvious a manner as possible. Make your emergency signals as loud, as bright (or dark), and/or as big as possible.

An international emergency distress signal is three of any specific signal: three blasts on a whistle, three fires arranged in an equilateral triangle, or three shots from a firearm.

Another signal is six whistle blasts in one minute, followed by a one-minute silence, followed by six blasts in one minute, and so on.

A gun or other loud report fired at approximately one-minute intervals continuously is another distress signal.

A continuously sounding fog horn, the word "Mayday" and "SOS" may be communicated by voice or in code. The International Code (table 15.1) for SOS is three short, three long, and three short (or dididit dahdahdah dididit). It may be repeated at appropriate intervals and may be made by an audible or visible signal (e.g. flashes of light, beeps on a horn, or tones on some type of transmitter).

Smoke is a very visible signal of location. Signal smoke bombs or grenades are typically orange. A smoky fire burning green wood or tires becomes quite visible. However, do not waste flares, smoke bombs, and other nonrenewable signals if no aircraft or other rescue vehicle or party is close by.

An X stamped out in snow or grass or sand in an open area is visible from the air and will mark your position. Make contrasts such as dark rock on light-colored sand, or white rocks on grass, or dark branches on snow to establish greater visibility. The larger the signal, the better. Standard military size for such marks is lettering eighteen feet long and three feet thick. The X signal made in this manner is one of the standard ground-to-air signals and means "can't proceed." It and others are presented in figure 15.1. For all these signals, the standard dimensions should be maintained if possible; the height of eighteen feet high is approximately six giant steps for an average adult and the three-foot width is about one giant step. Bright-colored clothing or other highly visible material, such as a "space blanket" may also be spread out to attract attention of rescue parties. A Marine distress signal is a three-foot-square international orange flag with a large black square and circles on it.

A mirror can send a signal for miles, even when atmospheric visibility is not at its best. Signal with a mirror often, and especially when you suspect a rescue party is near. During hazy conditions, your mirror flash may be seen by a rescue party on foot or in a vehicle (such as an airplane) before you see them. A signal mirror usually has a sighting hole in it. This can also be improvised by making a small hole in the reflective surface on the back of a mirror. Then just sight through the hole from the back of the mirror and aim at a target with the reflected light. Try it on visible objects for "target practice."

Another method of sighting recommended by the U.S. Air Force is to hold one hand out in front of you toward the object to be signaled and, holding the mirror with the other hand, reflect the sun's rays onto the outstretched hand. Now make a V-shape

TABLE 15.1
INTERNATIONAL MORSE CODE

A	• —	O	— — —	Query (?)	• • — — • •
B	— • • •	P	• — — •	Error	• • • • • • • •
C	— • — •	Q	— — • —		Numbers
D	— • •	R	• — •		
E	•	S	• • •	1	• — — — —
F	• • — •	T	—	2	• • — — —
G	— — •	U	• • —	3	• • • — —
H	• • • •	V	• • • —	4	• • • • —
I	• •	W	• — —	5	• • • • •
J	• — — —	X	— • • —	6	— • • • •
K	— • —	Y	— • — —	7	— — • • •
L	• — • •	Z	— — • •	8	— — — • •
M	— —	Period(.)	• — • — • —	9	— — — — •
N	— •	Comma (,)	— — • • — —	0	— — — — —

between your fingers and sight the target through the V while the reflected light is shining on the hand. Lower your hand and the light will be aimed at the target.

The U.S. Coast Guard approves of hand-held or aerial red flares, hand-held or floating orange devices (these are more effective during daylight and light wind than at night), and red meteors and parachute flares fired from launchers to give distress signals. Launchers are usually what would be called flare pistols. Kits sporting one or many of the above devices are available from several manufacturers, including Olin, Bristol Flares, Smith & Wesson, Heckler & Koch, Patrex, Kilgore, and Penguin. The cheapest and simplest devices are, of course, hand-held flares. Signal pistols that are of respectable quality are available for under $50.

While many distress signals are given in threes, an answer to a distress signal can be given in twos; for example, a distress signal of three shots should be answered with two shots.

SOURCES OF RADIO EQUIPMENT

Radio Shack
stores all over the United States

Heath Company
Benton Harbor, MI 49022

Amateur Electronic Supply, 4828 West Fond du Lac Ave., Milwaukee, WI 53216. Stores in Chicago, IL; Las Vegas, NV; Clearwater and Orlando, FL; Wickliffe, OH. Phone 800-558-0411.

Ham Radio Outlet. Stores in many cities in California and Atlanta, GA. Phone 800-854-6046.

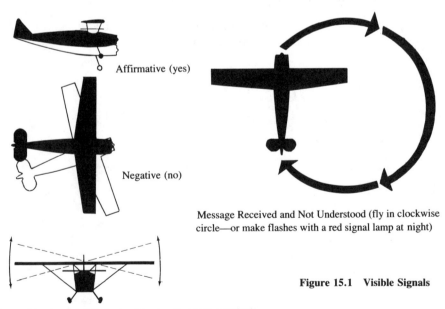

Affirmative (yes)

Negative (no)

Message Received and Not Understood (fly in clockwise circle—or make flashes with a red signal lamp at night)

Figure 15.1   Visible Signals

Message Received and Understood (rock wings back and forth or make flashes with a green signal lamp at night)

⊢	Wind Direction
L L	All Well
N	No
Y	Yes
⌐L	Not Understood
W	Require Mechanic
I	Require Doctor, Serious Injury
II	Require Medical Supplies
X	Unable to Proceed
F	Require Food & Water
≫	Require Firearms and Ammunition
□	Require Map and Compass
¦	Require Signal Lamp With Battery and Radio
K	Indicate Direction to Proceed
↑	Am Proceeding in This Direction
▷	Will Attempt Take Off
L⌐	Aircraft Seriously Damaged
△	Probably Safe to Land Here
L	Require Fuel and Oil
L L L	Operation Completed
╫	Have Found Only Some Personnel
X X	Not Able to Continue
<	Divided Into Two Groups—Proceeding in Directions Indicated
→•→	Information Received That Aircraft is in This Direction
N N	Nothing Found—Will Continue To Search

All OK Do Not Wait   Use Drop Message

Our Receiver is   Need Mechanical Help
Operating   or Parts—Long Delay

Negative (NO)   Affirmative (YES)

Can Proceed Shortly,   Do Not Attempt
Wait If Practicable   to Land Here

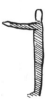

Land Here   Need Medical Assistance URGENTLY   Pick Us Up

HAM RADIO LICENSES

Call or write to your regional Federal Communications Commission Office (FCC). You may have to call your local Federal Information Center for the number or address, or if you know a ham radio operator he can probably help.

The address of the FCC western regional office in San Francisco is: FCC, 555 Battery St., Room 422, San Francisco, CA 94111.

Signal Flares

Nitro-Pak Survival Foods and Supplies, 325 West 600 South, Heber City, UT 84032, (801) 654-0099, (800) 866-4876.

Information

QST. The Official Journal of the American Radio Relay League, Newington, Connecticut. Official organ of the Canadian Radio Relay League. Devoted entirely to amateur radio—write to QST, 225 Main St., Newington, CT 06111.

CHAPTER 16

Miscellaneous Considerations

A few remaining items should also be considered in the collection of survival thoughts: transportation, barter, care of infants, methods of keeping some things cool, knots, and suggestions for some sanity-saving games are included here.

TRANSPORTATION

So very many of the survival situations we can envision could involve the leaving of our homes. Therefore, a few considerations on transportation should be made. The first item should be the Fourteen-day Emergency Kit. This kit in one form or another could be extremely important to rapid departure from home.

As with other survival items, transportation devices for John Q. Average American need to be selected to fit into the overall scheme of life. To buy an expensive transportation device for emergency preparedness only would be ridiculous for about ninety-eight percent of everybody. But if something not now being used were considered as a way of making life more efficient at the present as well as in a survival situation, it could then make sense to work toward it.

Keeping an automobile in working order and filled with fuel is an important aspect of leaving home. Some suggest that, as a protection against economic collapse, some regularly needed automobile repair parts be stored. This might include items on the order of tires, oil, filters, tune-up parts, and so forth. This may not strike your fancy, but at least keep the family automobile full of fuel and in good running order.

A family with four children and only a four-door sedan for transportation may wish to consider purchasing a trailer to use on trips or in an emergency to haul luggage and gear. On the other hand, if a family consisted of only a man and wife and he drove a pickup and she a sedan, there may be no need of a trailer. Some people may not own an automobile, making a trailer of no use to them (unless, of course, they were stout enough to pull it by hand). At any rate, a used trailer may be purchased for around $200 on up. One company makes a "survival trailer" which is sturdy and well built. Other places to shop are the classified ads of your local newspaper, auto wrecking yards, and small metal fabrication shops. Some of the larger department store chains also offer small trailers with small wheels, but trailers with large wheels are definitely recommended if you ever plan to take one over rough terrain.

Bicycles, motorcycles, and motor bikes represent some legitimate methods of getting around when other modes are no longer usable. In my own opinion, a bicycle is one of the great ways to get around if the distance is not too great. And using them is a

Figure 16.1   Transportation

good way to get some exercise. Used bikes are often inexpensive but still sound. A bicycle needs no fuel (it does need energy, however), and only a few spare parts could last a long time. Some master links for the chain, oil, tubes, tires, tire-repair equipment (including a hand air pump), and tools are the absolute minimum if you plan to keep it going. The small fuel-efficient motorcycles will run a long time on a gallon of fuel and are also a travel mode to consider.

Some other devices occasionally mentioned in discussions of survival transport are handcarts (some of the pioneers used them), wheelbarrows, travois, garden carts, tractors, all-terrain motorcycles, ultra-light aircraft, horses, mules, and llamas. Some of these may somehow fit into what you could consider in a survival situation as well as your normal life. Motor homes, travel trailers and campers may also be considered. Differing situations would certainly dictate different devices of transport.

A final idea to consider on transportation is that of keeping a good hand-operated tire pump and some tire patching materials on hand. A flat tire could be awfully difficult to fix without them.

BARTER

Bartering during periods of general distress would undoubtedly be much more common than at present. In practical terms, items for barter should be easily divided and tradable in small quantities. Some items to think about are food, hardware (shovels, picks, plastic sheeting, nails, tools, repair parts, etc.), sanitary items (toilet paper, diapers, soap, etc.), fuel, labor, ammunition, repair parts, spices (basics such as cinnamon), sewing items (material, needles, thread, scissors, etc.), gold, silver, and so on.

The items you can consider will vary with your situation, where you live, what your interests are and so on. The above-mentioned items are only meant to stimulate thought. Obviously, a farmer whose livestock have survived a disaster would be in a good position for barter, as would other small-business owners with stock remaining and people with specialized skills and equipment, either through their professions or their hobbies (mechanics, weavers, gardeners, carpenters, doctors, and so on).

A pertinent consideration in barter is protocol. Consider a trade carefully; value given should be value received. Something that seems of great worth one day may seem of little value after only a few ensuing days. Likewise, a seemingly poor trade may become valuable over time. So after an agreement has been reached, unless some startling new information shows the agreement to be drastically unsound and unless

further negotiations can be made, stick to the bargain; enjoy the fruits of the trade and allow the other party to do likewise. This is important because in tough situations people tend to be irritable and easily jump to the conclusion that they are "getting taken." It is probably best to keep things as low-key as possible in order to maintain a spirit of cooperation—especially if you may be associated with the other party, in one way or another, for some time to come.

CARE OF INFANTS

Much of the basic food used for emergency storage is not suitable for babies and without proper nutrition they would soon weaken and become ill. The best solution to infant nutrition is nursing.

The powdered and liquid canned formulas are also good, and if family finances of those with infants permit buying an emergency supply, it should be acquired. However, the formula is fairly expensive and specialized. A formula can be made by adding one-third cup plus two teaspoons of nonfat milk powder (instant is easiest to mix) to one and one-third cups of safe water and mixing it thoroughly. Then add one tablespoon of vegetable oil and two teaspoons of sugar. Mix this and serve it to the baby three times daily. If you have no bottle, do the feeding with a spoon.

Vitamin enrichment of the baby's diet can be achieved with multiple vitamins. Crush a multiple vitamin tablet to a powder and dissolve about one third of it in a small amount of liquid and feed it to the infant. This would be approximately suitable for one day's ration.

In the absence of anything else babies may be fed well-cooked, pureed (or thoroughly mashed) solid food. If it is still too coarse for very small infants, the food could also be strained through a porous cloth. A mixture of three parts cereal grain and one part legume (beans) is also nutritionally wholesome for the very young. Cook such a mixture until it is very soft and press it through a sieve or cloth before feeding it to the baby.

Sanitary conditions should be carefully maintained for infants. Where possible, use sterile bottles and utensils and germ-free water. Bottles and utensils can usually be made acceptably clean with plenty of hot, soapy water.

The other end of the matter can be summed up by saying: Keep a large number of disposable diapers and some premoistened wipes on hand. If that is not possible, and water is at a premium, shake the cloth diapers out and rinse them in a bucket of water and then wash them in hot, soapy water and/or boil them in water for a few minutes (not very long or you'll make mush of the diapers). Rinse and dry them. At all stages keep diapers away from flies.

KEEPING IT COOL

There are always foods that need to be kept cool to keep them fresh for a reasonable length of time. Cool drinks and foods can also be very refreshing and relaxing, especially in times of stress or when hard labor is in process.

Ice

Ice can be a means of keeping things cool in summer. A confined area with good insulation and a block of ice in it can make a good cooler to keep things from spoiling and to provide refreshment.

In the old days (and it was not really that long ago) people would go to the streams or lakes in the late spring (or whatever time of year it is just before the ice starts melting) and, with a saw and axe, cut blocks of ice and haul them to an ice house. The ice house was a ventilated structure providing shade over a pile or bin of sawdust or other such material. The ice was packed in the sawdust in rows and layers with sawdust surrounding each block. As the ice was needed it was retrieved from the sawdust. Using this method there was, in many areas, ice for nearly the entire summer.

Cool Places

A small enclosure that will provide shade and protection can be built and placed in a stream such that some water can flow through the enclosure. Food placed in containers in the enclosure will be cooled to the temperature of the stream water. The simplest form of such a device would be an item in a stream in the shade of a tree. Or, getting a little more complex, a small structure that provides protection, shade, and some flow-through of water can be built at a spring (springhouse) or on a stream. Also, a root cellar that is well insulated with a covering of dirt can keep food considerable cooler than it would be above ground.

KNOTS

A survival book should show how to tie a few knots. Three knots designed to join two ends of ropes (squareknot, surgeon's knot, sheet bend), three hitches used for lashing and tying down (clove hitch, timber hitch, taut-line hitch) and two loop knots (bowline, lariat loop) are shown (figure 16.2). *Everyone* should know how to tie a square knot, a bowline, and a clove hitch without referring to a book! The bowline is especially important as a rescue loop or to tie animals up. It will hold an animal or a person without tightening its loop. Practice up a little!

SANITY-SAVING GAMES

When I am watching a favorite sports team play or am playing ball or Scrabble or fishing with my sons and daughters, there is seldom room for worry about mowing the lawn, fixing the car, or other such trivia. Games can take the mind's attention away from troubles and worries and provide a welcome break. They can also pass the time enjoyably during a period of enforced confinement. There is a special need for games when children are involved.

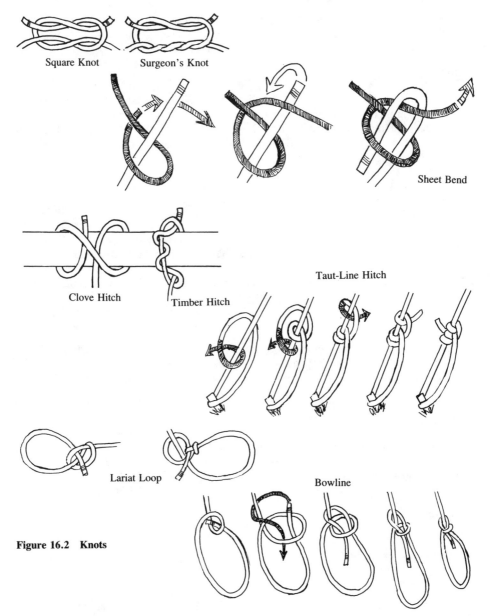

Square Knot   Surgeon's Knot

Sheet Bend

Clove Hitch   Timber Hitch

Taut-Line Hitch

Lariat Loop   Bowline

Figure 16.2   Knots

A word of caution is in order. In an already strained situation it is important to avoid cheating or being overly competitive. Play to lose sometimes and be gracious in winning or losing. Children might be offered good-sportsmanship awards.

A few suggestions of games to play follow.

Concentration

Take a deck of playing cards and spread them on a flat surface face down. Take turns turning over two cards at a time, trying to match them by number or some other preselected designation. If you select a matched pair, keep going until you miss. The one with the most pairs when all cards are taken is the winner.

Ball

Any number of games can be played with a ball inside or outside. Make up rules for whatever you want to play. Try to put a small ball through a hoop or hole indoors. Outdoors, play improvised touch football with any kind of ball, or baseball with almost any kind of ball and bat, and so on. Indoor balls can be made of foam rubber to prevent injury and damage.

Games of Skill

Throw coins at a mark; closest to the mark wins. Throw rings onto a stick or bottle. Stack miscellaneous items up, taking turns; the one who makes the stack fall loses. Make up definitions for unfamiliar words; the player who can guess which definition is the true one wins. Make up other games that challenge the various skills of the players—and do not forget old childhood stand-bys such as Twenty Questions, I Spy, Simon Says, and so on. A crossword puzzle book could provide hours of stimulating entertainment.

Prepackaged Games

Coloring books and crayons or colored pencils; board games such as chess, Scrabble, checkers, Monopoly; card games such as Uno, Old Maid, or Bridge; other games such as dominos, tic-tac-toe, and tiddlywinks can all help relieve tension and tedium.

Relay Races

With a little practice you can invent different relay races as fast as you can run them in an area as small as a room or as big as a football field. Do acrobatics en route, carry something in an awkward position, take some form of handicap, perform some task at the half-way point and/or anything else you can dream up that seems appropriate.

Charades

Depict names of movies, songs, animals. When this gets boring, draw pictures instead of acting. Choose teams or play individually.

Plays

Take turns making up and directing plays or musicals or take along a book of mystery plays and act them out.

These suggestions are brief and few, but should give enough ideas to get things moving in a good direction. Setting a specific, alloted time for games can provide a reward and give something to look forward to, but it is a good idea to creatively seize all opportunities to create and maintain stable and comfortable family life—especially in a disaster situation.

Push the stick into the ground at a level spot, so that it is straight up and down and casts a shadow. Mark the tip at the shadow with a small rock and wait 10-15 minutes until the shadow moves a few inches.

Mark the tip of the second shadow.

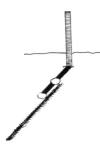

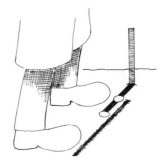

Draw a line from the first rock to the second rock and at least a foot past the second rock.

Stand with the toe of your left foot at the first rock and the toe of your right foot at the end of the line you drew.

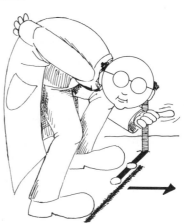

You are facing north.

**Figure 16.3   Finding North by
          Shadow Tip Method**

REFERENCES

Trapp, Jack. *How To Barter And Trade*. New York: Cornerstone Library, Inc., 1981.
Library, Inc., 1981.

The Official Boy Scout Handbook. Irving, Texas: Boy Scouts of America, 1979.

Brandreth, Gyles. *The Book of Solo Games*. New York: Peter Bedrick Books, 1983.

Heaton, Alma. *Family Recreation*. Provo, Utah: Published by the author, 1985.
Alma Heaton is a prolific producer of fun and fun books. This book is available by mail order from the author for $5 plus postage and handling. Send to Alma Heaton, 235 South 900 East, Provo, Utah 84601. Many other good books of this type are also available from Alma Heaton.

_____. *Double Fun: 100 Outdoor And Indoor Games*. Provo, Utah: Published by the author, 1974.
Available by mail order as *Family Recreation,* above.

SOURCES

Handcarts

Jim Sukoski, 330 West 100 South, Salt Lake City, UT 84101. Old fashioned carriage wheel maker.

Boulder Outdoor Survival School
P.O. Box 905, Rexburg, ID 83440 (208) 356-7446
and P.O. Box 345, Boulder, UT 84716 (801) 335-7404

Electric Power Generators

China Diesel Imports, 15749 Lyons Valley Rd., Jamul, CA 92035. Small diesel powered electric generators.

Solar Power

Real Goods, 3041 Guidiville Road, Ukiah, CA 95482. Alternative energy products including all kinds of solar products.

Books on how to do everything

Lindsay Publications, Inc.
P.O. Box 12
Bradley IL 60915-0012
(815) 468-3668

III
MEDICAL CARE

Survival medical care, as interpreted here, consists of first aid (Chapter 17) and use of herbs (Chapter 18). These two traditional aspects of self-help medical care do not really cover the same ground, although they do overlap. And they are potentially very important in a survival setting.

First Aid

One important aspect of reasonable preparedness is first aid. And adequate and proper first aid knowledge is probably the foremost feature of first aid preparedness. Although a good first aid kit and step-by-step procedure manuals are essentials, the more you already know about first aid and caring for the sick and injured the better. The easiest means to gain such experiences is through local first aid classes. Community first aid classes are becoming increasingly popular and are offered at minimal costs, and American Red Cross classes are likewise inexpensive and widely available. For further information call your nearby American Red Cross office or your local hospital.

After the immediate dangers of any medical emergency have been dealt with, always seek out the best medical help available.

MANUALS

A first aid manual may be simple or complex but will probably only be as good as your ability to use it. Purchase first aid manuals that cover most emergencies that could happen. First aid, according to the American Red Cross, is "immediate and temporary" care of the injured or suddenly ill. In an extended or large-scale difficulty such as a nuclear war, the basic procedure may, of necessity, have to suffice for longer than minimal time to get the victim to competent medical practitioners. Such medical services may be hard to find and may also be very busy when found. For this and other reasons it may be advisable to obtain more comprehensive first aid information, such as that found in the references listed at the end of this chapter. I highly recommend a careful look at this list of references. Remember, a book is like a tool; it is most useful when you know how to use it.

PROCEDURES

In any emergency rapid, calm, efficient efforts can minimize problems; and even in prolonged emergency situations, sticking with standard first aid care may be better than risking life-threatening procedures. Some very basic procedures are included here to help extend first aid to those in need when the following conditions exist.
- need for immediate lifesaving measures
- shock
- need for cardiopulmonary resuscitation (CPR) and artificial respiration
- choking (obstruction in the airway)
- bleeding
- poisoning (external, internal)
- psychological trauma
- bites

- dental pain
- sprains
- injuries due to heat and cold
- skin injuries (lacerations and blisters)
- fractures
- diarrhea
- emergency childbirth

Following are procedures for dealing with them.

Immediate Lifesaving Measures

Most injuries can be dealt with calmly and without hurry. In serious life-threatening injury, however, certain steps must be taken immediately to preserve life. *First,* open the victim's airway and restore his breathing and heartbeat if necessary (see Cardiopulmonary Resuscitation—CPR). *Next,* stop any bleeding (see Bleeding) and dress and bandage wounds to prevent infection. *Third,* treat the victim for poisoning; and *Fourth,* treat him for shock.

Shock

In any first aid emergency—TREAT FOR SHOCK! Shock may be immediate or delayed and is a life-threatening illness that can be caused by almost any traumatic injury—burns, fractures, hyperthermia, and so on. Injuries involving large fluid loss such as bleeding and burns are common causes of shock. Shock is a depression of the action of the nervous system and its control over body functions such as circulation and respiration and is characterized by weakness, rapid and weak pulse, paleness, and cool perspiration in the victim. The pupils of the eyes may be dilated and at the extreme the victim may also become incoherent.

Reassure and comfort the victim and have him/her lie down. Treat the causes of the shock (burns, fractures, bleeding, or whatever). Maintain normal body temperature. (Warm a victim with a depressed body temperature; cool a victim with an elevated temperature.) If no head injury is present, elevate the legs. Call for emergency help.

Mild fluids may be given if medical assistance is not readily available, as may be the case in an extended emergency. A saline solution made by mixing one teaspoon of salt and one-half teaspoon of baking soda in a quart of water may be used. If abdominal injuries are present, do not give fluids. If there is any question of the victim losing consciousness, do not give fluid because the victim may regurgitate and aspirate the vomitus.

Cardiopulmonary Resuscitation (CPR) and Artificial Respiration

CPR is a combination of artificial respiration and artificial circulation by means of external cardiac compression.

Artificial Respiration   Mouth-to-mouth is the preferred and normally most effective method of artificial respiration. If a victim is not breathing or his breathing is so shallow that it will not sustain life, begin artificial respiration immediately. Shake the victim to determine consciousness (obviously, use caution if fractures or other wounds are apparent or suspected) and then listen for signs of breathing. If the victim's heart has stopped beating, external cardiac compressions must also begin.

To begin artificial respiration, quickly remove any foreign material from the victim's mouth and tilt his head back by gently lifting the neck. Pinch the victim's nostrils together with one hand while continuing to hold the neck up with the other hand to maintain the airway, and give four rapid, successive breaths by placing your mouth over the victim's mouth.

Check the victim's carotid pulse to see if his heart is beating by placing your index and middle fingers at the side of the Adam's apple (larynx) between the muscles of the neck and the trachea. If no pulse is detectable, begin CPR immediately. If pulse is present, continue artificial respiration by breathing into the victim's mouth once about every five seconds. Check the carotid pulse periodically while giving artificial respiration.

The victim's chest should rise with each breath. If the air goes to his stomach (as seen by the stomach rising instead of the chest) turn the victim onto his side and press on his abdomen to push the air out. Turning the victim to the side should prevent the inhalation of any regurgitated matter into the lungs. If the victim regurgitates, clean the matter out of the victim's mouth and continue giving him artificial respiration. A drowning victim will almost always vomit as air replaces water in the lungs.

If the victim is a baby or young child, place your mouth over the nose and mouth rather than pinching the nostrils. Use puffs of air on an infant or young child rather than large breaths, and deliver them at a slightly faster rate (every three to four seconds).

Artificial respiration may be given mouth-to-nose if the victim's mouth is severely injured. Also, if artificial respiration is necessary for a person with a stoma (an opening in the neck to facilitate breathing) just blow into the stoma. If the stoma is open to the mouth and nose, as some are, it may be necessary to close off the nose and mouth with a free hand while using the stoma for respiration. Do not stop giving artificial respiration until the victim can breathe for himself or until he is pronounced dead by a physician.

Cardiac Compression   When the victim has no pulse (check the pulse as described above), artificial circulation must also be provided by the rescuer(s) without delay. The victim must be on a hard surface, and his legs may be elevated eight to ten inches if this can be done without injuring him further or delaying the administration of CPR.

Figure 17.1   CPR

Cardiopulmonary resuscitation (CPR) is a combination of cardiac compression and artificial respiration. Both of these skills should be practiced in a first aid class.

If you are alone, kneel by the victim's side and place the heel of your hand on the center of the chest one and one-half to two inches above the notch of the victim's sternum. Place your other hand on top of the first hand and, with arms straight and your shoulders directly above your hands, begin compressing the victim's chest one and one-half to two inches at the rate of about eighty times per minute. Keep your elbows straight and your fingers off the victim's chest; press only with the heel of your hand. Every fifteen compressions, stop and give two quick breaths to provide artificial respiration, then resume compressions.

If two rescuers are available, one should give artificial respiration while the other gives cardiac compressions. The compressions should be performed by one rescuer at a rate of sixty per minute, with a breath of artificial respiration given after every five compressions by the other rescuer. Recommended procedure is for the one doing compressions to count out loud, "One, one thousand; two, one thousand; three, one thousand; four, one thousand; five, breathe"—at which time the other rescuer gives the victim a breath before the next compression begins. The process is continuously repeated. The compressions are given on the numbers, and hence the breath is given between compressions after every fifth one.

CPR should not be interrupted for longer than five seconds.

In injuries where the heart or chest have been severely damaged, CPR may not be reasonable to pursue. Artificial respiration may, however, be possible in these cases. Obviously, the most competent medical help available should be sought in such serious difficulties.

CPR is not easy, and it would be difficult to perform it properly after only reading about it. Instructions here are only meant to renew what has already been learned. CPR should be learned under competent supervision with hands-on experience gained in the learning. Take a class on CPR!

Choking (Obstruction in the Airway)

When a person indicates an inability to speak or breathe, (the standard signal to alert others that you are choking is to wrap your throat with your fingers) it is likely that his airway is blocked. Airway blockage with food is common. If this occurs, ask the victim if he can speak. If he can, allow him to cough or gag on his own. If he cannot speak, stand behind the victim and wrap your arms around his chest. Place one closed fist in the victim's abdomen between the naval and the bottom of the ribcage and grab the fist with the other hand. Give a quick forceful thrust inward and upward into the victim's abdomen. Be careful of the placement of your fist in the victim's abdomen to avoid breaking the xyphoid tip at the bottom of the sternum (breastbone). If the victim is lying down, place him on his back and straddle his hips. Place the heel of your hand in the victim's abdomen, as described above, place your other hand on top of the first, and administer the inward and upward thrusts to the abdomen of the victim. In the case of a child, the force used should be reduced. In the case of an infant, place him face up on your extended upturned forearm and carefully compress the chest with the other hand. The chest compression can promote the forceful rush of air to clear the trachea of the infant without damaging the internal organs. In any case do not cease efforts to help; repeat the thrusts until they are successful. If the victim collapses to the floor, four thrusts in the abdomen may be alternated with four blows to the back followed by sweeping the mouth with a finger to remove any dislodged object.

If you are alone and find yourself choking, you can apply the thrusts to yourself by pressing the appropriate spot on your abdomen quickly and firmly against the arm of a chair or some other appropriate object. Repeat the thrust several times if necessary. Do not cease efforts until the obstruction is cleared.

Once the obstruction is dislodged, always seek medical help; the esophogus often swells following such trauma.

The above are descriptions and/or variations of what is sometimes refered to as the Heimlich Maneuver.

Bleeding

Extensive bleeding can cause death if not stopped promptly. External bleeding can be stopped by direct pressure; pressure at an appropriate location on the supplying artery; and, as a drastic last effort, by a tourniquet.

Direct Pressure   Direct pressure is the first step in controlling bleeding, and is applied by pressing a sanitary dressing directly to the wound. If there is no dressing available, use the bare hand. If blood soaks through a dressing *do not remove it;* add another dressing on top and continue the pressure.

Digital Pressure to Artery   Digital pressure to the artery supplying blood to the wounded area may be used in addition to the direct pressure if the direct pressure does not stop the bleeding. While continuing the direct pressure as described above, choose the pressure point between the heart and the wound that lies closest to the wound. There is particular need to know the brachial and femoral artery pressure points because of the frequency of injury to arms and legs. The purpose of using a pressure point is to press the artery between the fingers of the first aider and the victim's bone, thus slowing the flow of blood to the injured area. When the pressure point is being effectively applied the first aider can almost always feel the pulse. See figure 17.2 for the locations of the pressure points. A wounded limb may also be elevated to help reduce bleeding.

A nosebleed can usually be treated effectively by having the victim sit upright in a comfortable position and then squeezing the nostrils together. The pressure should be applied equally to both sides of the nose and should be hard enough to stop bleeding out of the nostrils or down the back of the throat. Continue the pressure for ten to fifteen minutes and then gradually release it.

Tourniquet   A tourniquet should be used only in extreme cases when direct pressure and pressure on the appropriate pressure point have failed to stop the bleeding and the victim's life is in danger, or in the case of traumatic amputation. The use of a tourniquet will very likely lead to the loss of the limb to which it is applied if it is left on for longer than a few minutes. Once a tourniquet has been applied it should not be removed or loosened until done by a physician.

If you do not have a specially designed tourniquet you can make one with any soft, strong, pliable material such as cloth or gauze. The band of material should be about two inches wide or wide enough so that it will remain at least one inch wide after it is tightened. A stick or other rigid material is needed to tighten the tourniquet. Place the band around the limb slightly above the wound (two inches to four inches). Tie a knot in the band, leaving it loose enough to put a stick into it. Insert the stick under the band and twist until just sufficient pressure to stop the bleeding is applied. Secure the end of the stick to the victim so that it will not come loose. Record the time the tourniquet was applied and seek medical assistance immediately. Again, never use a tourniquet unless life is threatened!

Treat a victim of severe bleeding for shock.

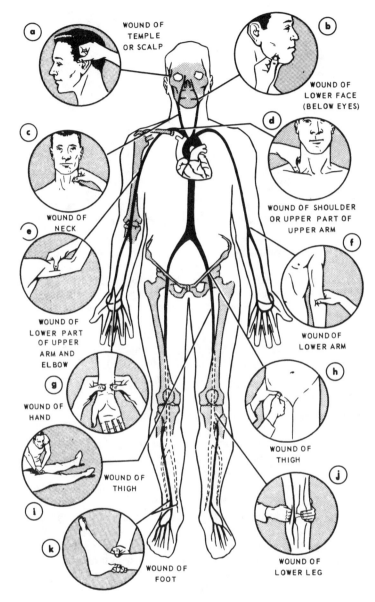

Figure 17.2   Pressure Points

Poisoning

In case of poisoning, either internal or external, if at all possible, call Poison Control before you do anything else. (Look up the number of your local Poison Control Center.)

External Poisoning   External poisoning due to contact with poison oak, poison

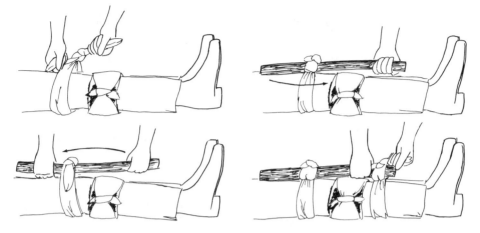

Figure 17.3   Application of A Tourniquet

A tourniquet should be applied only when a victim's life is in danger and after all other properly applied efforts to stop bleeding have been unsuccessful. There is a great probability that the limb will have to be amputated after a tourniquet has been applied for even a relatively short time.

ivy, or poison sumac can be painful. First symptoms may be the formation of small blisters; redness; itching; rash; and even fever, headache, and general discomfort. To treat these, remove the victim's contaminated clothing and gently cleanse the infected area, using copious amounts of water. Apply calamine lotion or 1% dibucaine ointment for topical relief.

Internal Poisoning   Internal poisoning can come from any of many sources. For general poisoning from ingested material, including poisonous plants and most other poisons except as noted below, dilute the poison by drinking one to two glasses of water and promote emesis (vomiting) with syrup of ipecac (about one-half ounce). If the initial dose does not induce vomiting, repeat the dose in fifteen or twenty minutes. (Vomiting may be promoted in some people by gagging with a finger or other reasonable instrument such as a spoon.)

In the case of poisoning by acid, such as phehol or hydrochloric acids, or alkali, such as lye or ammonia, the victim should be given one glass of milk or more if it can be given without promoting vomiting. Do not induce vomiting since these substances corrode body tissues with every contact and would cause as much damage coming up as they do going down. If milk is not available, milk of magnesia, egg white, or flour in water may be given in place of the milk to coat internal tissues.

In the case of poisoning with petroleum distillates, such as gasoline or kerosene, again, do not induce vomiting. Give the victim milk of magnesia, milk, or olive oil to slow down absorption of the poison. If necessary, start artificial respiration. Immediate medical help must be sought to eliminate stomach contents before brain and head toxicity occur.

Some cases of poisoning can also be treated by giving powdered charcoal which is made for the purpose. The charcoal is sometimes also given with a laxative. The charcoal absorbs some kinds of poison and the laxative can help speed elimination of the charcoal and poison. This treatment is sometimes used after the contents of the stomach have been removed by vomiting. Some substances for which charcoal is sometimes given as an antidote include asprin, arsenic compounds, aniline, atropine, cold and headache remedies, codeine, food poisoning, garden sprays, insecticides, mercury compounds, opium derivitives such as morphine, mushroom poisoning, muscle poisoning, nitrites, nitroglycerine, phenobarbital, sleeping pills, and strychnine. Give water followed by about fifty grams or about one-half cup mixed in some palatable liquid, or follow directions on the charcoal package.

Psychological Considerations

Catastrophic difficulties frequently bring about severe emotional reactions in the parties involved. These are often unpredictable but are normally only temporary.

If the person is not violent, the first thing to be done is to treat physical injuries first. A violent person may first need to be restrained to prevent injury to himself or others. Stay calm. Comfort the person; do not criticize him but on the other hand do not be overly solicitous. Avoid expressions such as "snap out of it" or "get hold of yourself." Avoid joking. Accept the person's feelings and try to reassure him. Make things as comfortable as possible. Involve the victim in meaningful but not overly taxing activity as soon as possible to help him release tensions and forget his troubles. Be patient with the disturbed. It may take a while, but they will probably return to normal. Do not give sedatives or tranquilizers; these will only delay his adjustment to the situation.

Bites and Stings

Insect bites, snake bites, and animal bites can all require first aid.

Insect bites and stings are usually from bees, scorpions, or spiders. *Bee, wasp, and hornet stings* should be treated by removing the stinger, washing with soap and water to remove loose venom, and applying a cold paste made with water and baking soda. Taking an antihistamine such as Benylin may help reduce swelling and discomfort. (If you know of a member of your family or party that is hypersensitive to any of these stings, it is good to be prepared to treat for anaphylactic shock. See *Respiratory Aids* in this chapter and consult your physician. Proper usage of such supplies include making sure when and where not to use them.)

Scorpion stings should be treated by applying ice or other available cold material to the wound. Keep the wounded area below the rest of the body. Keep the victim warm and quiet. Apply cold baking soda paste to the wound.

For *spider bites* (especially black widow bites) victims should be treated for shock; kept quiet and warm with legs slightly elevated. Heat applied to the abdomen may help relieve the abdominal cramps which frequently occur. Black widow venom in particular can drastically slow the digestive process, and victims should not ingest food until the effects of the bite have somewhat abated.

Ticks are small arachnids; flat, round, have eight legs, and are usually some shade of brown. When in infested areas, make frequent checks (at least daily) on yourself and on each other in your group. If a tick is found, apply alcohol, fingernail polish, or heavy mineral oil over the area where the tick has entered. This should make the tick back out. The oil may need to stay on as long as half an hour before the tick runs out of air. Touching the exposed end of the tick with a heated needle or the heated sharp point of a knife may also make it back out, but do not kill the tick by doing so. This only increases the difficulty in removing it. If none of these methods work, use some splinter forceps to remove the beast. Remove by pulling straight back until you observe the skin being stretched, then twist back and forth while increasing the pull until the tick pops out.

Take care not to leave the head or any other part of the tick in the skin, and avoid handling the whole tick or parts of it to reduce your exposure. Scrub the area with soap or disinfectant. Apply triple antibiotic ointment and a bandage. If an abscess forms, open it and allow the pus to drain.

A strong solution of DEET can act as a tick and chigger repellant.

Snakebites can be dangerous because of poisons and infections. Treat a poisonous bite (and treat all bites as poisonous unless you know for certain they are non-poisonous) by (1) calming the victim and restricting all physical activity to prevent the spread of the venom; (2) placing a lightly constrictive bandage two to four inches above the wound; (3) cold packing the wound area; (4) immobilizing the wound area at about the same level as the heart; (5) sending for medical help. If such help is more than three to four hours away, you may consider using a snakebite kit to make an incision through each fang mark so that the incisions are parallel and go with the "axis" (length) of the body. (Never cut across the length of the body because of possible muscle damage.) Each incision should be just through the skin and about three-eighths of an inch long. After the incisions are made, use the suction cup provided in the kit to apply suction for about twenty minutes but no longer than thirty minutes. Suction may also be done with the mouth if no suction cup is available and if there are no open sores in the mouth of the one applying the suction. If this is done by mouth, the material removed from the wound should, of course, be spit out each time suction is done. Follow the directions in the snakebite kit.

In many areas the authoritative local first aid protocol does not include the incision-and-suction procedure at all. There is some question about its effectiveness, and infections frequently develop at the site of the incisions. Another problem is that, for cosmetic reasons and because of the difficulty of making incisions on bony extremities filled with blood vessels, it would be undesirable to make incisions in the areas of the body that are most frequently bitten (hands, face, and legs) unless it were certain that the procedure would make a substantial contribution to normal recovery. And in most cases that certainty does not exist. Check on the local snakebite protocol in your area, and check for information on suggested procedures for treating bites from the kinds of snakes common to your area.

Be sure to cleanse a snakebite very thoroughly even if the bite is from a nonpoisonous snake. One of the greatest dangers from snakebites is infection. People have died from infections from nonpoisonous snakebites.

It is interesting also to note that a substantial percentage of the persons bitten by poisonous snakes are not envenomated. Still, it is wisest to treat all bites as if they were venomous. For an important update on snakebite please see page 303.

Animal bites should be cleansed immediately with water (running water if possible) and then scrubbed with soap and water and/or a disinfecting cleanser such as phisoderm or Hibiclens to disinfect the area. After about twenty-four hours, (some recommend doing it immediately), the wound may be soaked in warm water or warm compresses or a poultice may be applied to help stimulate drainage four times daily.

Rabies can be contracted from the bites of rabid animals of several species, including dogs, coyotes, foxes, skunks, squirrels, wildcats, and some species of bats. Even the blood or saliva of an infected animal can transmit the disease, as can the environs of an infected cave. If bitten by a rabid animal, you must receive medical help immediately. Unattended rabies victims generally die. If it is possible to safely retrieve the body of the rabid animal for tests, do so, and take it with you to the doctor. Avoid handling it directly.

Dental Pain

An infected tooth certainly makes its presence felt, but locating it is sometimes difficult. Have the victim bite on a piece of folded paper or small stick one tooth at a time or tap each one lightly until the bad one is found. Dry the tooth and clean any loose filling or other foreign material from it. Place a drop or two of oil of cloves (eugenol) in the cavity and give the patient an appropriate pain medication. Some kits are available that provide temporary filling material. Zinc oxide powder can be made into a stiff paste with oil of cloves and used to fill a dry cavity for a temporary filling.

Gum ailments can often be treated by removing food particles, cleaning, applying heat, and flossing. Irritated wisdom teeth can be temporarily treated by cleaning, flossing, and swabbing with salt water to relieve added irritation from food particles.

Sprains

A sprained joint may be accompanied by swelling and discoloration. If it is possible, it should be determined if there are any fractures present; if so, the joint should be immobilized. Elevate the sprained joint and apply cold packs to reduce the swelling for the first half-hour or so and intermittently for twenty-four to forty-eight hours. Excessive swelling can cause considerable damage. After this period, heat may help relieve pain and promote healing, although most recent studies show that cold applications do more good. Do not rush into strenuous activity with a sprained joint—baby it along. (A first aid axiom for sprains is ICE: apply Ice, Compress the sprain with an elastic bandage but maintain circulation, and Elevate it.)

Temperature Injuries (Heat and Cold)

Extremes of temperature create some notable first aid problems: burns, hypothermia, hyperthermia, and frostbite.

Burns   Thermal burns can be characterized according to severity. There are first-, second-, and third-degree burns. The main concept for immediate treatment is to cool it. Immerse the burned area in cold water for twenty to forty minutes. If immersion is not possible, then use cold compresses. Rings should be removed if any substantial amount of body area has been burned or if the fingers, hands, or arms are burned. The rings act as a tourniquet as the body swells. The same is true for other tight-fitting items of clothing that cannot be loosened or easily removed if swelling continues, such as cowboy boots.

First-degree burns produce redness, pain, and some swelling; sunburn is a good example. The pain of first-degree burns may be treated topically with aloe vera juice or gel, 1% dibucaine ointment, or benzocaine.

Second-degree burns are as first-degree burns but also form blisters. To treat first- and second-degree burns that do not cover a large portion of the body, flush the area with or immerse it in cold water as noted above to completely cool the burned area. Be very careful not to contaminate any blister openings with unsterile water. Gently dry any blistered areas and apply a dry sterile dressing. Leave intact blisters alone but remove loose skin. Apply antibiotic ointment (Polysporin is preferred for this) and non-stick gauze to areas where the skin is open in second-degree burns. Change the dressing every two days. Healing time should be ten to twenty days. Crusting may be soaked loose after about five days.

Any burn that is larger than the victim's hand or burns in critical areas such as the face, diaper area, or hands should be treated by competent medical help if available.

Those who are burned over large portions of limbs or trunk should be treated for shock.

Third-degree burns involve the destruction of the skin, including discoloration and charring. In many cases, third-degree burns may not be as immediately painful as second-degree burns because of the destruction of nerve tissue.

Third-degree burns should be treated by cutting any clothing away from the burned area (do not attempt to pull it away) and covering the burned area with a thick, dry, sterile cloth dressing. Do not allow adjacent burned body surfaces, such as fingers or groin area to touch each other without a dressing on each surface. Otherwise the surfaces may stick together. It is also a good idea to keep the burned area elevated. Remove all watches, rings and other metallics from burned areas because they retain heat. A third-degree burn the size of an adult thumbnail or larger is likely to require skin grafting for proper healing.

A victim of third-degree burns or extensive first- and second-degree burns should be treated for shock. This includes the restoration of body fluids. Use a solution of one-half teaspoon of salt and one-fourth teaspoon of baking soda in one quart of water. The water may be sweetened and flavored to promote palatability, but it is a good idea to avoid potassium-rich fruit juices such as orange or apple during the first day or two. Some commercial preparations are made for the restoration of body fluids. Allow the patient to drink all the fluid possible.

Hyperthermia   High temperatures and/or high effort can lead to heat exhaustion or heat stroke.

Heat exhaustion is shock caused by heat and heavy fluid loss. Symptoms include weakness, nausea, excessive sweating, dizziness, headaches, and muscle cramps. Treat for shock: have the victim lie down in a cool area, normalize his body temperature by fanning him or covering him if he becomes cold, loosen his clothing, slightly elevate his feet, and give him copious amounts of water. The water given may contain one-half to one teaspoon of salt per quart. Do not give stimulant fluids such as coffee or tea.

Heatstroke (sunstroke) occurs when the processes that regulate body temperature (such as sweating) completely break down. The victim will stop sweating (making the skin feel hot and dry), and may begin with headache, dizziness, nausea, and mental confusion before collapsing and possibly losing consciousness. This constitutes an *extreme life-threatening emergency*. The victim's body temperature must be immediately lowered or death or permanent injury will occur. Move the victim into the shade and immerse him in cold water if possible. If this is not possible, remove all heavy clothing and pour water on the victim or apply water with cloths to promote cooling. Fan the victim. Once the victim's body temperature is lowered, medical care should be sought. The temperature may rise again or continue to be stable, but the person will be ill and unable to perform normally for an unpredictable length of time.

Hypothermia   Hypothermia is the condition of having a depressed body temperature. Many have died from the effects of hypothermia simply because of the subtlety of its onset. At a body temperature of about 96 degrees the victim experiences shivering and decreased dexterity. At 86 degrees his skin turns blue and his muscular responses and mental faculties become seriously degraded. At 82 degrees he exhibits mental incoherence and depressed pulse and respiration, and he finds muscular activity nearly impossible. Death is imminent. At about 78 degrees or above death results.

Beware of wet and cold together. Much of the subtlety of hypothermia is that it can occur even when it does not seem "that cold." Be aware also that eating snow or ice lowers the body temperature.

A victim of moderate hypothermia should be sheltered and warmed as quickly as possible. Placing him in a tub of very warm water (about 110 degrees farenheit) is suitable. If there is no warm water, replace wet clothes with dry clothes and warm him with a suitable source of heat (which may include warm liquids). If these are not possible, it is recommended first aid practice to place the dry victim in a warm, dry sleeping bag with a dry rescuer. *A victim alone in a sleeping bag will not warm up unless there is a good source of heat.*

A victim of severe hypothermia who is in shock can actually be worsened by warming the skin such as by placing him in warm water as described above. The *center* of the body must be warmed. If the victim is unconscious this can only be done with medical equipment.

If none of the above are available, victims should huddle together, build a fire, do calesthenics or isometrics to increase blood flow, drink warm liquids (*not* alcohol—this eventually acts as a depressant on the system), stay sheltered as much as possible, and/or do anything else that will safely increase body temperature. Sleep should be avoided at all costs until body temperature is restored to normal.

Frostbite   Frozen skin and subcutaneous tissues are frostbite. In frostbite, ice crystals formed in the interstitial fluids. If rubbed, these crystals rupture cell membranes. If these crystals form, thaw, and refreeze, they recrystalize in larger form and cause greater tissue damage. Frostbitten skin becomes white, stiff, and numb. Members of a group should watch each other for signs of frostbite.

Frostbite should be treated by rapidly thawing the frostbitten area. If only the skin is frostbitten, it may be warmed by placing warm skin next to it—for example, placing a warm bare hand on a frostbitten cheek or a frostbitten hand in the opposite armpit or a frostbitten foot on the warm abdomen of a compassionate companion.

Deep frostbite should be treated by thawing the affected area in warm water (about 110 degrees farenheit) only until the area is pink or red. The process is painful. If it is necessary for the victim to walk to a safe area, do not attempt to thaw frozen feet until the walk is completed. Once a deeply frozen foot is thawed, the victim will be unable to walk. In severe cases, it is better not to cut away dead or blistered areas (some of which will turn black), but to allow these areas to drop off on their own. It may take several months for deeply frostbitten parts to completely heal.

Skin Injury

The skin is the most frequently injured part of the body. Lacerations, blisters, and scrapes are all common.

Lacerations   Stop the bleeding from *lacerations* by direct pressure. After bleeding has stopped, the area should be scrubbed clean and any foreign matter removed. Use Hibiclens Surgical Scrub, Phisoderm, or a povidone-iodine prep pad to scrub and disinfect the area. The scrubbing will probably restart the bleeding, which must be stopped again as before. Dry the skin, push the edges together, and while holding them together apply butterfly bandages (see figure 17.4). If you do not have any commercially produced butterfly bandages or Steri-Strips, make some with adhesive tape by pinching the adhesive in the center of the strips together from the sides and then applying them to the wound. This will avoid tape contact on the wound as described. A

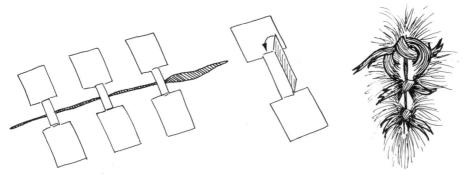

Figure 17.4   Butterfly Bandage   Figure 17.5   Closing Scalp Wounds With Hair

dry dressing should be applied and kept dry. A light coating of triple antibiotic ointment should also be applied directly to the wound to help prevent infection. Scalp wounds can often be closed by tying locks of hair from opposite sides of the wound together (see figure 17.5).

If a laceration is too large to be held together with butterfly bandages the wound should be sutured (stitched). Treat this carefully; you might have greater problems from infection arising from the materials or instruments than from the wound itself. And, even if you let the wound go, remember that even the worst scar can be dealt with later. The only reason for a layman to suture is if there is no other way to keep the wound closed. For limbs and trunk a good suture material to use is 3-0 Ethilon with a curved FS-1 needle. Use four stitches per inch. For wounds of the face a 5-0 Ethilon suture with an FS-2 needle can be used (six stitches per inch). For wounds inside the mouth a recommended material is 3-0 plain gut suture with an FS-2 needle. If a ruptured blood vessel should need to be tied off in a wound, clamp the vessel with the needle holder or a hemostat and tie it off with the plain gut material, using a square knot.

The hands of the one doing the suturing, the wound area, and the instruments (needle holder) must be thoroughly cleansed with disinfectant. Use either Hibiclens surgical scrub or povidone-iodine. If sterile gloves are available, scrub hands thoroughly with soap and then use the gloves.

Needles and suture material come packed in sterile packages. They, the needle holders, and the disinfectants mentioned are all available without prescription from mail-order houses and/or first aid supply houses. If stitching is done soon after an injury, while the area is still numb from trauma, it will probably not be much more painful than are the injections which are used to deaden wounds before suturing. The standard 2% Xylocaine used for deadening and a syringe for its application are prescription items.

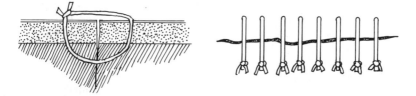

Figure 17.6   Proper Placement of Sutures

Stitches must be placed evenly to prevent excessive scarring and puckering of the skin (see figure 17.6). Figures 17.7 a-f illustrate the process. Hold the needle in the holder so that you can twist your wrist to perforate the skin on one side of the wound and push the needle steadily through to the opposite side and up and out. Pull the suture material through until only about an inch remains on the side of the original perforation. (a) Hold the needle with the free hand, loop the long needle-end around the holder, and take hold of the short end with the holder. (b) Pull the wound closed by pulling the long end and tightening the knot. (c) Repeat this process by looping the long end around the needle holder in the opposite direction from the first loop and then grabbing the short end and pulling tight (d, e and f). Repeat once more with a loop made in the same manner as the first one. If a good square knot does not result, do not worry: any secure knot should work just fine.

It is normal for bleeding to occur where the needle perforates the skin. Use direct pressure to stop the bleeding.

If infection develops, get medical help if it is available. Otherwise, soak the area with hot wet packs for a few minutes at a time several times a day. A sanitary poultice may also be applied. Abscesses may have to be perforated and gently relieved of their contents. Do not open an abscess, however, until it has come to a head and the head softens. Do not squeeze an abscess or boil.

Blisters   A friction blister should be cleansed with a disinfectant. A blister may be left intact until it breaks on its own or may be punctured at its outside edge with a sharp sterile instrument. Then apply a light coating of triple antibiotic ointment and cover it with a sterile dressing. Use tape, plastic strips, or moleskin to protect the blistered area and take the pressure off the area.

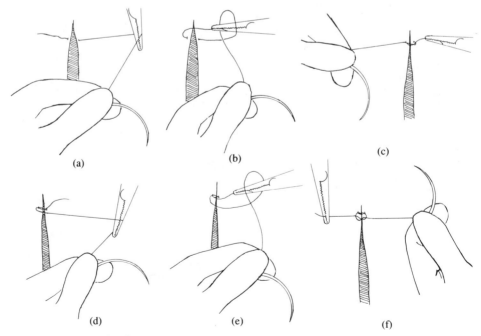

(a)   (b)   (c)   (d)   (e)   (f)

Figure 17.7   Suturing Steps

Scrapes   Wash scrapes with mild soap and gauze pad. Then treat it much the same as a second-degree burn: apply antibiotic ointment and a non-stick gauze pad. Change dressing every other day. If the sore is too painful to wash with soap and water, spray with an anesthetic spray, such as Bactine or Solarcaine, and then proceed with other care.

An alternative is to wash the sore and allow it to air dry. This may produce slightly worse scarring and longer healing time.

Fractures

Bone fractures must be cared for properly to prevent additional damage to the broken bones and surrounding tissues and to prevent infection. Fractures are classified as open or closed. An open, or compound, fracture is characterized by pieces of bone protruding through the skin (figure 17.8). In a closed, or simple, fracture the bone may or may not be deformed but it is not protruding (figure 17.9).

The general rule is to immobilize fractures where they are until medical treatment can be obtained; and, if it is not certain that the bone is fractured, treat as though it is. This can be done with any easily portable, rigid device and wrappings, using plenty of padding, and should include immobilization of the joints above and below the fracture. Wire splints and pneumatic splints (which are blown up like a balloon) are widely available from first aid suppliers. If the break is open and medical help is readily available, immobilize the break and obtain help. If no medical help is available for one or two days, the open wound should be gently and carefully cleaned with one of the standard disinfectants or, if nothing else is available, rinsed by pouring over it a solution of one teaspoon of salt in a quart of water. The protruding fragments or bone ends may be more correctly aligned by applying traction to compound limb fractures. Carefully pull both sides of the break (above and below), allowing the bone ends to slip below the surface of the skin. Cleanse the wound and apply triple antibiotic ointment. Do not attempt to close the puncture site (with sutures or butterfly bandages); it should be left open and allowed to heal.

If a limb is deformed and the circulation drastically restricted, as seen by a blueness of the limb and lack of pulse at pressure points below the break, then it may also be necessary to straighten the limb before splinting it to restore circulation. Apply slight traction, as described above, while straightening the deformity. There is danger in this because of the possibility of further damaging tissues with the sharp bone fragments, so proceed cautiously. (The victim's yells at this point would probably induce due caution.)

Special caution should be used in caring for possible neck and back fractures. Any rapid or improper movement can cause permanent damage or even death. Do not attempt traction or any other remedy; seek competent medical care.

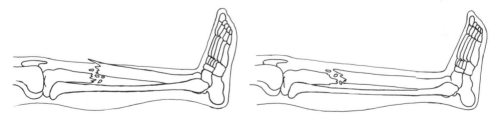

Figure 17.8   Open Fracture   **Figure 17.9   Closed Fracture**

Immobilizing and elevating fractures can help relieve pain. This helps reduce swelling, which can cause pain and additional damage. For pain in fractures, Percogesic is recommended because it also contains a muscle relaxant.

Diarrhea

In some cases diarrhea can be a serious malady. When it occurs, the victim should stop eating solid food and start a diet of clear fluids, such as broths, jello water, juices, and so on. Milk should be avoided. Fluids could also include a sugared salt solution made with one and one-half tablespoons of sugar and one teaspoon of salt in a quart of water. Kaopectate also helps. While the diarrhea continues, pay particular attention to sanitation and hygiene to prevent spread of the disease.

An expedient remedy for diarrhea is small amounts of a tea made of hardwood bark boiled in water for one to two hours.

Emergency Childbirth

Some things just do not necessarily wait for convenience, and childbirth is certainly one of them. The best advice on obstetrical emergencies is to avoid them if at all possible. Try to have a trained medical practitioner of the highest order you can find present, but if this is not possible then you must do what you can to help. Delivery can be normal or it can have complications.

Normal Delivery   The onset of childbirth is normally signaled by the beginning of labor or the contracting of the uterus (womb). This process gradually forces the cervix (opening of the uterus) to open wide enough for the baby to be pushed out. The appearance of a watery, bloodstained discharge occurs with the beginning of serious labor; and—usually later but sometimes even before labor begins—the ''water breaks'' when the sack of water surrounding the baby breaks and discharges a pint or more of clear fluid.

The mother should relax during early labor and take large breaths during contractions to try to minimize them. As the contractions grow more intense and closer together (every two to three minutes), the mother will have the uncontrollable urge to push or ''bear down,'' much like in a bowel movement. She should not work too hard but allow the process to continue naturally. The intensity of the contractions can be reduced if the mother pants (quick breaths) during the contractions. This aids in making a smooth, gentle delivery and is easier on the infant.

When serious labor begins the mother's inner thighs, vaginal area, and rectal area should, if possible, be sponged or cleaned with warm soapy water. No water or other material should be allowed to enter the vagina. Wash from front to back. If possible, place a sterile sheet under the mother and sterile towels on her thighs. The person attending in the birth should scrub his hands continuously for several minutes with soap and water and if possible wear sterile gloves.

In addition, preparations should be made to receive the baby. If no hospital or medical care is available within an hour or so, preparations should be made to cut the umbilical cord with sterile ties and sterile blade or scissors. If medical care will be available shortly, then it is normally not necessary to cut the cord. Suture material may be used to tie the cord, or cloth or tape made sterile by soaking it in 70 percent isopropyl alcohol (common rubbing alcohol) for fifteen to twenty minutes or by boiling it in water for five to ten minutes. A blanket, diaper, shirt, and warm place should be prepared for the delivery. If there is the remotest possibility that there could be a mixup of which baby belongs to whom (as with more than one birth occurring at the same location), then make identifying tags with adhesive tape and attach them to the baby's ankle and the mother's wrist.

When the birth is imminent the mother should assume the supine or a semi-sitting position with legs spread apart. When the infant's head begins showing, the mother should again try to slow the delivery process by panting. This will help her avoid tearing. Talk to the mother; let her know what you are doing and try to gain her confidence.

The person attending should support the baby's head as it comes out and then support the rest of the baby as it is delivered. Do not pull the baby out in any way.

As the baby's head is delivered it will normally rotate to the side to allow the body to assume better position for delivery. Support the baby's head with both hands while this takes place. If the baby's face is still covered with membrane break the membrane and pull it away so the baby can breathe. Continue to support the baby's head and body as it is delivered.

As soon as the infant is delivered hold it on your forearm with its head lower than its feet and clean out its airway with a suction bulb. Hold the baby at the level of the mother or below until the cord stops pulsating.

If the infant does not cry spontaneously rub the feet or rub and pat the back. If the baby does not cry within 2 to 3 minutes use gentle artificial respiration.

After the cord stops pulsating, it may be tied and cut. Make the first tie about 4 inches from the baby and the other about 2 inches farther away (closer to the mother). Use sterile material and secure square knots to make the ties. Then cut between the ties.

At this point the baby may be put to the mother's breast to suckle. This will help promote the contracting of the mother's uterus.

The afterbirth or placenta should deliver within a few minutes after the baby is born. Do not pull on the cord to hasten the process. Gently massage the mother's abdomen to help the uterus contract and expel the afterbirth. The massaging of the mother's abdomen should continue every few minutes for about an hour after the placenta has been delivered.

The baby should be dried and kept warm. There is no hurry to clean off the waxy material (vernix) that covers the baby's skin. If an area of properly controlled temperature is not available, the baby can be kept warm by keeping it next to the mother.

Delivery With Complications   Before and during delivery there are infrequently complications to pregnancy and childbirth. The serious complications need competent medical help (a doctor) to deal with them. Some of the possible complications are mentioned here to help make it known when help is needed.

SUPINE HYPOTENSIVE SYNDROME.   If a woman who is near term complains of dizziness when lying on her back, the heavy fetus and placenta may be pressing on her inferior vena cava, thereby slowing blood circulation. This may be especially pronounced if the woman is low in blood volume. The woman should lie on her side to prevent further hypotension. Be alert to other problems if these symptoms occur; there is also a possibility of internal hemorrhage.

ECTOPIC PREGNANCY.   In rare cases the fertilized egg has implanted outside the uterus in the abdominal cavity, in a fallopian tube, on an ovary, on the outside wall of the uterus, or even outside the cervix. Some of the symptoms are internal bleeding, spotting, weakness, elevated pulse, unusual pains, and shock. If it becomes obvious or suspected that one of these conditions exist, treat the mother for shock and obtain medical assistance.

PLACENTA PREVIA.   Occasionally the placenta is attached to the lower part of the uterus and grows in a position such that it covers or partially covers the opening that the baby must pass through for normal delivery. This condition is normally accompanied by painless vaginal bleeding during the later stages of the pregnancy with the attendant decrease in the mother's health. These are emergencies endangering the lives of both the mother and baby and require expert medical assistance. If the placenta tears during labor or before, the circulation between mother and baby is impaired or stopped and the mother could bleed to death. If this occurs, treat the mother for shock and get expert help immediately. If the placenta has grown over the opening, it should be diagnosed through competent medical care prior to labor and delivery and steps taken to protect the mother and child. However, no vaginal exam should be done while bleeding is occurring as this would increase the hemorrhaging.

SPONTANEOUS ABORTION.   Some pregnancies end prematurely in a spontaneous abortion of the fetus. Usually this is nature's method of removing an unviable growth. Treat the mother for shock and obtain medical assistance. Save the tissue to be examined by a doctor to aid in determining if there is any undelivered tissue.

PREMATURE BIRTH.   A premature but normal delivery is common and is not the same as spontaneous abortion. With proper care, a premature infant born after about the sixth month of pregnancy usually stands a good chance of living. Extreme care must be taken with a premature infant to avoid disease and to keep its body temperature at the proper level.

STILLBORN BABY.   Sometimes an infant is born dead. Such a fetus usually has large blisters all over the body, has a very soft head, and will smell rotten. No efforts to revive such a fetus need be made. Reassure and comfort the mother and treat her with normal post-delivery care.

BLEEDING.   Some amount of *bleeding* is normal during childbirth (about a total of one-half to one cup of blood). Gently massaging the mother's abdomen and allowing the newborn infant to suckle at the mother's breast help accelerate the mother's natural protective responses. (Any significant bleeding before the baby is out is a danger signal.) In cases of postpartum hemorrhage (severe bleeding after delivery), the uterus may have ruptured or there may be other causes. A ruptured uterus is indicated by a sudden relaxation of the uterus during a strong contraction. In these cases, elevate the mother's hips, treat her for shock, and get help!

Abruptive placenta may also be a cause of bleeding. This is a condition created when the placenta prematurely separates from the uterine wall, causing severe abdominal pain, rigid abdomen, and bleeding. Treat the mother by keeping her warm, elevating her legs, and obtaining medical help. There is an immediate risk to the baby.

BREECH BIRTH.   Occasionally the baby will be positioned buttocks first (breech) instead of head first. In these cases, allow the baby's buttocks and trunk to be expelled to the umbilical cord while supporting the baby at the back. If the head takes longer than two to three minutes to deliver after the body has entered the birth canal, it is necessary to prevent the baby from suffocating if the cord is compressed. If the baby is facing down, reach into the birth canal with the middle and index finger of the same

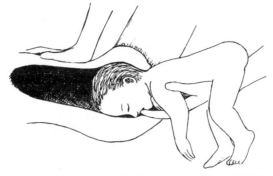

Figure 17.10   Mauriceau or Wiegand-Martin Maneuver in a Breech Birth when Head Does Not Deliver Spontaneously

Fingers of left hand inserted into infants' mouth or over mandible; right hand exerting pressure on head from above.

arm supporting the baby (see figure 17.10). While creating an air channel with the middle finger, gently open the baby's mouth with the index finger and hold the airway open until the head is delivered.

LIMB PRESENTATION.   Rarely a limb (arm or leg) is the first body part to be forced through the birth canal. This can mean that the infant is positioned so that it would be difficult or impossible for the baby to be born without changing its position. This situation is dangerous to both the mother and the infant; medical assistance is needed immediately.

CORD AROUND NECK.   Occasionally the cord is looped around the baby's neck. The dangers are compressing the cord or strangling the baby. As the baby is delivered, if it becomes obvious that this is the case, try carefully to remove the cord from around the baby's neck. If the cord is tight around the neck and removal is not possible, the cord may be tied at any two points about two or three inches apart and cut between the ties.

PROLAPSED UMBILICAL CORD.   Rarely, the umbilical cord will prolapse, or present itself first during labor. In this case elevate the mother's hips, very gently push the infant back into the mother, place a sterile towel or other dressing around the exposed cord, and immediately obtain medical assistance.

INVERTED UTERUS.   The uterus can (rarely) invert during delivery when the organ is turned inside out and expelled with the baby and/or placenta. It must then be physically replaced to its proper position. This is best done by a competent medical practitioner.

MECONIUM PASSAGE.   Occasionally a greenish fluid is expelled with the infant. This is from the baby's bowels and is a sign of stress. Clean all of this material from the baby's airway as soon as possible.

SUPPLIES

Some of the items that make up a good, basic, nonprescription first aid supply are listed below. Of course, not all items listed may be necessary; but they are given as examples to consider. A checklist of supplies is included. Prescription items may be added to your kit according to availability and need. Always anticipate your needs of special medications.

Antiseptics and Topicals

- germ-killing preparations (Hibiclens, Phisoderm, povidone-iodine prep pads—such as Betadine, alcohol, hydrogen peroxide)
- triple antibiotic ointment (such as Neosporin)
- Aloe-Gel (for minor burns and skin irritations)
- calamine lotion (for itching, contact with poison plants, etc.)
- dibucaine ointment (1%) (such as nupercainal—a topical anesthetic which will give temporary relief from minor burns, bites, itching, and hemorrhoids)
- antifungal cream (a 1% tolnaftate such as Tinactin or Desenex is very effective)
- skin care items such as hand lotion, something for chapped lips, and zinc oxide cream for baby's diaper rash

Gastrointestinal Upset Aids

- antacid (such as Mylanta or Di-Gel)
- laxative (such as Bisacodyl)
- something to check diarrhea (such as Bacid capsules—which contain lactobacillus acidophilus to restore a friendly gastric culture and a medium to hold it—or Kaopectate)
- enema
- meclizine (to control vomiting)

Dressings

- plastic strip bandages
- rolled gauze (two or three inches wide)
- multi-ply gauze pads (sterile four-inch-square pads—feminine napkins, and disposable diapers also make good large dressings)
- non-stick gauze pads (such as Telfa)
- elastic bandage
- tape to secure dressings (one inch adhesive and paper tapes)
- a large triangular bandage

Closure Devices

- butterfly bandages
- suturing kit (only if you feel that you can use it and that there is a strong possibility you will need it—otherwise, forget about the stitches), including needle, needle holders, and a general combination of sutures such as 3-0 ethilon (coarse), 5-0 ethilon (fine), 3-0 plain gut.

Respiratory Aids

- antihistamine, such as chlorpheniramine (come in many over-the-counter remedies—often in the form of the brand name Chlor-Trimeton or Benadryl)—for hayfever, other allergies, colds, etc.
- vasoconstrictor and decongestant, such as pseudophidrine (as in Sudafed or Actifed)—to relieve congestion
- anaphylactic shock kit (if a member of your party is known to be hypersensitive to insect stings). This kit may include epinephrine and benadryl, or chlorpheniramine. Epinephrine is a prescription drug, but if it is known to be needed get a doctor to give you a prescription and instructions on using it. These precautions are much cheaper than emergency hospital visits and can be life saving. The procedure for using epinephrine is dangerous and should only be carried out in strict compliance with medical advice. A kit called "Anakit" is made by Hollister Stier Laboratories and has the necessities for such treatment. Chlorpheniramine and Benylin syrup may be somewhat helpful in cases of anaphylactic shock if nothing else is available.

Analgesics

- buffered aspirin
- acetaminophen (as in Tylenol)
- Percogesic
- gargle
- ear drops
- anesthetic throat lozenges such as Sucrets

Toothache Remedies

- oil of cloves (Eugenol)
- toothache gel (comprised of 0.05 percent banzalkonium chloride in a gel base)
- emergency dental kit (which would include at least one of the above plus temporary filling material and more—examples are "Mini EDK" from Kaufman's West, 504 Yale S.E., Albequerque, NM 87106; and "Dental Emergency Kit" from Early Winters, 110 Prefonatine Place South, Seattle, WA 98104-9977).

Ear Care

- Schein Otic Drops (softens wax and can aid treatment of mild outer ear infections)
- Triple antibiotic ointment (may be carefully applied to outer surfaces with cotton swabs)

Snakebite Kit

- knife
- suction cup
- constrictive bandage

- antiseptic
- instructions—see also Snakebite Update on page 303
- Cutter Snakebits Kits are compact, inexpensive, and available almost everywhere

Remedies for Poisoning

- syrup of ipecac (to induce vomiting for removal of poisons from the stomach)
- powdered activated charcoal (may be used to absorb poison—use U.S.P. grade)

Ophthalmics

- soothing eye drops (such as Visine or Murine)
- yellow oxide of mercury (1%) (for minor infections)

Emergency Birth Kit

- two umbilical clamps (or some sterile cloth tapes or suture material)
- sterile gloves
- flannel receiving blanket
- two disposable diapers
- suction bulb
- three sanitary pads (large)
- sterile scalpel or sterile scissors to cut umbilical cord
- soap
- four-inch-square sterile gauze dressings (2)
- adhesive tape to label baby and hold dressing over cord, and pencil

Miscellaneous

- table salt (to use as gargle and restoration of body fluids)
- baking soda (to use in restoration of body fluids and as a topical paste for bites)
- sugar cubes (for insulin shock)
- insect repellent
- sun-screening lotion
- potassium iodide (prophylactic against radioactive iodine)

Basic Equipment

- splinter forceps (fine point—for splinter removal)
- needle
- bandage scissors
- disposable scalpel (No. 10 or 11, sterile)
- thermometer
- elastic bandage
- hot water bottle

- safety pins
- medicine dropper
- matches
- basin—rigid or collapsable plastic
- plastic sacks, plastic wrap (for containers, poultices, etc.)

Other Desirable Equipment

- tissue scissors
- tissue forceps
- sterile gloves
- wire and/or inflatable vinyl splints
- sterile cotton-tipped applicators
- penlight and batteries
- rubber tourniquet
- measuring vessel
- material for sling (triangular bandage)
- bulb syringe

The following items need not be purchased unless you know how to use them or intend to learn how to use them.

- Kelley hemostat
- Stethoscope
- Sphygmomanometer and cuff (for blood pressure readings)
- plastic airways (oro-pharyngeal airways)

FIRST AID SUPPLY CHECKLIST

Antiseptics and topicals

- ☐ Germ Killing cleanser
- ☐ Antiseptic prep pads
- ☐ Antibiotic ointment
- ☐ Irritation ointment
- ☐ Antifungal
- ☐ Minor burn preparation
- ☐ Skin care items
- ☐ Insect repellent

Gastrointestinal upset aids

- ☐ Laxative
- ☐ Antacid
- ☐ Anti-diarrhea agent
- ☐ Anti-vomiting agent

Dressings

- ☐ Plastic strips
- ☐ Rolled gauze
- ☐ Sterile gauze pads (4'' x 4'' or other pads)
- ☐ Adhesive tapes
- ☐ Elastic bandage
- ☐ Triangular bandage

Closure Devices

- [ ] Butterfly bandages
- [ ] "Steri-strips"
- [ ] Sutures

Respiratory Aids

- [ ] Antihistamine
- [ ] Decongestant

Analgesics

- [ ] Buffered Aspirin
- [ ] Acetaminophen (Tylenol)
- [ ] Percogesic
- [ ] Ear drops
- [ ] Gargle and/or throat lozenges
- [ ] Toothache Remedies
- [ ] Oil of cloves (Eugenol), toothache gel
- [ ] Emergency dental kit

Ear Care

- [ ] Ear Drops
- [ ] Triple Antibiotic Ointment

Snakebite

- [ ] Kit and instructions

Poisoning

- [ ] Emetic—Syrup of ipecac
- [ ] Powdered charcoal (USP only—this is not burned toast)

Opthalmics

- [ ] Soothing eye drops
- [ ] Oxide of mercury (1%) for minor infections

Emergency birth kit

- [ ] Umbilical clamps (2)
- [ ] Disposable diapers (2)
- [ ] Receiving blanket
- [ ] Large sanitary napkins (3)
- [ ] Suction bulb
- [ ] Sterile gloves
- [ ] Germ killing cleanser

Container

Several manufacturers are packaging first aid kits in rugged, lightweight, well-designed packs that facilitate carrying and use. There are both rigid and soft cases available. These are ideal. Fishing tackle boxes and other containers can also be used satisfactorily to hold first aid kits.

SELECTED REFERENCES AND NOTES

The American National Red Cross. *Advanced First Aid and Emergency Care*. Garden City, New York: Doubleday & Company, Inc., 1973.

Barnhart, Edward R., publisher, and others. *Physicians Desk Reference for Non-Prescription Drugs*. Oradell, New Jersey: Medical Economics Company, 1984.

Berkow, Robert, M.D., and others. *The Merck Manual of Diagnosis and Therapy*. 13th ed. West Point, Pennsylvania: Merck, Sharpe, and Dohme Research Laboratory, 1977.
This is quite technical, but if you can read it, it can be very helpful.

Dickson, Murray. *Where There Is No Dentist*. Palo Alto, California: The Hesperian Foundation, 1983.

Forgey, William, M.D. *Wilderness Medicine*. Revised edition. Maryville, Indiana: Indiana Camp Supply, 1982.
A very practical book. The author simplifies things a great deal.

Fries, James F., M.D., and Donald M. Vickery, M.D. *Take Care of Yourself: A Consumer's Guide to Medical Care*. Reading, Massachusetts: Addison-Wesley Publishing Company, 1981.

Glass, Thomas G., Jr., M.D. *Snakebite First Aid*. San Antonio, Texas: Glass Publishing Co., 1981.
A good treatment of the subject. Available by writing to Glass Publishing Co., 8711 Village Dr., San Antonio, Texas 78217.

Hafen, Brent Q., and Keith J. Karren. *Prehospital Emergency Care and Crisis Intervention*. 2nd ed. Englewood, Colorado: Morton Publishing Company, 1983.
A very good Emergency Medical Technicians manual. Also look for other EMT manuals. Call your local EMT station (police or fire department) and ask their advice. Get the text and workbook for more complete training. One publisher of EMT manuals is the Reston Publishing Company of Reston, Virginia.

Kothe, William C., ed. *Medical Handbook for Unconventional Operations*. Boulder, Colorado: Paladin Press, 1968.
Includes some help on dental care and obstetrics.

Rothenberg, Robert E., M.D., F.A.C.S. *The New American Medical Dictionary and Health Manual*. New York: The New American Library, Inc., 1975.
Since so many of the manuals have technical terms, this could help a great deal.

Sehnert, Keith. *How To Be Your Own Doctor (Sometimes)*. New York: Grosset and Dunlap, Inc., 1975.

Thygerson, Alton L. *The First Aid Book*. Englewood Cliffs, New Jersey: Prentice-Hall, Inc., 1982.
A good all-around first aid book.

U.S. Army Special Forces Medical Handbook. Boulder, Colorado: Paladin Press, 1982.
Somewhat technical, but full of useful information, including obstetrics, some animal care, and some primitive methods.

U.S. Department of Defense. *Emergency War Surgery*. Washington, D.C.: U.S. Government Printing Office, 1975.
Technical. Treatment of war wounds.

U.S. Public Health Service. *Laboratory Section of the Packaged Disaster Hospital*. Revised edition. Publication No. 1071-F-4. call no. FS3.302:F-4.[Washington, D.C.: U.S. Government Printing Office, 1966].
How to do some vital lab work such as blood and urine tests, and blood typing with minimal equipment.

U.S. Public Health Service. *Therapeutic Guide for Pharmaceuticals in the packaged Disaster Hospital*. Publication No. 1071-C-1, call no. FS 2.302:C-1. [Washington,

D.C.: U.S. Government Printing Office, 1965].
Gives doses, dangers, etc., on many drugs and medications.

U.S. Public Health Service and the War Shipping Administration. *The Ship's Medicine Chest and First Aid at Sea.* Revised edition. Miscellaneous Publication, No. 9. [Washington, D.C.: U.S. Government Printing Office, 1955].
A wealth of information on emergency care, although it is a bit outdated.

Werner, David. *Where There Is No Doctor.* Palo Alto, California: The Hesperian Foundation, 1977.
The title says it all.

Wilkerson, James A., M.D., ed. *Medicine for Mountaineering.* 2d ed. Seattle, Washington: The Mountaineers, 1976.

SOURCES

First aid supplies are widely available, but it may be difficult to find satisfactory kits and some desired items. Two good mail order sources are Indiana Camp Supply and NitroPak.

SNAKEBITE UPDATE

The practice of using high voltage direct current (DC) shocks to neutralize the effects of snakebites, insect bites, and other animal bites is increasing dramatically. There is a growing body of evidence from many parts of the world that this remedy, when correctly applied, is amazingly effective. Although not currently approved by the FDA and other official agencies, research is currently under way which may lead to such "official" approvals. In the meantime lives and limbs as well as pain and suffering for both humans and pet animals are being saved by use of this method.

Good quality stun guns (used for self-defense) such as those made by Nova Technologies, are the easiest tools of delivery for the shocks. Some of the "cheap imitations" are apparently less effective. Bites of smaller critters, such as insect stings, are usually near the surface of the skin and are usually treated by spanning the bite area with the two probes of the stun gun. Snakebites are deeper and are usually treated by placing the probe in the fang mark while extending the other probe with a wire to a few inches above (closer to the body) the fang mark and applying the shock. While the shock is unpleasant, it is most certainly less unpleasant than losing a limb or dying.

One missionary doctor, Doctor Ronald Guderian, working in Ecuador has been very successful with this "shocking cure."

The above information is meant to inform only and should not serve as instructions for emergency care.

Stay in touch with local medical practitioners for complete instructions and approved protocols using this treatment.

An interesting, two-part related article appeared in *Outdoor Life*, published by Times Mirror Magazines, Inc., titled "A Shocking Cure" (June, 1988, p. 64-) and "A Shocking Cure for Snakebite" (July, 1988, p. 46-).

CHAPTER 18

Herbs

Modern medicine has done some wonderful things for us, but it may also have taken us away from some of the independent uses of nature that we might do well to exercise. In the modern system, we want to take a pill or have a shot or a surgery and be instantly healed from our maladies. However, some of our health problems could probably be treated effectively by herbal remedies, and knowledge of these remedies could serve as a valuable tool in a survival situation.

The use of herbal remedies is a somewhat forgotten but reviving practice. It takes, in many cases, common sense, persistence, and learning to be able to use herbs with skill and effectiveness. It may be a little more difficult to take charge of your own body and be your own best doctor for some maladies than it is to go to a doctor and let him decide what to do. This is in no way intended to mean that one should not consult with the best authorities available in modern medicine when the situation merits it; and sometimes, a person becomes so ill in health that he needs someone to take charge of him.

In a nutshell, herbal use is believed to be an effective nutritive resource for survival and in other situations as well. My wife and I have used herbal remedies in our home for years while raising a fairly large family. We have taken classes, read books, discussed them, tried, tested, and sifted through what we have learned. In some cases we have kept looking for better remedies. Indeed, there are times when it is better to go to the doctor and times when it is better to save your money. In many cases of illness we have found some of the most effective and reliable remedies to be simple, inexpensive, and locally grown.

There isn't room in this book for a comprehensive treatment of this subject, but we have condensed our findings to a few basic, common herbs and ways of preparing and using them. The uses discussed are recorded in more than one source and/or derived from experience, but they are certainly not all documented by controlled study. And, of course, no claims whatsoever are made for the effect of any of these remedies on you personally. They are merely reported uses, and you must assume the liability for your use of any of them. Be cautious and alert and test the things you try gradually to observe their effect on you, as you should do with any medication. A few reported side-effects are also mentioned.

GLOSSARY

In describing the therapeutic values of medicines or herbs, it is simplest to use certain descriptive terms. Since these terms may not be widely familiar, a glossary is provided here to define them.

Alteratives gradually help to revive and restore bodily functions.
Anodynes relieve pain.

Anthelmintics kill or expel intestinal worms.

Antiemetic helps stop vomiting.

Antiepileptic helps relieve seizures.

Antiseptics inhibit growth of or kill bacteria.

Antispasmodics prevent or relieve cramps and spasms.

Aphrodisiacs stimulate sexual power. They are also sometimes used to help regulate male or female glandular imbalances.

Aromatics have a smell and/or taste which helps make other medicines more palatable when used in combination with them. They may also be used as stimulants.

Astringents cause a contraction of tissue.

Cardiacs affect the heart.

Cathartics aid in the evacuation of the bowels.

Demulcents help reduce inflammation and provide a protective coating.

Diaphoretics promote perspiration.

Diuretics reduce water retention and increase urine flow.

Emetics induce vomiting.

Emmenagogues prompt menstrual flow.

Expectorants facilitate the loosening and discharge of mucus from the respiratory system.

Febrifuges reduce fevers.

Laxatives promote defecation.

Nervines relax the nervous system.

Parturients promote labor in childbirth.

Pectorals relieve chest infections.

Purgatives are powerful laxatives recommended for careful use with adults only.

Sedatives quiet and reduce tenstion.

Stimulants generally speed up some of many body functions.

Styptics reduce bleeding. They help tissues and blood vessels at the area of application to contract.

Sudorfics promote perspiration.

Vermifuges promote elimination of worms.

METHODS OF USE

Herbs are generally used as oils, tinctures, poultices, baths, teas, or powders. The roots, leaves, stems, seed pods and/or seeds, flowers, multiple parts, or whole plants are used.

Three common ways in which herbs are used are as teas or infusions, decoctions, and fomentations. Teas/infusions are generally made from the more delicate parts of the plants. Decoctions are usually made from roots, seeds, bark, and so on, where a more rigorous treatment of the plants is necessary to get the nutrients into solution.

Tea or Infusion

Teas can be not only nourishing and/or settling, but the process of drinking them can also be comforting. To make a tea, heat an appropriate amount of water to boiling; add to the water the herbs, plants, leaves, blossoms, or whatever is to be used; and remove it from the heat. Cover, if possible to prevent medicinal entities from being ''steamed'' away, and let it stand for five to fifteen minutes. Strain the tea through a cloth or other filtering device and it is ready to drink. In some cases a little honey or other sweetener may make the drink more pleasant. The amount of material to add to the water varies with use and the herb used, but about one teaspoon of crushed, dry leaves per cup of water is a good place to start.

Decoction

In some cases the roots, seeds, bark, branches, and other tougher portions of an herb are used and the nutritional or medicinal entities may not be as easy to get into a water solution as is the case with leaves and flowers; or it may be desirable to remove larger quantities or entities that are more difficult to remove from stems, flowers, leaves, and more delicate portions. In these cases a decoction is usually made. This is done by placing the herbs in cool water, bringing it to a boil, and then covering and simmering it for twenty to thirty minutes or longer.

Fomentation

A fomentation is used by locally applying a cloth dampened with a tea or decoction to the desired body area. The pieces of herbs used to make the tea may generally be included in the application.

HERBS TO GROW AND USE

A few herbs that are easy to grow, time-tested, and useful are listed first. A more comprehensive list of herbs and briefly stated uses follows the first group.

Alfalfa

Alfalfa has been found to be a good source of calcium, iron, and vitamin A. It can be eaten as a green, blended into a green drink, or steeped as a tea. The young, tender leaves are usually preferred. Also, alfalfa seed produces a sprout that is a favorite of many (see instructions on sprouting). Alfalfa seems to grow almost everywhere.

Aloe Vera

Aloe vera is fairly easy to grow from root division (planting a small portion of live root). The leaves may be cut open and the juice used to soothe and heal burns, stings, bites, and other minor wounds. It acts as an astringent on open wounds. When taken internally, the juice has a laxative or purgative effect. It may be grown indoors in full sunlight and needs infrequent waterings.

Camomile

The sun never sets on the use of camomile. It is used the world over as a treatment for colds, stomach upset, and colitis; as a sedative, and a general tonic; as an anodyne during the menstrual period; and topically as a gargle, an eyewash, and an application for eczema and inflammation. It is used almost exclusively as an infusion or tea. The blossoms or flowers are steeped in hot water for about ten or fifteen minutes in a covered vessel in the standard manner. One or two teaspoonfuls of the blossoms per cup of water is a good place to start for a satisfactory dose.

Figure 18.1   Alfalfa

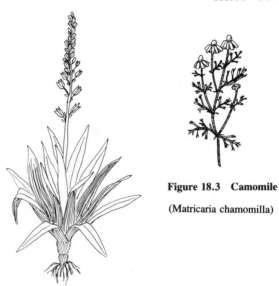

Figure 18.3   Camomile

(Matricaria chamomilla)

Figure 18.2   Aloe Vera

There are two common varieties of camomile: *Anthemis nobilis* (Roman camomile) and *Matricaria chamomilla* (German camomile). The *Anthemis* is easier to grow because it is perennial and can be started by root division. The *Matricaria* must be grown from seed. If you want to have a good patch of camomile, give each healthy plant one and one-half to two feet of space on every side. Pick the fully bloomed flowers, dry them away from direct sunlight, and store them for use.

Cayenne (Capsicum)

Cayenne grows in the form of hot red peppers that are good for flavorings and are prized by herbalists. This herb has been used as a gargle (when added to a liquid), a stimulant, an internal body cleanser, a relaxant, an ulcer healer (some swear by; others against it), and a general tonic. Cayenne is high in its content of several minerals and vitamins A and C.

Cayenne is a warm-weather plant with a fairly long growing season (four to four and one-half months). Start it indoors from seed and move it outside when the danger of frost is past. The Long Red Cayenne is a good variety to plant; sow it in fertile soil. When it is ready to harvest, cut the mature pepper stem off about an inch or so above the pepper and dry it, or else dry the entire plant. Grind the well-dried peppers into a fine powder for use.

Comfrey

Comfrey is an excellent nutritive resource for both man and animal, has been used effectively as a medicinal plant for thousands of years, and is very easy to grow. The plant is reported to contain calcium, potassium, phosphorus, and trace minerals; the amino acids tryptophan, lysine, isolencine, and methionine; and some vitamin B12 (uncommon in vegetation). The leaves are also sources of vitamins A and C (Bricklin, *The Practical Encyclopedia of Natural Healing*, p. 231) and are high in chlorophyll. In

Figure 18.5   Comfrey

Figure 18.4   Capsicum

(Symphytum officinale)

addition, comfrey produces allantoin, which can aid in cell proliferation (Ibid, p. 232). This may help to explain why in days of yore comfrey earned the nickname "knitbone" and "knitback" because of its seeming ability to help tissues to knit together.

Comfrey leaves are used in salads, blended (with a blender) into green drinks with fruit juices such as pineapple, and so on. Infusions or teas are nutritionally beneficial, soothing to inflamed mucous membranes, and good for making poultices. Comfrey root tea is a longstanding pectoral remedy in folk medicine, good for relieving chest ailments including whooping cough. It is also said to be an expectorant. However, there are more problems with using *comfrey root* than good and I recommend not using comfrey roots at all.

A poultice can be made of comfrey by mixing the leaves with boiling water, cooling them and applying them between layers of gauze or thin cloth to the desired area. (An open wound should be well cleaned before the poultice is applied and the leaves should be clean.) A strong decoction can be applied with a piece of bread or an absorbent cloth. The poultices have produced some amazing recorded results. They are said to be good for bruises, sprains, bursitis, boils, burns, gangrene, and other wounds. An ointment can also be made by mixing a strong comfrey decoction with some petroleum jelly or other skin salve.

The allantoin in comfrey remains concentrated in the roots (especially the smaller lateral roots) in the winter and gradually concentrates more in the leaves, buds, and shoots as the plant reaches full growth later in the year.

Comfrey plants can be started easily from root cuttings from existing plants. (These can be ordered from many seed catalogs or nurseries.) They are difficult to grow from seed. Because they are very hardy and prolific, the root cuttings should be planted in a permanent location, about eighteen inches apart under three or four inches of soil. The leaves may be cut every month or so to within two or three inches of the ground and used or dried and stored for future use.

Garlic

Garlic has been called "nature's antibiotic," and because of its use as such in Russia it has sometimes been called "Russian Penicillin." During the Great Plague of Europe in the seventeenth century, garlic was credited with protecting many from the disease. It reportedly induces sweating, acts as a diuretic, cleanses the stomach, aids digestion,

Figure 18.6   Garlic

(Allium sativum)

Figure 18.7   Horseradish

(Armoracia lapathifolia)

Figure 18.8   Oak

(Quercus)

loosens congestion, removes intestinal worms, opens blood vessels and reduces blood pressure, aids in kidney and bladder functions, and is antiseptic (in fact, it was used as an antiseptic during World War I). In addition, when chewed, garlic cloves are said to reduce toothache. They also make good poultices for snake and insect bites.

Garlic is a wonderful flavoring in soups and meat dishes, but it is much easier to take garlic medicinally in capsule form (if available). Other methods of taking the raw juice include mixing it with honey, boiling garlic cloves gently in milk, or just squeezing the juice and taking a teaspoonful of it or eating a piece of the raw bulb. Poultices can be made from the crushed plant or a cloth can be saturated with the juice.

Garlic bulbs can be split into about eight to twelve cloves. A clove can be planted early in the spring to produce a late summer or early fall harvest. Plant them pointed end up about two inches deep and about six inches apart. Hang the harvested plants in a warm place to dry.

Horseradish

Horseradish is another commonly eaten food that is also considered to be a medicinal herb and a good source of vitamin C. Reportedly horseradish root can be used as a diuretic, a treatment for colds, and a pain-relieving compress. The leaves have reportedly been used as an antiseptic and as a healing compress; it is also claimed that chewing the leaves helps to combat food poisoning.

A common medicinal recipe calls for about one ounce of freshly chopped horseradish root to be mixed with about one-half ounce of bruised mustard seed, added to one pint of boiling water, and left to steep for three or four hours. The steeped mixture is strained and given in doses of six to eight tablespoons full per day in three or four equal portions.

Horseradish is a perennial and can be propagated by dividing or cutting roots. It does not grow fast, but it is hardy. Plant good-sized root pieces (a few inches long) early in the year (January or Februrary, if soil can be worked), with the bottom of the root a foot deep and the pieces a foot apart. Harvest the roots in the fall, or leave them in the

Figure 18.9 Rosemary	Figure 18.10 Sage
(Rosmarinus officinalis)	(Salvia officinalis)

ground under a covering of straw and dig them up as needed. Store them in a cool, damp area or in damp sand.

Oak Bark

The bark of many oak varieties—and white oak in particular—is very good when used as a poultice to draw and remove infections. It is easiest to use when powdered. It has also been used effectively to soothe red and swollen gums.

Peppermint and Spearmint

Tasty teas of spearmint and peppermint have been used to relieve headache and upset stomach. Mints are said to be useful as antispasmodics, diaphoretics, aromatics, and stimulants. Mint leaves can also be used as an addition to salads. Peppermint is stronger than spearmint. Make mint teas in the standard manner, using liberal amounts of leaves.

Mints are started from cuttings or divided roots. They are very prolific and like moist areas. They also cross-pollinate, so it may be desirable to keep the various species separated. To harvest them, strip the leaves when they have reached their full size and the bottom leaves have begun yellowing. Dry the leaves out of direct sunlight and store them in a tight container. Cut the stripped stems off an inch or two above the ground to prevent plant disease. (It may be easier to cut the stems before removing the leaves, hang them on a string to dry, and then strip the dried leaves and store them.) (See figure 8.8.)

Red Raspberry

Red raspberry plants are easy to grow and have delicious fruit and leaves which, when made into a tea, are soothing to the digestive system and are good for diarrhea and vomiting. The tea is also said to provide nutrients that strengthen the walls of the uterus and other organs in preparation for childbirth.

Rosemary

Rosemary is a good herb to add spicy flavor to meat dishes and also a traditional remedy. The young leaves, tops, and flowers, made into a tea or infusion, are reportedly good for colic, colds, nervousness, headaches, and depression, as well as making a hearty tonic. In some cultures, dried powdered rosemary is sprinkled on the newly cut umbilical cords of infants as an antiseptic and astringent.

Rosemary is an evergreen shrub and is not fast growing as many herbs are. It is also not as hardy as some plants and should be protected from extreme cold by burlap, mulch, or a greenhouse. It may be grown from seed, but it can be grown faster from divided root.

Sage

It is claimed that sage is effective for quickening the mental faculties and moderating extremes (either high or low) of sexual appetite. In addition, it is reportedly useful as a gargle; a nervine or a poultice to help heal sores, ulcers, and other open wounds. It is also astringent and is reported to be good for sore throats and coughs.

Use one-half to one ounce of sage per quart of water and add lemon or orange juice to the infusion for a bracing, therapeutic drink. Sage is also used for flavoring meat and other dishes.

Sage is perennial but can be started easily from seed. It is hardy and is said to be a good garden companion to rosemary. It may take two years to get a full crop. Harvest the higher leaves and use them fresh, or dry and store them.

OTHER HERBS

Following is a list of some other commonly used herbs and their reported therapeutic values:

- Balm of Gilead—stimulant, expectorant, general tonic.
- Barberry—bark of roots: antiseptic, purgative, tonic, liver problems; bark: high in vitamin C, febrifuge.
- Black cohosh—astringent, alterative, diuretic, emenagogue, expectorant; used to aid blood circulation, seizures, convulsions; supplies the female hormone estrogen.
- Black walnut—used in treatment of ringworm, tuberculosis, internal parasites.
- Blessed thistle—galactagogue (promotes the flow of milk from nursing mothers), vermifuge; helps relieve menstrual distress.
- Blue cohosh—demulcent, emmenagogue, nervine, antispasmodic; stimulates rapid birth of baby ready to be born.
- Buchu—diuretic, diaphoretic, stimulant, febrifuge; good for prostate, urinary organs.
- Brigham tea—blood purifier; used in treating kidney problems, sinus problems, asthma. Use sparingly.
- Burdock root—diaphoretic, alterative, diuretic, tonic; used for arthritis.
- Carrot—roots: vermifuge, ememenegogue, treatment for arthritis, dropsy; blossoms (made into tea): used for dropsy.
- Chaparral—blood purifier, liver cleanser; used for leg cramps, boils, acne, arthritis, kidneys. Use cautiously.

- Chickweed—high in vitamin C.
- Damiana—aphrodisiac; used for adjusting hormone imbalance, alleviating female "hot flashes."
- Corn—silk: demulcent, diuretic.
- Dandelion—root: diuretic, laxative, tonic; good for liver and kidneys.
- Dong quai—provides female hormone estrogen.
- Echinacea—blood cleanser.
- Eyebright—mild astringent, tonic; used as an eyewash.
- Fennel—seed: used to relieve gas pains, colic, cramps; used for seasoning.
- Fenugreek—reportedly good for lungs. Seeds can be used for poultices.
- Fo-ti—reportedly good for endocrine system, brain.
- Ginger—used for colds, lung problems; good spice.
- Ginseng—stimulant, tonic.
- Golden seal—root: alterative, tonic, laxative; used for infections, flu, disorders of internal organs; helps control secretions, aids digestion. Diabetics or hypoglycemics must use with caution.
- Gotu kola—reportedly good for high blood pressure, mental fatigue; general tonic.
- Hawthorne—used for cardiac problems; tonic.
- Hops—anodyne, nervine, tonic.
- Horse nettle—anodyne, antispasmodic; used to treat epilepsy, hysteria.
- Horsetail grass—used to relieve kidney stones, glandular disorders, eye, and nose and throat problems.
- Horehound—diaphoretic, expectorant, tonic, vermifuge.
- Huckleberry—used to stimulate pancreas in diabetics.
- Hyssop—carminative, diaphoretic, pectoral, tonic.
- Irish moss—demulcent, tonic; also has nutritive value.
- Jerusalem artichoke—source of minerals; used as an aid to diabetics.
- Jojoba—nut: helps with skin and scalp problems.
- Juniper—berries: carminative, diuretic, stimulant.
- Kelp—source of minerals; used to aid thyroid, other glands.
- Lady's Slipper—anodyne, antispasmodic, nervine.
- Licorice—root: aphrodisiac; also used for heart and lung and circulatory problems, hypoglycemia.
- Lobelia—antispasmodic, emetic, expectorant, stimulant, relaxant. A strong herb; should be used carefully.
- Mandrake—alterative, carminative, cathartic.
- Marigold—stimulant, antispasmodic, diaphoretic.
- Marshmallow—demulcent, emollient; also used for kidney and bladder problems.
- Milkweed—used to help expel gallstones.
- Mullein—astringent, emollient, pectoral, demulcent.
- Myrrh gum—antiseptic, tonic, stimulant.
- Nettle—astringent, diuretic, tonic.
- Oregon grape—blood purifier, tonic.
- Parsley—diuretic; high in vitamin A.
- Passionflower—relaxant; used for high blood pressure, nervousness.
- Peach—bark: diuretic.
- Pennyroyal—carminative, diaphoretic, emmenagogue, stimulant.
- Peruvian bark—astringent, febrifuge, tonic.
- Plantain—alterative, diuretic; leaves: poultice for stings, bites, sores to relieve pain and promote healing.

- Pleurisy root—antispasmodic, diaphoretic, expectorant.
- Poke root—alterative, cathartic, emetic, tonic; high in iron.
- Psyllium—seed: demulcent, diuretic, vermifuge.
- Red clover—sedative, alterative, depuritive.
- Rue—emmenagogue, antispasmodic, stimulant.
- Saint John's wort—sedative, astringent, expectorant.
- Sarsaparilla—alterative; used to balance hormone systems (especially male).
- Saw palmetto—used to improve function of male and female sex glands.
- Skullcap—antispasmodic, nervine, tonic.
- Senna—cathartic, laxative, vermifuge.
- Skunk cabbage—diaphoretic, antispasmodic, expectorant.
- Slippery elm—demulcent, emollient, pectoral; aid for weak digestion, poultice.
- Squaw vine—parturient, astringent, tonic, diuretic.
- Tansy—emmenagogue, anthelmintic, tonic.
- Thyme—antiseptic, carminative, tonic.
- Uva ursi—astringent, diuretic, mucelagenous; used for kidney problems and to strengthen organs.
- Valerian root—nervine, antispasmodic, tonic.
- Vervain—emetic, nervine, tonic.
- Wild cherry—pectoral, astringent, tonic.
- Wintergreen—astringent, stimulant.
- Witch hazel—astringent, tonic.
- Wood betony—alterative, nervine; also used to strengthen heart when there are heart problems.
- Wormwood—anthelmintic.
- Yarrow—source of potassium; used to reduce fevers; leaves: chewing them is said to reduce toothache.
- Yellow dock—root: cardiac; high in iron.

SOME SIDE EFFECTS

Some of the side effects that have been reported from herbal use are listed below, along with the herbs responsible.

EFFECT	HERB
Allergic reaction	Chamomile, goldenrod, marigold, yarrow
Cancer	Sassafras (has been banned by the FDA)
Poisoning	Mistletoe, pokeweed, Indian tobacco
Nervous system reactions	Catnip, lobelia, burdock, hydrangea, juniper, wormwood, nutmeg, jimsonwood
Sensitivity to sun	Saint-John's-wort
Cardiovascular disturbance	Foxglove (source of digitalis), licorice
Diarrhea (this may or may not be detrimental. In some cases diarrhea is a means of body "housecleaning.")	Aloe leaves, buckthorn bark, senna, dock root

ITEMS OF INTEREST

As a matter of interest, herbs are catagorized by some into "ABC" catagories. "A" stands for aromatic, activating, energizing herbs that reportedly work on the circulatory, respiratory, urinary, reproductive, and nervous systems; examples are cayenne, garlic, ginger, and mint.

"B" stands for bitter herbs that build, rejuvenate, and "bless," such as goldenseal, lobelia, hops, horehound, cascara sagrada, and yellow dock. These reportedly affect the lymphatic system and lower intestines, among other areas.

"C" stands for cleansing astringent herbs that are said to act on skin and membranes around nerves, stomach, and muscular and skeletal systems. Comfrey, slippery elm, and flax are representative of this category.

As another matter of interest, certain herbs are reputed to be folk cures for cancer. Some of them have had compounds isolated from them that have been found to be active against some forms of cancer, but no specifics will be given here. (I make no claims whatsoever for these. If you have cancer, obtain the best and most well-informed help you can find!) These herbs include absinthe, aronica, atriplex, beet, black walnut, borage, calendula, celery, chaparral, chicory, chive, Concord grapes, colocynth, crimson clover, crown vetch, cucumber, cumin, flax, garlic, hot pepper, licorice, onion, peanut, poke root, safflower, salvia, stinging nettle, tamarind, tansy, tea, tomato, vitamin A, blue violet, red clover, yellow dock, parsley, laetrile juice, apricot seeds, cherry seed, buckthorne, and burdock.

SELECTED REFERENCES AND NOTES

There are many books on herbs and a long list could be made. Here are just a few.

Bricklin. Mark. *The Practical Encyclopedia of Natural Healing*. Emmaus, Pennsylvania: Rodale Press, Inc., 1983.
A very sensible treatise on the subject.

Coon, Nelson. *Using Plants for Healing*. New York: Hearthside Press, Inc., 1963.
Very good indexing of plants and descriptive terms.

Gordon, Monteen and Velma Kieth. *The How To Herb Book*. Pleasant Grove, Utah: Mayfield Publications, 1984.

Hylton, William H., ed. *The Rodale Herb Book*. Emmaus, Pennsylvania: Rodale Press, Inc., 1976.
A very good treatment on growing, harvesting, and using herbs, as well as sources for common herbs.

Kadens, Joseph. *Encyclopedia of Medicinal Herbs*. New York: Arco, 1970.

Kloss, Jethro. *Back to Eden*. Loma Linda, California: Back to Eden Books, 1982.
An old standby, not overly sophisticated in some places, but full of sincere and useful information. First Published in about 1939.

Krochmal, Arnold, and Connie Krochmal. *A Field Guide to Medicinal Plants*. New York: The New York Times Book Co., Inc., 1973.

Malstrom, Stan D. *Own Your Own Body*. New Canaan, Connecticut: Keats Publishing, Inc., 1977.
A fairly comprehensive book about how the body works, giving many ideas on how to fix and renew it. Well arranged.

McGrath, William A. *Common Herbs for Common Illnesses*. Provo, Utah: Nulife Publishing, 1977.

Moulton, LeArta. *Herb Walk*. Provo, Utah: The Gluten Company, 1979.
A great book if you want to be able to identify herbs and other growing plants. Chock-full of good pictures. Also videos: P.O. Box 482, Provo, UT 84603.

South, J. Allan. *The Sense of Survival*. Orem, Utah: Timpanogos Publishers, 1985.
Remarkably informative and entertaining, *tour de force*. Delightfully illustrated.

EPILOGUE

As a young man I was able to gain some valuable experiences doing farm work, keeping machinery running with "spit and baling wire," and living on very little. My parents, when they were young, had much less than that; and the previous generation had even less. Behind these simple statements of fact lie the memories of so many details of a life at once simpler and more difficult: the regular trips to the spring for water, the mail-order catalogue sitting in the outdoor privy, the Saturday night bath (whether needed or not), the wood chopping and coal hauling, the basic foods, the ingenuity and the simple delights, the struggles and the frustrations, the faith and the hopes and the prayers.

If a bus crashes and some are killed and some injured, with none completely unscathed, it is called a tragedy. If a nuclear war occurred some would be killed, some would be injured, no one would be unscathed, and it would be a tragedy magnified countless times over.

In this or any other imaginable difficulty that lies ahead for humanity, the devices that now fill our everyday lives will tend to become less important than the inner workings of the human spirit and the justice of governments it created. Even in a situation of tremendous destruction, there would be an incredible salvage yard of machines and technologies with which to rebuild. But if we had not the fortitude, the sense of cooperation and ingenuity, and the hope and will to survive when disaster strikes we may find the times of difficulty much harder to bear. In such times we will find that service and uprightness amid the sufferings and the disorder become more important than the physical comforts we may have lost.

Man stands now with the ability to bring about unimaginable destruction, and, as history has proved many times over, nature is ever able to outdo man. Until humankind can overcome its own selfishness, distrust, deceit, and aggression and continue to reach up and place greater value on principles than on possessions, on orderliness than on pleasures, on peace than on gaining attention or power, it will ever stand in danger from itself.

As for me and my house, we are determined to face what comes with courage, with a twinkle in the eye, and with faith that there is an eternal healing to come that will justly and powerfully reckon with it all. May we avoid all the trouble we can and be prepared, both inwardly and outwardly, for that which we cannot avoid.

If we are prepared, we shall not fear.

Appendix A — Fallout Shelters

CONCRETE BLOCK
SHELTER—BASEMENT
LOCATION

General Information

This compact basement shelter will provide low-cost protection from the effects of radioactive fallout. Its purpose is to provide adequate protection for the minimum cost in an existing basement. In addition to the low cost, materials should be readily available, and the labor time will be short.

Technical Summary

This shelter has about 50 square feet of area, 300 cubic feet of space and will provide shelter for five persons.

The materials required to build this shelter are obtainable at local concrete block plants and/or lumber yards.

Natural ventilation is provided by the entranceway and the air vents in the shelter wall.

Estimated construction time for the basic shelter is less than 44 man-hours.

MATERIALS LIST

Item	Actual Number Required
Masonry:	
4'' x 8'' x 16'' solid concrete masonry units or	296 blocks
2¼ x 4'' x 8'' solid bricks	1776 bricks
4'' x 8'' x 16'' hollow concrete masonry units	7 blocks
Lumber: ("Construction" or "No. 1" grade or better)	
posts 2 x 4 x 5'4''	6
joists 2 x 4 x 5'4''	7
beams 2 x 4 x10'5½''	2
frame 2 x 8 x 5'4⅜''	2
header 2 x 8 x 2'3''	2
plywood 1'4'' x 6'9¼'' x ¾'' (utility B-C grade)	4 pieces
plywood 1'4'' x 4'3¾'' x ¾'' (utility B-C grade)	4 pieces
Hardware:	
8d nails	2 pounds
10d nails	2 pounds
⅜'' bolt size multiple-expanding machine bolt anchors	18
⅜'' x 3½'' square-head unfinished anchor bolts	18
Mortar (prepared dry-mix bags)	9 bags

317

Special tools:

¾'' star drill for ¾'' x 2-7/8'' anchor bolts

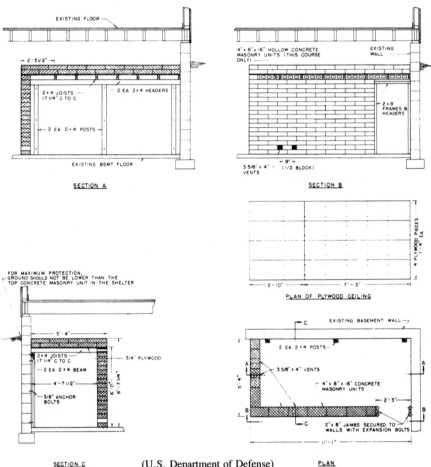

(U.S. Department of Defense)

Construction Sequence

1. Lay out guidelines with chalk on basement floor for shelter walls.
2. Lay first course of 4'' x 8'' x 16'' solid blocks in a full bed of mortar to make the walls 8'' thick. Vary the thickness of mortar bed if basement floor is not level.
3. Set door frame in place and continue to lay wall blocks. Be sure to leave the 4'' spaces for air vents as shown on the drawing.
4. Continue this procedure until the walls have been laid up to a height of 5'8'' (17 courses). This height can be increased, if the basement headroom permits and provided the shelter roof remains below the outside ground level.
5. Fasten posts and door frame to the basement wall using two expansion anchors and bolts for each. Be certain the posts rest on the floor.
6. Nail two 2 x 4 boards together to make the wall beam. Nail the beam on top of the posts and secure with expansion anchors and bolts to the wall.

7. Place wood joists in position and secure with nails.
8. Place the 4'' x 8'' x 16'' hollow blocks between joists as shown on the drawing. The holes in the blocks will afford ventilation.
9. Put several 3/4'' pieces of plywood on the joists as shown and nail them to the joists with 8d nails.
10. Lay two layers of solid 4'' x 8'' x 16'' blocks flat on top of the plywood; stagger the joints. Mortar is not required in the ceiling.
11. Continue procedures 9 and 10 until the roof is completed.
12. Additional blocks stored in the shelter are for stacking in the entryway after occupancy.

BELOWGROUND CORRUGATED STEEL CULVERT SHELTER

General Information

This shelter is designed to provide low-cost protection from the effects of radioactive fallout. Its principal advantages are that most of the structure is generally available as a prefabricated unit ready for lowering into an excavation and that it requires only simple connections and covering to complete the installation.

Technical Summary

Space and Occupancy—This shelter has 32 square feet of area and about 120 cubic feet of space (including the entranceway). It could provide space for three persons. The addition of a 4-foot length would provide for one more person.

Availability and Cost of materials—This type of shelter is avaiable from steel culvert fabricators or their sales outlets in most population centers. This prefabricated shelter, including ventilation system, plastic wrap, and sandbags is designed to be sold for $150 or less, excluding delivery and installation.

Fallout Protection Factor—When the entranceway is properly shielded as shown in the drawings, the protection factor should be greater than 500.

Blast Protection—This shelter could be expected to withstand a limited blast overpressure of 5 pounds per square inch.

Ventilation—A sheet metal intake vent 3 inches in diameter is provided together with a manual airblower for more than three persons. Air is vented through the sandbag closure at the entrance.

Installation Time—One man working with hand excavation tools should be able to complete the excavation in less than 2 man-days. Two men will be needed to roll the shelter structure into the excavation from the point at which the shelter has been delivered. If lifting rather than rolling is necessary to transport the structure, four men will be required. Time for this phase will vary upward from 1 hour depending on distance of the move. It will then take one man 4 working days to complete the covering and installation phases.

Structural Life Expectancy—The estimated life of this galvanized steel shelter will be at least 10 years under most soil conditions. Under normal conditions highway culverts of similar material have been known to last indefinitely with little maintenance.

CONSTRUCTION SEQUENCE*

1. Select well-drained site. The total area required, including the mounding, will be approximately 15' x 20'.
2. Use stakes to mark the corners of the area, and excavate. The hole required for the main shell is 5' x 9' x 2' deep, and the entrance requires an additional 2½' x 4' x 6''.
3. Line hole with plastic film wrap.
4. Lower galvanized steel shelter into place on supporting wood strips.
5. Assemble and install the vent pipe.
6. Cover shelter with plastic wrap.
7. Backfill and mound. Be sure the shelter is covered by at least 2 feet of packed earth. Depth may be checked with a wire probe. The mound should be covered with grass as soon as possible by sodding or seeding to prevent the protective soil from being eroded.
8. Place small sandbags inside the shelter. These are used to fill the entrance completely after the shelter is occupied.
9. 1-inch boards may be used on 2'' x 4'' blocks to provide a floor.

*This is a generalized construction sequence for a prefabricated steel culvert shelter. Detailed instructions are provided with the construction kit.

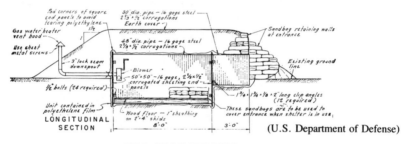

(U.S. Department of Defense)

BILL OF MATERIALS
(To shelter 3 persons)

Item	Quantity
Prefabricated steel culvert shelter (with bolts and clips supplied, if unit is not spot welded).*	1.
Galvanized steel lock-seam downspout	6 feet.
Elbow for steel lock-seam downspout	1 foot.
Ventcap (gas water-heater type)	1.
Intake air blower (optional for 3 persons or less)	1.
Scrap lumber	9 board feet.
6 mil. polyethylene film (20' width)	30 feet.
Sandbags (to hold 75 to 100 pounds each)	18.
Sandbags (to hold 15 to 20 pounds each)	30.
Flyscreen 7'' x 7'', for ventpipe	1.
Entranceway insect screen 36'' x 36''	1.
Soil or sand (for shelter cover)	5 tons.

*Fabricators should treat spot-welded areas with bitumastic compound or other approved waterproofing material.

BELOWGROUND NEW CONSTRUCTION CLAY MASONRY SHELTER

General Information

This shelter will provide protection against the effects of radioactive fallout. It can also protect from limited blast overpressures. The shelter is located belowground outside a house but is reached from the basement. Its principal advantages are in flexibility of shape and design to conform to the house design and in the use of materials that tie in with the new construction of a house. Because of the headroom and interior space the shelter can be used for other purposes.

Technical Summary

Space and Occupancy—The shelter in this design has over 70 square feet of area and 420 cubic feet of space. It will provide occupancy for six persons.

Availability and Cost of Materials—Structural clay masonry units, brick, and structural tile are available in concrete-block plants and lumberyards. Cost of the materials and equipment for the basic shelter is estimated at $300 to $350. Labor cost should run approximately $250 to $300 when performed as part of new house construction.

Fallout Protection Factor—The protection factor for a shelter of this type is over 1000.

Blast Protection—This shelter has a structural blast resistance of 5-pounds-per-square-inch overpressure.

Ventilation—Ventilation equipment and pipe are required. A hand-operated blower should be specified to furnish at least 20 cubic feet of air per minute. The air is exhausted through a separate ventpipe.

Construction Time—A home-construction project that includes this shelter will not require additional trades or crafts not already on the project. The time for construction of this shelter could increase normal house construction time by a few days.

Structural life Expectancy—Assuming normal construction practices, this structure, with a minimum of maintenance, should last more than 30 years.

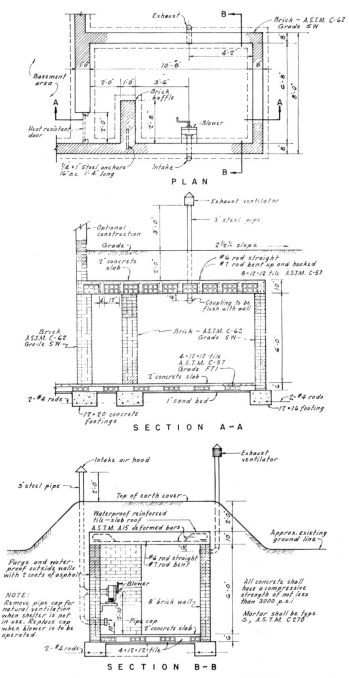

(U.S. Department of Defense)

CONSTRUCTION SEQUENCE

No construction sequence is given for this shelter because the work would probably be supervised by a contractor familiar with new construction.

BILL OF MATERIALS

Item	Quantity
Roof:	
8'' x 12'' x 12'' structural clay tile ASTM—C57—grade FTI	72 pieces.
Steel reinforcing, No. 6 deformed bars 7'6'' length, ASTM—A—15—Straight.	10 pieces.
Steel reinforcing, No. 7 deformed bars 10' length, bent up and hooked ASTM-A-15	10 pieces.
Concrete, minimum 3,000 pounds per square inch	1.5 cubic yards.
Walls:	
Brick, standard size (2-L'' x 4'' x 8'') ASTM—C62—grade SW.	3,800 pieces.
Anchors (¼'' x 1'' x 4'') steel	4.
Mortar (1¼-3¾ cement-lime-sand)	65 cubic feet.
Floor:	
Tile (4'' x 12'' x 12'') structural clay ASTM—C57—grade FTI.	96 pieces.
Concrete, minimum 3,000 pounds per square inch	0.7 cubic yard.
Footings:	
Concrete, minimum 3,000 pounds per square inch	1.0 cubic yard.
Steel reinforcing, No. 4 reinforcing bars ASTM—A15	68 linear feet.
Miscellaneous:	
Parge 1¼-3¾ mortar ASTM—C270—Type M	8 cubic feet.
Asphalt	5 gallons
Blower (at least 20-cubic-feet-per-minute rating)	1.
Mounting bracket, blower	1.
Intake and exhaust ventpipe, 3'' steel (sufficient for both intake and vent pipes).	16 linear feet.
Fittings:	
Ells 3'' steel	2.
Tees 3'' steed	1.
Ventpipe cap	1.
Flyscreen 7'' x 7'' (for vent and intake pipes)	2.

THESE ARE PLANS FOR EXPEDIENT FALLOUT SHELTERS

SAVE THESE PLANS—THEY MAY SAVE YOUR LIFE

General Information

Without protection, untold numbers of Americans would die needlessly in the event of a nuclear attack. The expedient shelters illustrated in the following pages provide protection to occupants from the deadly radiation of radioactive fallout generated by a nuclear detonation—their use can save the lives of millions of Americans.

Even though the illustrated shelters are very austere, there are a number of things that can be done to improve their habitability after they have been built. With the use of a little ingenuity and effort, the shelters can be made more comfortable. Some of the things that can be done are:

●Construct seats, hammocks, or bunks.

●Cover the floor with boards, pine boughs or logs and drape sheets or material over the earth walls.

●Provide safe, dependable light.

●For hot weather, construct the expedient air ventilation pump.

●For cooking, construct the expedient cook stove for use in the entryway. In cold weather, seal the entrance and use the stove for heating the shelter area. *Be sure ventilation is provided whenever the stove is used.*

●Store shelter supplies in entryway for more living space. Cover all open containers. Radiation will not damage these supplies.

Humans must have water and food to live. When people are to live in a shelter for a week or two, sufficient food and supplies must be provided for the occupants. The minimum necessities are:

●Water—minimum requirements (dependent upon temperature—less in cold weater, more in warmer) will be from one quart to one gallon per person per day. Storage can be accomplished by using disinfected metal or plastic trash cans or boxes lined with strong polyethylene film or strong plastic bags. For purity, eight drops (one teaspoon) of a 5¼% chlorine solution (e.g., Clorox) should be mixed into each 5 gallons of water.

●Food—all food should require no refrigeration and should be brought to the shelter in airtight tins or bottles. Under shelter conditions, people will require about half as much food as usual. Foods should have a high nutritional value and a minimal amount of bulk (i.e., canned meats—fruits—vegetables, dried cereals, hard candy, etc.)

●Sanitation—a metal container with a tight-fitting lid for use as a toilet with which plastic bags can be used. Toilet paper, soap, towels, sanitary items and a quantity of strong plastic bags will be needed.

●Medical supplies—a well-stocked first-aid kit comparable to what is usually kept at home. Take special medicines for infants and others and a good first-aid handbook.

●Clothing and bedding—several changes of clean clothing, especially socks and underclothing—dependent upon the weather, blankets, pillows and sleeping bags may also be needed.

●Portable radio—lastly, but hardly least important, a portable radio with fresh and extra batteries. Radio station broadcasts will advise you when it is safe to abandon the shelter and also provide you with other important emergency information.

SMALL-POLE SHELTER
(Capacity 12 Persons)

(Taken from *Expedient Shelter Handbook* authored by G. A. Cristy and C. H. Kearny for the Defense Civil Preparedness Agency at the Oak Ridge National Laboratory (1974). The book is available from the National Technical Information Service, U.S. Department of Commerce, Springfield, VA 22161.)

1. This expedient shelter has a protection factor (PF) greater than 1000 if covered with 3 ft. of earth—that is, an occupant of this shelter would receive less than 1/1000th of the radiation from fallout that he would receive outside in the open. Furthermore, this shelter would give good protection against thermal radiation and the fires that large nuclear explosions, even 30 to 40 miles away, could ignite. Calculations and a blast test have shown that in most soils this shelter can survive and save its occupants from blast overpressures over 30 pounds per square inch. Such a pressure would destroy most buildings.

2. For 12 persons to successfully live for days in this shelter, the benches and bunks must be built with the dimensions and spacings given in the pictorial view. Then 3 persons out of the 12 can sleep in shifts on the overhead bunks, while the remaining persons can sit with plenty of head room. On each side of the shelter there should be 9 feet of benches and 9 feet of overhead bunks.

3. *Study both of the two drawings and read all of these instructions before beginning work.* THEN CHECK OFF EACH STEP WHEN COMPLETED.

4. Materials required for building a 12-person small-pole shelter:

a. *Green* Poles—with no pole having a smaller diameter *top* (= small end) than the minimum diameter specified for its use by the drawings. Table 1 lists the required number and sizes.

b. Rainproofing Materials

   (1) Preferably one 100-ft. roll, 12-ft.-wide, 6 1/n mil polyethylene, or at least 200 sq. ft. of 6-mil polyethylene, or at least 100 sq. ft. of other waterproof plastic, plastic table cloths, shower curtains, and/or linoleum rugs.

   (2) Three hundred feet of sticks, 1/n/to 1½ in. in diameter, of any lengths (for drains).

c. Nails, Wire, and/or Cord

   Ten pounds of 40-penny nails plus 4 lbs. of 16-penny nails are ideal. However, 7 lbs of 16-penny nails can serve. Nails are useful but *not essential*—for you can use *wire, cord,* or *rope* instead of nails to hold crossbraces, etc., in place *until earth pressure tightens all parts of this shelter and holds it together.*

d. Boards for Benches and Overhead Bunks

   (Boards are desirable, but not essential.)

   (1) 2'' x 4'' boards—70 feet for frames (or use 3-in. diameter poles).

   (2) 1'' x 8'' boards—100 feet (or use 1-in. to 2-in. diameter poles).

5. To cut and haul poles more easily and safely, study the instruction sheet, *How To Cut and Haul Logs and Poles More Easily,* at the end of these instructions.

6. Desirable tools for building a 12-person, small-pole shelter:

Tools	Quantity	Tools	Quantity
Axe, long handle	2	Hammer, claw	2
Saw, bow, 28-inch	2	File, 10-inch	1
Saw, cross-cut, 2-man	1	Tape, steel, 10-ft	1
Pick	2		
Shovel, long handle	3		

(Also useful: a 50-ft. steel tape and 2 hatchets)

7. To save time and work, sharpen all tools, and keep them sharp.

8. Wear gloves from the start—even tough hands can blister after hours of digging and chopping. Blisters are painful, seriously slow the work, and could result in dangerous infections.

9. Before staking out the outlines of the excavation, check for rock depth—by driving down a 6-ft., sharpened small pipe or rod. To avoid groundwater problems, avoid low ground.

10. To help drain the floor, locate the shelter so that the original ground level at the entrance is about 12 inches lower than the original ground level at the far end of the shelter.

11. Stake out the trench for the entire shelter. Even in very firm ground, at *the surface* make the excavation 9 ft., 8 in. wide and 18 ft. long (3 ft. longer than the entire length of the wooden shelter). The sloping sides of the excavation are necessary, even in firm earth, to provide adequate space for backfilling and tamping.

12. Check the squareness of the staked trench outline by making its diagonals equal.

13. Clear all brush, tall grass, etc., off the ground to a distance of 10 ft., all around the staked location—so that later you can easily shovel loose earth back onto the roof.

14. If excavating in unstable ground, excavate with appropriately less steep sides.

15. When digging the trench for the shelter, use a stick 7 ft., 8 in. long (the minimum bottom width) to repeatedly check excavation width.

16. When digging with a shovel, pile the earth dug from near ground level about 10 ft. from the edges of the excavation. Then earth dug from 5 or 6 feet below ground level can easily be piled on the surface only 1 to 5 ft. from the edge of the excavation.

17. Finish the bottom of the excavation so that it slopes 1/2 inch vertically per foot of length toward the entrance, and also slopes toward the central drain ditch—in which will be placed sticks covered with porous fabric, to serve like a crushed-rock drain leading to a sump.

18. While some persons are excavating, others should be cutting *green* poles and hauling them to the site. Cut poles having *tops with diameters* (not including bark) *no smaller than the diameters specified for each type of pole.*

19. For ease in handling the wall and roof poles, select poles with *top* diameters no more than 50% larger than the specified minimum diameters.

20. Sort the poles by size and lay all poles of each size together, near the excavation.

21. Before a shelter is completed, some workers should provide for essential forced ventilation in warm or hot weather by building an expedient Kearny Air Pump (KAP). Carefully follow the instructions given in Appendix B.

22. Before the excavation is completed start building the ladder-like, horizontal parts of the shelter frame. Construct these parts on smooth ground near the excavation. Place two straight poles, each 10 ft-6 in. long (with small-end diameters of 3½ to 4 in.), on smooth ground, parallel and 6 ft., 2 in. apart. Securely hold these poles so that their outer sides are exactly 6 ft., 2 in. apart, by driving into the ground two pairs of stakes so that they just touch the outsides of the two longitudinal poles. Each of the four stakes should be located about one foot from the end of a pole. To keep the 10 ft., 6 in. poles from being rotated, during the next step, nail two boards or small poles across them perpendicularly, as temporary braces about 4 feet apart.

Then with an axe or hatchet, slightly flatten the inner sides of the two poles at the spots where the ends of the 6 cross braces will be nailed. Next saw each cross brace pole to the length required to fit snugly into its place, and toenail each cross brace in place, preferably with two 40-penny nails in each of its ends.

23. Place the lower, ladder-like part of the shelter frame of the main room on the floor of the excavation.

24. Build the frame of the main room. Secure the vertical poles in their vertical positions by nailing, wiring, or tying temporary diagonal brace-boards (or small-diameter, temporary brace-poles) to their *inner* sides and to the *inner* sides of the longitudinal frame poles on the bottom of the excavation. To keep the vertical frame poles exactly 6 ft., 2 in. apart until the ladder-like upper horizontal part of the frame is secured in position, nail a temporary horizontal brace across the vertical frame poles, about a foot below their tops.

25. To support the ladder-like upper horizontal part of the frame, nail blocks to the inner sides of the vertical frame poles (as shown in the upper right-hand corner of the pictorial view). If you have big nails, use a block about 3 inches thick and 6 inches long, best cut from a *green,* 4-in.-diameter pole.

26. In the finished shelter, do *not* leave vertical poles under the upper longitudinal poles; to do so would seriously reduce the usable space for benches, bunks, and occupants.

27. While some workers are building the frame of the main room, other workers should be making the rectangular entrance frames and then the complete entrance. To keep the rectangular entrance frames square during construction and backfilling, nail a temporary diagonal brace across each one.

28. Once all frame poles of the main room are in place, put the vertical wall poles between the frame poles, touching each other, until all walls are completed. When placing the wall poles, keep them vertical by alternately putting a butt and a top end uppermost. Wall poles can be held in position by backfilling and tamping about a foot of earth against their lower ends, or they can be wired in position until backfilled.

29. Be sure to use the two 5-in. diameter poles, each 6 ft., 2 in. long, by placing one next to the top and the other next to the bottom of the doorway to the main room. Use braces, each 2 ft., 3 in. long, to hold the top and bottom of this doorway apart.

30. To prevent dirt from coming through the larger cracks between wall poles, cover the cracks with plastic, rugs, roofing, scrap lumber, sticks, or even cardboard.

31. After all horizontal bracing and vertical wall poles are in place, begin backfilling, putting earth between the walls and trench sides. Pay particular attention to the order of filling. The earth fill behind all the walls must be brought up quite evenly, so that the earth fill behind one side is no more than 12 inches higher at any one time than the earth on the opposite side. *Tamp earth fill in 6-inch layers.* Do not use a mechanical tamper. A log makes a good tamper.

TABLE 1

Length	Minimum Diameter (= Diameter of Top)	Number of Poles Required	Width[c]
6 ft, 2 in.	5 in.	2	---
3 ft, 1 in.	5 in.	6	---
10 ft, 8 in.	4½ in.	-	9 ft.
10 ft, 6 in.[b]	4½ in.	2	---
7 ft, 2 in.	4½ in.	-	43 ft.
5 ft, 6 in.[a],[b]	4½ in.	8	---
2 ft, 3 in.	4½ in.	8	---
7 ft, 2 in.	4 in.	-	12 ft
6 ft, 3 in.	3½ in.	4	---
5 ft, 6 in.[a],[b]	3½ in.	10	---
3 ft, 10 in.	3½ in.	-	15 ft
10 ft, 6 in.[b]	3½ in.	2	---
2 ft, 2 in.	3½ in.	6	---

[a]To be cut to fit for crossbraces.

[b]These poles for the horizontal bracing should have tops with diameters no more than about 15% larger than the minimum diameters specified.

[c]Distance measured across a single layer of poles when a sufficient number of poles are laid on the ground side by side and *touching*, so as to cover a rectangular area.

NOTE: The above list does not include flooring materials, to be placed between the bottom crossbraces. Scrap boards, poles, sticks, etc., will serve.

32. Next, lay the roof poles side by side, touching each other. Cover at least the larger cracks with plastic, roofing, boards, or sticks to keep dirt from falling through. If the earth is sandy, cover the whole roof with some material (such as bed sheets or even several thicknesses of newspaper) to keep sand from running through cracks.

33. **CAUTION:** Do not try to rainproof this flat roof, and then simply cover it with earth—because, if you do, water will seep straight through the loose earth cover, puddle on the flat roofing material, and leak through the joints between pieces of roofing material or through small holes in the roofing material.

34. Mound earth over the shelter, piling it about 15 inches deep along the center line of the roof and sloping it toward the sides of the roof, so that the earth is only about 2 inches deep over the ends of the roof poles. (Preparatory to mounding earth onto the roof, place gradestakes in position, so you will be able to know the locations and depths of roof poles as you cover them.) Continue these slopes to two side drainage ditches (that are to be buried later, as shown in the drawing). Smooth this mounded earth with a rake, and remove any sticks or rocks likely to puncture the rainproofing roofing material to be laid on it later.

35. Place rainproofing material (preferably plastic film, such as 6-mil polyethylene—or roofing, plastic shower curtains and table cloths, or canvas) on top of the smooth, mounded earth, as shown in sections of the drawings.

36. Place the rest of the earth cover over the shelter, being sure that the corners of the shelter have *at least 2½ feet of earth over them*. Mound the dirt, smoothing its surface so that water will tend to run off to the surface drainage ditches which you should dig on the two sides.

NOTE: If enough rainproofing material is available, the buried drain ditches can be omitted by continuing the buried-roof rainproofing material, sloping it all the way out to the surface drain ditches around the outermost edges of the mounded earth cover.

37.   To install a 36-in. long x 24-in. wide KAP (Kearny Air Pump)—to keep a crowded shelter from **dangerous** overheating in warm weather— narrow the ends of the overhead bunks so that the aisle between them is about 28 inches wide for a distance of 38 inches from the doorway.

38.   Build the benches and overhead bunks. If available, use boards. If not, use small, straight poles. On each side, build a row of benches and bunks 9 ft. long, centered in the shelter. In order to use the shelter space to the greatest advantage, make the heights and widths of the benches and bunks the same as the thoroughly tested heights (14 in., and 4 ft.-5in.) and widths (16 in., and 24 in.) given by the drawings. Also be sure to space their vertical supports 3 ft. apart—so two men can sit between each pair of vertical supports.

39.   Place a fly or canopy—*open on all sides*—over the entrance, to minimize the entry of sand-like descending fallout particles. The plan and elevation drawing shows a "plastic fly cover," its ridgepole, and one of its two vertical pole supports. Such a fly also can be made of canvas or shower curtains with wires or cords connected to the fly corners and to 4 stakes.

40.   To improve the floor, lay small poles (or sticks covered with scrap boards) between the lower brace poles, so that the floor is approximately level.

41.   If the water table is very near the surface and gravity drainage by ditching is not practical, this shelter should be built in a shallower excavation, or even on the surface. However, the mounded earth should slope not more steeply than 2:1, and mechanized earth-moving equipment probably would be necessary to cover it adequately within 48 hours of a typical civilian group's beginning construction.

42.   *Expedient Ventilation—Cooling.* (If pushed for time, those workers who are only going to work on the shelter itself need not read this section before beginning their work.)

Install a 24'' x 36'' KAP near the top of the doorway. (See Appendix B for instructions.) To enable the KAP to efficiently pump fresh air from the outdoors all the way through the shelter, block the lower half of the KAP doorway with a quickly removable covering, such as a plastic-covered frame made of sticks. Be sure to connect the KAP's pull-cord only *11 inches* below its hinge line. This prevents excessive arm motions which would cause unnecessary fatigue.

In windy or cold weather, control the natural flow of air through the shelter by hanging adjustable curtains in the doorways at both ends, and/or by making and using trapdoors on the tops of the vertical entryway. For an adjustable curtain, use a piece of plastic with a supporting stick connected to its upper edge—so that you can provide different sized openings in the doorway *above the top of the adjustable curtain.*

Even in the coldest weather, in order to occupy this crowded underground shelter for hours without getting headaches, or worse, from breathing too much exhaled carbon dioxide, it is necessary that about 3 cubic feet per minute of air from the outdoors should flow through the shelter for each shelter occupant. An airflow of 36 cfm (3 x 12) is enough for 12 persons. This very slow-moving but essential airflow can be checked by repeatedly dropping a dry piece of toilet paper measuring only ¼ in. and ½ in. Drop this small piece of paper in the center of the shelter, from a height of 7 ft, being careful that no one breathes toward it. If on the average this paper lands on the floor about 1½ inches off the vertical and consistently in one direction, then about 36 cfm is flowing through the shelter.

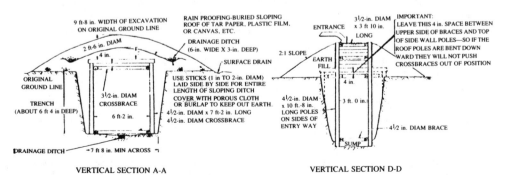

VERTICAL SECTION A-A   VERTICAL SECTION D-D

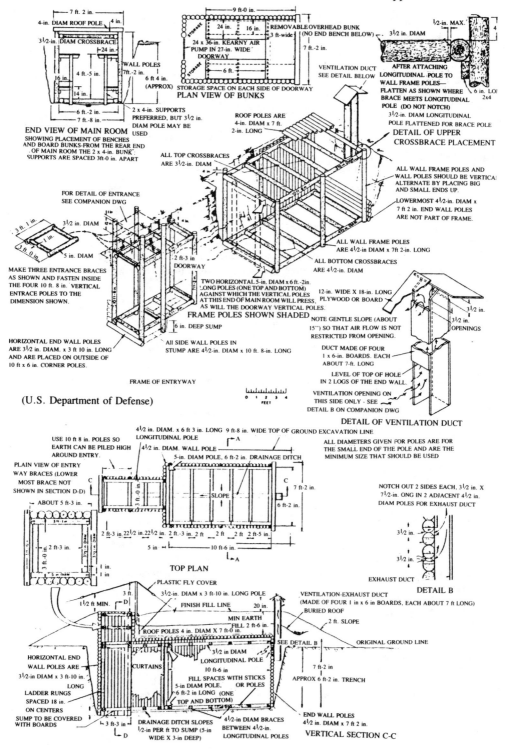

- 7 ft. 2 in.
- 4-in. DIAM ROOF POLE
- 4 in.
- 3½-in. DIAM CROSSBRACE
- 24 in.
- WALL POLES
- 7 ft.-2 in.
- 16 in.
- 4 ft.-5 in.
- 6 ft 4 in. (APPROX)
- 14 in.
- 6 ft.-2 in.
- 7 ft.-8 in.
- 2 x 4-in. SUPPORTS PREFERRED, BUT 3½ DIAM POLE MAY BE USED

END VIEW OF MAIN ROOM
SHOWING PLACEMENT OF BENCHES AND BOARD BUNKS-FROM THE REAR END OF MAIN ROOM THE 2 x 4-in. BUNK SUPPORTS ARE SPACED 3ft-0 in. APART

- 9 ft-0 in.
- STORAGE
- 24 in.
- 16 in.
- REMOVABLE OVERHEAD BUNK (NO END BENCH BELOW)
- 3 ft-wide
- 24 x 36-in. KEARNY AIR PUMP IN 27-in. WIDE DOORWAY
- 7 ft.-2 in.
- 6 ft.
- STORAGE SPACE ON EACH SIDE OF DOORWAY
PLAN VIEW OF BUNKS

- VENTILATION DUCT SEE DETAIL BELOW
- 3½ in. DIAM

- ½-in. MAX.
- AFTER ATTACHING LONGITUDINAL POLE TO WALL FRAME POLES- FLATTEN AS SHOWN WHERE BRACE MEETS LONGITUDINAL POLE (DO NOT NOTCH)
- 6 in. LO 2x4
- 3½-in. DIAM LONGITUDINAL POLE FLATTENED FOR BRACE POLE

DETAIL OF UPPER CROSSBRACE PLACEMENT

- ROOF POLES ARE 4-in. DIAM x 7 ft. 2-in. LONG
- ALL TOP CROSSBRACES ARE 3½-in. DIAM

- ALL WALL FRAME POLES AND WALL POLES SHOULD BE VERTICAL ALTERNATE BY PLACING BIG AND SMALL ENDS UP.
- LOWERMOST 4½-in. DIAM x 7 ft 2 in. END WALL POLES ARE NOT PART OF FRAME.
- ALL WALL FRAME POLES ARE 4½-in DIAM x 7ft 2-in. LONG

- FOR DETAIL OF ENTRANCE SEE COMPANION DWG

- 3½ in. DIAM
- 3 ft. 1 in.
- 1 in.
- 3 ft-0 in.
- 5 in. DIAM
- 2 ft-3 in. DOORWAY

- ALL BOTTOM CROSSBRACES ARE 4½-in. DIAM

MAKE THREE ENTRANCE BRACES AS SHOWN AND FASTEN INSIDE THE FOUR 10 ft. 8 in. VERTICAL ENTRACE POLES TO THE DIMENSION SHOWN.

- TWO HORIZONTAL 5-in. DIAM x 6 ft.-2in. LONG POLES (ONE TOP AND BOTTOM) AGAINST WHICH THE VERTICAL POLES AT THIS END OF MAIN ROOM WILL PRESS, AS WILL THE DOORWAY VERTICAL POLES.
FRAME POLES SHOWN SHADED

- 12-in. WIDE X 18-in. LONG PLYWOOD OR BOARD
- 6 in. DEEP SUMP
- NOTE GENTLE SLOPE (ABOUT 15") SO THAT AIR FLOW IS NOT RESTRICTED FROM OPENING.

- 3½ in.
- 3½ in. OPENINGS

- HORIZONTAL END WALL POLES ARE 3½ in. DIAM. x 3 ft 10 in. LONG AND ARE PLACED ON OUTSIDE OF 10 ft x 6 in. CORNER POLES.
- ALL SIDE WALL POLES IN STUMP ARE 4½-in. DIAM x 10 ft. 8-in. LONG

- DUCT MADE OF FOUR 1 x 6-in. BOARDS. EACH ABOUT 7-ft. LONG
- LEVEL OF TOP OF HOLE IN 2 LOGS OF THE END WALL.

FRAME OF ENTRYWAY

- VENTILATION OPENING ON THIS SIDE ONLY - SEE DETAIL B ON COMPANION DWG

(U.S. Department of Defense)

FEET 0 1 2 3 4

DETAIL OF VENTILATION DUCT

- USE 10 ft 8 in. POLES SO EARTH CAN BE PILED HIGH AROUND ENTRY.
- 4½ in. DIAM. x 6 ft 3 in. LONG LONGITUDINAL POLE
- 9 ft-8 in. WIDE TOP OF GROUND EXCAVATION LINE
- 4½ in. DIAM. WALL POLE
- 5-in. DIAM POLE, 6 ft-2 in. DRAINAGE DITCH

- ALL DIAMETERS GIVEN FOR POLES ARE FOR THE SMALL END OF THE POLE AND ARE THE MINIMUM SIZE THAT SHOULD BE USED

PLAIN VIEW OF ENTRY WAY BRACES (LOWER MOST BRACE NOT SHOWN IN SECTION D-D)

- ABOUT 5 in.-3 in.
- 3 ft-0 in.
- SLOPE
- 7 ft-2 in.
- 6 ft-2 in.

- NOTCH OUT 2 SIDES EACH, 3½ in. X 7½-in. ONG IN 2 ADJACENT 4½ in. DIAM POLES FOR EXHAUST DUCT

- 2 ft-3 in.
- 2 ft-3 in. 22½ in. 22½ in. 2 ft.-3 in. 2 ft 2 ft 2 ft 2 ft 5 in.
- 5 in.
- 10 ft-6 in.

TOP PLAN

- 3½ in.
- 3½ in.
- EXHAUST DUCT

DETAIL B

- PLASTIC FLY COVER
- 1½ ft MIN.
- 3 ft.
- 3½-in. DIAM x 3 ft-10 in. LONG POLE
- FINISH FILL LINE
- 20 in.
- VENTILATION-EXHAUST DUCT (MADE OF FOUR 1 in x 6 in BOARDS, EACH ABOUT 7 ft LONG)
- BURIED ROOF
- 2 ft. SLOPE
- MIN EARTH FILL. 2 ft-6 in.
- ROOF POLES 4 in. DIAM X 7 ft-0 in.
- SEE DETAIL B
- ORIGINAL GROUND LINE

- HORIZONTAL END WALL POLES ARE 3½-in DIAM x 3 ft-10 in. LONG
- LADDER RUNGS SPACED 18 in. ON CENTERS
- SUMP TO BE COVERED WITH BOARDS
- CURTAINS
- 3½ in DIAM LONGITUDINAL POLE 10 ft-6 in
- FILL SPACES WITH STICKS OR POLES 5-in DIAM POLE, 6 ft-2 in LONG (ONE TOP AND BOTTOM)
- 7 ft-2 in.
- APPROX 6 ft-2 in. TRENCH
- 3 ft-3 in.
- DRAINAGE DITCH SLOPES ½-in PER ft TO SUMP (5-in WIDE X 3-in DEEP)
- 4½-in DIAM BRACES BETWEEN 4½-in. LONGITUDINAL POLES
- END WALL POLES 4½-in. DIAM x 7 ft 2 in.

VERTICAL SECTION C-C

Smoking produces carbon monoxide, which causes severe headaches under ventilation conditions which, though austere, are adequate when no one smokes.

43.   When building *a shelter for more than 12 persons,* increase the illustrated 10 ft., 6 in. length of the main room by one ft. for every person beyond 12. Furthermore, to assure adequate cooling—ventilation (especially in warm or hot weather) for Small-Pole Shelters sized for 12 to 24 persons, the ventilation duct should be replaced by a second full-sized entryway.

44.   If more than 24 persons are to be sheltered, build two or more separate shelters.

HOW TO CUT AND HAUL LOGS AND POLES MORE EASILY

1.   Sharpen your tools before getting to work—no matter how much the rush.

2.   When sawing green trees, oil your saw with kerosene or diesel fuel. If you lack these, use motor oil, grease, or even soap.

3.   When felling a small tree, to cut it off square, to keep your saw from being pinched, and to help make the tree fall in the desired direction: (A) First, saw the tree about one-third through on the side toward which you want it to fall. (B) Then, while sawing the opposite side, have another person push on the tree with a 10-ft. push-pole, by pressing the end of the push-pole against the tree about 10 feet above the ground. A push-pole with a forked end or a big nail on its end is best.

4.   After a tree is felled, trim off all limbs and knots so that the log or pole is smooth, and will require no additional smoothing when you are preparing to drag it, or use it to build your shelter.

5.   To speed up the measuring of poles and logs and to cut them the right length, make and use a measuring stick.

When dragging poles, it is usually best first to cut them exactly two or three times the length of the final poles to be used in the shelter. To drag several poles (or a log) *by hand,* cut a 3½ ft. long stick (2 to 2½ inches in diameter), tie a short piece of one-quarter-inch (or stronger) rope to its center; make a lasso-like loop at the free end of the rope, so that when it is looped around the log and two men are pulling the log, the end of the log is raised about 6 inches above the ground. The loop should be tightened around the log about 2 ft. from its end, so that the end of the log cannot strike the backs of the legs of the two men pulling it.
CAUTION: If you drag a log down a steep hill, one man should tie a rope to the rear end of the log, and then follow the dragger, always keeping ready to act as the brake.

7.   Avoid carrying logs on your shoulders: You can injure yourself severely if you trip, and you will certainly tire yourself more than if you drag them.

8.   When you get your poles or logs to the location where you are building the shelter, cut them to the desired minimum diameters and specified lengths, and place together all of one specified type. Be sure that the diameter of the *small end* of each pole of one type is at least as large as the minimum diameter specified for its type.

DOOR COVERED TRENCH SHELTER

General Information

This shelter is designed for areas where there is a shortage of small trees and/or building materials. The depth to ground water and rock must also be below the bottom of the trench. In addition, the earth must be sufficiently firm and stable so that the trench walls will not collapse. The shelter (3-person capacity) can be constructed by 3 people working an approximate total of 12 hours each. Read and study all instructions before beginning to build.

STEP 1   Select a reasonably level site. Lay out the shelter as illustrated by laying doors side by side to determine the shelter length. Door knobs should be removed.

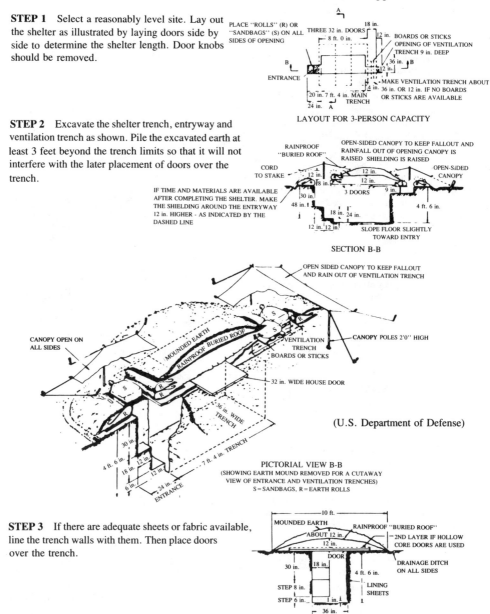

PLACE "ROLLS" (R) OR "SANDBAGS" (S) ON ALL SIDES OF OPENING

THREE 32 in. DOORS
18 in.
8 ft. 0 in.
12 in. BOARDS OR STICKS OPENING OF VENTILATION TRENCH 9 in. DEEP
36 in. B
12 in.
B
ENTRANCE
MAKE VENTILATION TRENCH ABOUT 36 in. OR 12 in. IF NO BOARDS OR STICKS ARE AVAILABLE
4 in.
20 in. 7 ft. 4 in. MAIN TRENCH
24 in. A

LAYOUT FOR 3-PERSON CAPACITY

STEP 2   Excavate the shelter trench, entryway and ventilation trench as shown. Pile the excavated earth at least 3 feet beyond the trench limits so that it will not interfere with the later placement of doors over the trench.

RAINPROOF "BURIED ROOF"
OPEN-SIDED CANOPY TO KEEP FALLOUT AND RAINFALL OUT OF OPENING CANOPY IS RAISED SHIELDING IS RAISED
CORD TO STAKE
12 in.
12 in.
OPEN-SIDED CANOPY
8 in.
3 DOORS
9 in.

IF TIME AND MATERIALS ARE AVAILABLE AFTER COMPLETING THE SHELTER. MAKE THE SHIELDING AROUND THE ENTRYWAY 12 in. HIGHER - AS INDICATED BY THE DASHED LINE
30 in.
48 in.
18 in. 24 in.
4 ft. 6 in.
12 in. 12 in.
SLOPE FLOOR SLIGHTLY TOWARD ENTRY

SECTION B-B

OPEN SIDED CANOPY TO KEEP FALLOUT AND RAIN OUT OF VENTILATION TRENCH

CANOPY OPEN ON ALL SIDES

MOUNDED EARTH
RAINPROOF BURIED ROOF
VENTILATION TRENCH
BOARDS OR STICKS
CANOPY POLES 2'0" HIGH
32 in. WIDE HOUSE DOOR
36 in. WIDE TRENCH

(U.S. Department of Defense)

30 in.
4 ft. 6 in. 12 in.
18 in. 12 in.
6 in.
24 in.
ENTRANCE
7 ft. 4 in. TRENCH

PICTORIAL VIEW B-B
(SHOWING EARTH MOUND REMOVED FOR A CUTAWAY VIEW OF ENTRANCE AND VENTILATION TRENCHES)
S = SANDBAGS, R = EARTH ROLLS

STEP 3   If there are adequate sheets or fabric available, line the trench walls with them. Then place doors over the trench.

10 ft.
MOUNDED EARTH
ABOUT 12 in.
RAINPROOF "BURIED ROOF"
2ND LAYER IF HOLLOW CORE DOORS ARE USED
12 in.
DOOR
30 in. 18 in.
4 ft. 6 in.
DRAINAGE DITCH ON ALL SIDES
STEP 8 in.
LINING SHEETS
STEP 6 in. 1 in.
36 in.

SECTION A-A

STEP 4   In order to hold in place an adequate amount of earth on top of the doors, construct earth "rolls" around the entryway as shown. The "rolls" will keep the earth fill in place. See how to make an earth roll.

STEP 5   Place earth fill and the waterproofing material over the doors. Place sandbags as shown on the illustrations.

STEP 6   Construct shallow drainage ditches on all sides and place canopies over the openings.

Tools and Materials

1. Doors (interior solid or hollow-core)—1 full size (32'' minimum width) for each person. If doors measure less than 32'' in width, use a combination of doors to provide the minimum width per person. **If doors are hollow core—use two layers.**
2. Pick and/or mattock.
3. Long-handled shovels and square bladed shovel.
4. Rainproofing material—(e.g., plastic sheeting, canvas plastic table covers, etc.) at least 25 square feet per person plus 2 pieces about 6 ft. by 6 ft. for use as canopies.
5. One bedsheet or the equivalent of 50 sq. ft. of cloth or plastic per person to line trench and make earth-filled rolls.
6. Two pillowcases per person to use as sandbags.
7. String or cord to tie canopies and sandbags.
8. Knife.
9. Several boards about 3 feet long.
10. Measuring tape and/or ruler.
11. Work gloves for each worker.
12. Hammer and hand saw.

How To Make An Earth Roll

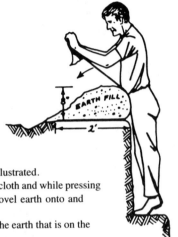

1. Select a piece of cloth or plastic at least as strong as a new bed sheet, 2 ft. wider than the side of the opening to be protected, and 5 ft. in length.
2. Place 2 ft. of the length of the cloth on the ground, as illustrated.
3. While using both hands to hold up 3 ft. of the length of the cloth and while pressing against the cloth with your body, have another person shovel earth onto and against the cloth.
4. While still pulling on the cloth, place the upper part over the earth that is on the lower part of the cloth.
5. Cover the upper edge of the cloth, forming an earth-filled "hook" in this edge.

TILT-UP DOORS AND EARTH

General Information

Read and study all instructions before starting to build. The location selected for this shelter should be level or gently sloping down and away from the masonry wall. A three-person shelter can be constructed by three people working a total of 6 hours.

STEP 1   Lay out the trench and earth notch widths, as dimensioned on the section below, adjacent to a masonry wall. Determine the length of trench and notch by allowing one door width of length per person to be sheltered.

STEP 2   Excavate trench and earth notch. Place excavated earth outside shelter limits for later use.

STEP 3   Remove door knobs from all doors. Place double layer of doors in notch and against wall as shown in sketch. Nail 1'' × 8'' board to door edges at entrance to serve as earth stop, *after* attaching plastic entrance cover as shown, *or* build retaining wall or sandbags in lieu of board. Place one door on edge lengthwise as the end closure.

STEP 4   Place one end of the rolled up waterproofing material under the top edge of the doors before earth fill is placed. Begin placement of earth fill on doors. Cover the earth fill with waterproofing material, securing it with earth at top and bottom to prevent it from blowing away.

STEP 5   Construct entrance—fill "sandbag pillowcases" with earth taken from the trench and stack to dimensions shown after doors are in place. Plastic or polyethylene (waterproofing material) entrance cover should be in place before earth fill is put on the doors.

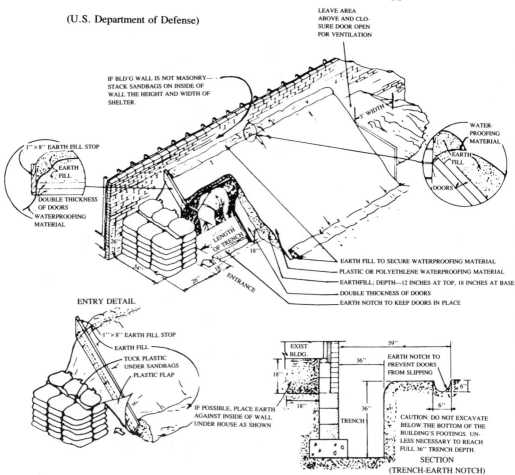

(U.S. Department of Defense)

LEAVE AREA ABOVE AND CLO-SURE DOOR OPEN FOR VENTILATION

IF BLD'G WALL IS NOT MASONRY—STACK SANDBAGS ON INSIDE OF WALL THE HEIGHT AND WIDTH OF SHELTER.

3' WIDTH

WATER-PROOFING MATERIAL

1" × 8" EARTH FILL STOP

EARTH FILL

EARTH FILL

DOORS

DOUBLE THICKNESS OF DOORS
WATERPROOFING MATERIAL

26"

LENGTH OF TRENCH

18"

54"

26"

18"

18"

ENTRANCE

EARTH FILL TO SECURE WATERPROOFING MATERIAL
PLASTIC OR POLYETHLENE WATERPROOFING MATERIAL
EARTHFILL; DEPTH—12 INCHES AT TOP, 18 INCHES AT BASE
DOUBLE THICKNESS OF DOORS
EARTH NOTCH TO KEEP DOORS IN PLACE

ENTRY DETAIL

1" × 8" EARTH FILL STOP
EARTH FILL
TUCK PLASTIC UNDER SANDBAGS
PLASTIC FLAP

IF POSSIBLE, PLACE EARTH AGAINST INSIDE OF WALL UNDER HOUSE AS SHOWN

EXIST BLDG.

18"

18"

36"

TRENCH

59"

36"

EARTH NOTCH TO PREVENT DOORS FROM SLIPPING

6"

6"

CAUTION: DO NOT EXCAVATE BELOW THE BOTTOM OF THE BUILDING'S FOOTINGS. UN-LESS NECESSARY TO REACH FULL 36" TRENCH DEPTH.

SECTION
(TRENCH-EARTH NOTCH)

Tools and Materials

1. Tools: pick, shovel, hammer, saw, screwdriver, knife, yardstick.
2. Sandbags pillowcases or plastic garbage bags—at least 39.
3. Lumber: 1" × 8" piece 7' long (or 20 more sandbags) for earth-fill stop at entrance edge of doors.
4. Rope or cord to tie sandbags.
5. Doors: two layers for length of shelter plus one for end closure. (Example: 7 doors for 3 person shelter).
6. Nails: 8 penny (2½" long), about 10 to nail earth stop to door edges at entrance.
7. Plastic or polyethlene (waterproofing material) to cover double layer of doors plus entrance.
8. Work gloves for each worker.

Other plans are available from U.S. Army AG Publication Center, Civil Preparedness Section, 2800 Eastern Blvd. (Middle River), Baltimore, MD., 21220

Appendix B

EXPEDIENT FALLOUT SHELTER
AIR VENTILATION PUMP

All expedient shelters are designed to provide for some natural ventilation. In very hot weather, additional ventilation may be required to provide a livable temperature. Construction of an air pump that can provide additional ventilation is illustrated below.

STUDY ALL INSTRUCTIONS BEFORE STARTING CONSTRUCTION

STEP 1   AIR PUMP

The air pump operates by being swung like a pendulum. It is hinged at the top of its swinging frame. It is swung by pulling an attached cord. The flaps are free to also swing and when they are in the closed position, air is pushed through the opening that the pump is attached to.

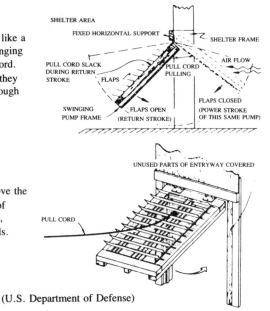

To obtain maximum efficiency and move the largest amount of air, the unused portions of the entryway should be covered with wood, plastic, cloth, stiff paper or similar materials.

(U.S. Department of Defense)

STEP 2   MATERIALS AND TOOLS NEEDED TO CONSTRUCT AN AIR PUMP

(Materials sized for a 36-inch by 29-inch pump)
Lumber sizes can be altered, depending on availability.

*A. LUMBER

SIZE	QUANTITY	SIZE	QUANTITY
1''x 2''x 36''	2	1''x 1''x 32''	1
1''x 2''x 32''	2	1''x 2''x 29''	2
1''x 1''x 36''	1	1''x 4''x 36''	1

B. One pair ordinary door or cabinet butt hinges, or metal strap hinges, or improvised hinges made of leather, woven straps, cords or four hook & eye screws which can be joined to form two hinges.

C. 24 nails about 2'' long, plus screws for hinges.

334

*D.   Polyethylene flim, 3 to 4 mils thick, of plastic dropcloth, or raincoat-type fabric, or strong heavy paper—10 rectangular-shaped pieces, 30'' x 5½''.

*E.   30' of smooth, straight wire for use as flap pivot wires—(about as thick as coat-hanger wire) or cut from 10 wire coat hangers, or 35' of nylon string (coat-hanger wire thickness).

*F.   30 small staples, or small nails, or 60 tacks to attach flap pivot wires to wood frame.

*G.   30' of ¾'' to 1'' wide pressure-sensitive waterproof tape that does not stretch, or use needle and thread to sew hem tunnels to the flaps.

*H.   For flap stops, 150 ft. of light string, strong thread, or thin smooth wire. 90 tacks or small nails to attach flap stops to the wood frame, or flap stops can be tied to the frame.

I.   10 feet of cord for the pull cord.

J.   Desirable tools: hammer, saw, wirecutter-pliers, screwdriver, small drill, scissors, knife, yardstick, and pencil.

* Items must be sized or adjusted to fit opening into which airpump is to be placed.

STEP 3   HOW TO CONSTRUCT THE AIR PUMP

A. Cut lumber and assemble frame as shown.

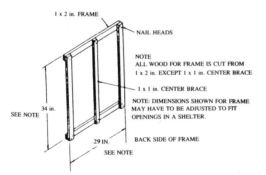

(U.S. Department of Defense)

B. Complete frame and attach hinges. If drill is not available to drill screw holes to attach hinges, use a nail to make the holes.

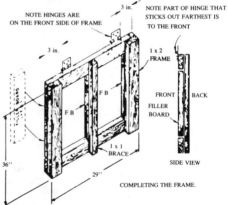

C. Cut 10 rectangular strips 30'' long by 5½'' wide for use as flaps. Hem flaps as shown. Use pressure-sensitive tape or sew hem shut to form hem tunnel.

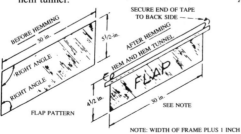

After hem is made, cut notches in flaps as shown. Avoid cutting tape that holds hem.

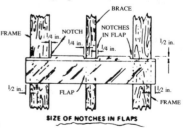

Insert 10 pieces of straight wire (pivot wires) into flap hem as shown. Flaps should swing freely. String can be used if wire not available (wire coat-hanger thickness).

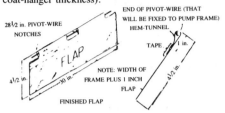

(U.S. Department of Defense)

D. Mark pump frame for pivot wire and flap stop locations.

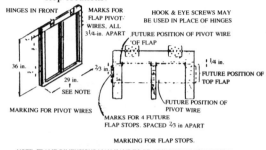

NOTE: FRAME DIMENSIONS MAY HAVE TO BE ADJUSTED TO FIT OPENING IN SHELTER

E. Attach flap stops (strings or wires) to the pump frame at the marked locations. 4 flap stops are needed between adjacent pivot wires.

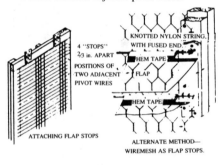

F. Starting from the bottom—staple, nail, tack or tie the flap pivot-wires with flaps in their marked positions. Attach hinges to horizontal support board. Attach pullcord to center brace.

STEP 4   TYPICAL INSTALLATION OF AIR PUMP

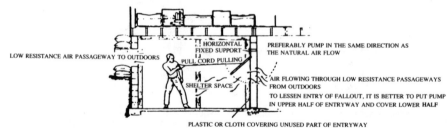

(U.S. Department of Defense)